Recent Titles in This Series

K–Theory and Algebraic Geometry: Connections with Quadratic Forms and Division Algebras

Proceedings of Symposia in
PURE MATHEMATICS

Volume 58, Part 1

K–Theory and Algebraic Geometry: Connections with Quadratic Forms and Division Algebras

Summer Research Institute on
Quadratic Forms and Division Algebras
July 6–24, 1992
University of California, Santa Barbara

Bill Jacob
Alex Rosenberg
Editors

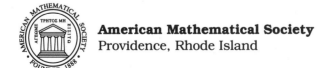

American Mathematical Society
Providence, Rhode Island

PROCEEDINGS OF THE SUMMER RESEARCH INSTITUTE
ON QUADRATIC FORMS AND DIVISION ALGEBRAS
HELD AT THE UNIVERSITY OF CALIFORNIA
SANTA BARBARA
JULY 6–24, 1992

with support from the National Science Foundation
Grant DMS-9122671

1991 *Mathematics Subject Classification.* Primary 12–06, 14–06, 16–06, 19–06.

Library of Congress Cataloging-in-Publication Data
Summer Research Institute on Quadratic Forms and Division Algebras (1992: University of California, Santa Barbara, Calif.)
K-theory and algebraic geometry : connections with quadratic forms and division algebras / Summer Research Institute on Quadratic Forms and Division Algebras, July 6–24, 1992, University of California, Santa Barbara; Bill Jacob, Alex Rosenberg, editors.
  p. cm. — (Proceedings of symposia in pure mathematics; v. 58, pts. 1–2)
  Includes bibliographical references.
  ISBN 0-8218-1498-2 (set),–ISBN 0-8218-0339-5 (pt. 1),–ISBN 0-8218-0340-9 (pt. 2)
  1. K-theory—Congresses. 2. Geometry, Algebraic—Congresses. I. Jacob, Bill. II. Rosenberg, Alex, 1926- . III. Title. IV. Series.
QA612.33.S86   1992
514′.23—dc20   94-34832
                                                                                          CIP

Contents

PART 1

PART 2

Preface

During the decade of the 1980s profound connections were discovered relating modern algebraic geometry and algebraic K-theory to arithmetic problems. Indeed, the phrase "arithmetic algebraic geometry" was popularized during that time and is now used by many to denote an entire branch of 20th century number theory. In addition, these same developments in algebraic geometry and K-theory greatly influenced research into the arithmetic of fields in general, in particular the algebraic theory of quadratic forms and the theory of finite-dimensional division algebras. The purpose of the 1992 AMS Summer Research Institute was to provide the research community with both a broad overview of the tools from algebraic geometry and K-theory that proved to be the most powerful in solving problems in the theory of quadratic forms and division algebras, as well as provide a forum for exposition of recent research.

The three week institute had six week-long speakers: three in K-theory, R. Swan, A. A. Suslin, and A. S. Merkurjev; and three in algebraic geometry, J.-L. Colliot-Thèléne, W. Raskind, and D. Saltman. A substantial portion of their lectures are reproduced in their expository articles in this volume. The editors hope these articles will help introduce young researchers to these important tools. In addition the institute had a series of individual research lectures, many of which can be found here. The organizers would like to thank the NSF for financial support, as well as the AMS and conference coordinator Wayne Drady for their efforts in administering the institute.

Bill Jacob
Alex Rosenberg
Santa Barbara, March 1994

Proceedings of Symposia in Pure Mathematics
Volume **58**.1 (1995)

Birational Invariants, Purity
and the Gersten Conjecture

J.-L. COLLIOT-THÉLÈNE

Contents

1991 *Mathematics Subject Classification.* Primary 14F20; Secondary 14C25, 14M20.
This paper is in final form and no version of it will be submitted for publication elsewhere.

§0. Introduction

In the last ten years, classical fields such as the algebraic theory of quadratic forms and the theory of central simple algebras have witnessed an intrusion of the cohomological machinery of modern algebraic geometry. Merkur'ev's theorem on the e_3 invariant, one way or another, relates to étale cohomology; so does the general Merkur'ev/Suslin theorem on the norm residue symbol. One of the purposes of the 1992 Summer Institute was to introduce workers in the two mentioned fields to the tools of modern algebraic geometry.

The organizers asked me to lecture on étale cohomology. Excellent introductions to étale cohomology exist, notably the set of lectures by Deligne in [SGA4 1/2]. There was no point in trying to duplicate such lectures. I thus decided to centre my lectures on unramified cohomology. This notion, although not novel by itself, has attracted attention only in recent times; it lies half-way between the scheme-theoretic point of view and the birational point of view. The notion itself may be defined for other functors than cohomology (other functors of interest will be discussed in §2.2).

Let k be a field and F a functor from the category of k-algebras to the category of abelian groups. Let X be an irreducible (reduced) algebraic variety X, and let $k(X)$ be its function field. Let $A \subset k(X)$ run through the rank one discrete valuation rings which contain k and whose field of fractions is $k(X)$. By definition, the unramified subgroup $F_{\mathrm{nr}}(k(X)/k) \subset F(k(X))$ is the group of elements in $F(k(X))$ which lie in the image of $F(A)$ for each A as above.

In order to demonstrate the utility of this notion, in §1, we shall start from scratch and establish the well known fact that an elliptic curve is not birational to the projective line, the tool being unramified H^1 with $\mathbf{Z}/2$ coefficients, presented in an even more down to earth form. Copying this proof for unramified H^i for higher integers i leads, for $i = 2$, to another approach to the celebrated Artin-Mumford examples of nonrational unirational varieties [Ar/Mu72]. One may go further along this way (see §4.2).

As one may observe immediately, the notion of unramified cohomology only depends on the function field $k(X)/k$ of the variety X. Indeed, one place where this notion has been extremely successful is Saltman's paper [Sa84], where the author shows that some function fields (invariant fields of a linear action of a finite group) are not purely transcendental over the ground field, thus settling Noether's problem over an algebraically closed field in the negative. Saltman's invariant is the unramified Brauer group $\mathrm{Br}_{nr}(k(X)/k)$, which may be shown to be equal to the (cohomological) Brauer group $\mathrm{Br}(X)$ of a smooth projective model X (assuming char.$(k) = 0$). In concrete cases, it is unclear how to construct such a model for a given function field. A key aspect of Saltman's paper is that the unramified point of view enables one to dispense with the construction of an explicit model, and even with the existence of such a model.

Various recent works have been devoted to extensions of these ideas, a number of them with the goal of detecting the nonrationality of some unirational varieties (see §4.2.4 and §4.2.5 below).

One aim of these lectures is to reconcile the birational point of view on unramified cohomology with the scheme-theoretic point of view. Among the reasons for doing so, let me mention:

– the study of the functorial behaviour of unramified cohomology with respect to arbitrary (not necessarily dominating) morphisms of varieties;

– the attempts to use unramified cohomology to introduce some equivalence relation on the rational points, or on the zero-cycles on a variety defined over a non-algebraically closed field (see e.g. [CT/Pa90]); the idea (a special case of the functoriality problem) is to try to evaluate an unramified cohomology class at an arbitrary point (some condition such as smoothness of the variety is essential at that point);

– the attempts to try to control the size of unramified cohomology groups over an algebraically closed ground field.

It will be demonstrated (§2) that good functorial behaviour of the unramified functor F_{nr} (and some other similar functors) attached to a functor F requires some basic properties, namely the specialization property, the injectivity property and the codimension one purity property (see Definition 2.1.4) for regular local rings.

The study of these properties, particularly the codimension one purity for étale cohomology, will lead us to a review of two fundamental properties of étale cohomology for varieties over a field (§3). The first property is cohomological purity, which is due to Artin ([SGA4]), and which is described in the standard textbooks on étale cohomology. The second property is the Gersten conjecture for étale cohomology, established by Bloch and Ogus [Bl/Og74], which is not studied in the standard textbooks. Although the circle of ideas around it is well known to a few experts, to many people it still seems shrouded in mystery. In §3, some motivation for this conjecture will be given together with its formal statement (§3.5). A guided tour through the literature will then be offered, but no proofs will be given. However, in §5, an alternate approach to some results of Gersten type ([Oj80], [CT/Oj92]) will be described in some detail.

No survey paper would be satisfactory without a couple of new results, or at least a couple of results which are in the air. While preparing these lectures, I thus came across

– finiteness results for unramified H^3-cohomology (§4.3), some of which had already been obtained by L. Barbieri-Viale [BV92b];

–a rigidity theorem for unramified cohomology (§4.4), also considered by Jannsen (unpublished);

– a new proof of a purity theorem of Markus Rost (Theorem 5.3.1).

The present notes are the outcome of the five lectures I gave at Santa Barbara. I have made no attempt at writing a systematic treatise, and I hope that the written text retains some of the spontaneity of the oral lectures.

Notation. Given an abelian group A and a positive integer n, we write $_nA$ for the subgroup of elements killed by n, and we write A/n for the quotient A/nA.

Given an integral domain A, we let $\mathrm{qf}(A)$ denote its quotient field.

In this paper, discrete valuation rings will be of rank one unless otherwise mentioned.

We let $\mathbf{A}_k^n$, resp., $\mathbf{P}_k^n$ denote n-dimensional affine, resp., projective, space over a field k.

If X is an irreducible variety, and p is a positive integer, we let X^p denote the set of all codimension p points of X, i.e., the set of points $M \in X$ whose local ring $\mathscr{O}_{X,M}$ is of dimension p.

§1. An exercise in elementary algebraic geometry

Let $\mathbf{C}(x)$ be the field of rational fractions in one variable over the complex field $\mathbf{C}$ and let $L = \mathbf{C}(x)(\sqrt{x(x-1)(x+1)})$ be the quadratic extension of $\mathbf{C}(x)$ obtained by adjoining the square root of $x(x-1)(x+1)$. This is a function field of transcendence degree one over $\mathbf{C}$, and it is none other than the field of rational functions of the elliptic curve E given by the affine equation

$$y^2 = x(x-1)(x+1).$$

Certainly one of the first statements in algebraic geometry is that E is not birational to the affine line $\mathbf{A}_\mathbf{C}^1$, in other words that L cannot be generated over $\mathbf{C}$ by a single element. There are several ways to see this. As a preparation to the notion of unramified cohomology, we give the following elementary proof, which only uses the notion of a discrete valuation (of rank one, with value group $\mathbf{Z}$).

PROPOSITION 1.1. *Let $L/\mathbf{C}$ be the field $L = \mathbf{C}(x)(\sqrt{x(x-1)(x+1)})$. Then L is not $\mathbf{C}$-isomorphic to the rational field $\mathbf{C}(t)$.*

PROOF. (a) Write $L = \mathbf{C}(x)(y)$ with

$$y^2 = x(x-1)(x+1) \in L.$$

Consider the element $x \in L$. We claim that for any discrete valuation v on the field L, the valuation $v(x)$ of x is *even*.

Case 1. If $v(x) = 0$, the result is clear.

Case 2. If $v(x) > 0$, then $v(x-1) = \inf(v(x), v(1)) = v(1) = 0$ and similarly $v(x+1) = 0$. From the identity

$$y^2 = x(x-1)(x+1)$$

in L, we deduce $2v(y) = v(x)$ and $v(x)$ is even.

Case 3. If $v(x) < 0$, then $v(x-1) = \inf(v(x), v(1)) = v(x)$ and similarly $v(x+1) = v(x)$; hence from the above identity we deduce $2v(y) = 3v(x)$; hence $v(x)$ is even.

(b) We claim that x is not a square in L.

Given any quadratic field extension $K \subset L = K(\sqrt{a})$ with char.$(K) \neq 2$, a straightforward computation yields a short exact sequence

$$1 \to \mathbf{Z}/2 \to K^*/K^{*2} \to L^*/L^{*2},$$

where the map $\mathbf{Z}/2 \to K^*/K^{*2}$ sends 1 to the class of a. Applying this remark to the quadratic extension of fields

$$K = \mathbf{C}(x) \subset L = \mathbf{C}(x, \sqrt{x(x-1)(x+1)}),$$

we see that if x were a square in L, either x would be a square in $\mathbf{C}(x)$ or $x(x-1)(x+1)/x = (x-1)(x+1)$ would be a square. But for each of $x \in \mathbf{C}(x)$ and $(x-1)(x+1) \in \mathbf{C}(x)$ we may find a valuation of the field $\mathbf{C}(x)$ which takes the value 1 on either of these elements. Hence, they are not squares, and x is not a square in L.

(c) In the rational function field $\mathbf{C}(t)$ in one variable over $\mathbf{C}$, any nonzero element, all valuations of which are even, is a square. Indeed, any $z \in \mathbf{C}(t)^*$ after suitable multiplication by a square in $\mathbf{C}(t)$ may be rewritten as a polynomial in $\mathbf{C}[t]$ with simple roots $z = \prod_{i \in I}(t - e_i)$. If I is not empty, i.e., if $z \neq 1$, then for any $i \in I$, we may consider the valuation v_i associated to the prime ideal $(t - e_i)$, and $v_i(z) = 1$ is not even.

(d) Putting (a), (b), (c) together, we find that the field L is not $\mathbf{C}$-isomorphic to the field $\mathbf{C}(t)$. $\square$

EXERCISE 1.1.1. Let p and q be two coprime integers, and let e_i, $i = 1, \ldots, q$ be distinct elements of $\mathbf{C}$. Let L be the function field $L = \mathbf{C}(x)(y)$, with $y^p = \prod_{i=1, \ldots, q}(x - e_i)$. Arguing with elements all valuations of which are divisible by p, show that L is not purely transcendental over $\mathbf{C}$.

Let us slightly formalize the proof of Proposition 1.1. Given a field k, char$(k) \neq 2$, and a function field F/k (by function field F/k we mean a field F finitely generated over the ground field k), we consider the group

$$Q(F/k) = \{\alpha \in F^*, \forall v \text{ trivial on } k, v(\alpha) \in 2\mathbf{Z}\}/F^{*2}.$$

Let Ω be the set of discrete valuation rings A (of rank one) with $k \subset A$ and with quotient field K. The group $Q(F/k)$ may also be defined as:

$$Q(F/k) = \{\alpha \in F^*/F^{*2}, \forall A \in \Omega, \alpha \in \mathrm{Im}(A^*/A^{*2})\}.$$

Our proof of the nonrationality of $L = \mathbf{C}(x)(y)$ fell into the following parts:

(a) We showed that if $F = \mathbf{C}(t)$, then $Q(F) = 0$. As the reader will easily check, this result may be extended. Namely, if $t_1, \ldots, t_n$ are independent variables, the natural map $k^*/k^{*2} \to Q(k(t_1, \ldots, t_n)/k)$ is an isomorphism. Even more generally, if K/k is a function field, the inclusion $K \subset K(t_1, \ldots, t_n)$ induces an isomorphism $Q(K/k) \simeq Q(K(t_1, \ldots, t_n)/k)$.

(b) We produced a nontrivial element in $Q(L)$.

In the next section, we shall generalize the formalism above to functors F other than $F(A) = A^*/A^{*2}$. In most cases, the analogue of (a) will be easy to check. However the computation of the analogue of $Q(L)$ will often turn out to be tricky. Before getting to this, we might comfort ourselves with the remark that in the case at hand, a few more elementary arguments enable one to compute the exact value of $Q(L)$.

If k is algebraically closed, the group k^* is divisible, hence any discrete valuation on F is trivial on k. In that case, we simply write $Q(F) = Q(F/k)$. In the above proof, we showed $Q(\mathbf{C}(t)) = 0$ and $Q(L) \neq 0$ for $L = \mathbf{C}(x, \sqrt{x(x-1)(x+1)})$. We shall go further and actually compute $Q(L)$.

First note that given any extension L/K of function fields over a field k, there is an induced map $Q(K/k) \to Q(L/k)$ (for any discrete valuation v on L, either v is trivial on K, or it induces a discrete valuation on K). When the extension L/K is finite and separable, there is a map the other way round.

LEMMA 1.2. *Let L/K be a finite extension of function fields over k. Assume that L/K is separable. Then the norm map $N_{L/K}: L^*/L^{*2} \to K^*/K^{*2}$ induces a homomorphism $N_{L/K}: Q(L/k) \to Q(K/k)$.*

PROOF. Given any discrete valuation ring $A \subset K$ with $k \subset A$ and $\mathrm{qf}(A) = K$, the integral closure B of A in L is a semilocal Dedekind ring ([Se68], I, §4). We thus have an exact sequence

$$0 \to B^* \to L^* \to \bigoplus_{i=1,\dots,m} \mathbf{Z} \to 0,$$

where the right hand side map is given by the valuations v_i, $i = 1, \dots, m$ at the finitely many maximal ideals of B. If all valuations $v_i(\beta)$ are even, then this sequence shows that β may actually be written $\beta = \gamma.\delta^2$ with $\gamma \in B^*$ and $\delta \in L^*$. Now the norm map $N_{L/K}: L^* \to K^*$ sends B^* into A^*, and we conclude that $N_{L/K}(\beta)$ is the product of a unit in A^* by a square in K^*, i.e. that $v_A(N_{L/K}(\beta))$ is even. $\square$

PROPOSITION 1.3. *With notation as in Proposition 1.1, we have $Q(L) = \mathbf{Z}/2 \oplus \mathbf{Z}/2$.*

PROOF. Given a quadratic field extension $K \subset L = K(\sqrt{a})$ with $\mathrm{char}.(K) \neq 2$, the exact sequence mentioned in the proof of Proposition 1.1 may be extended to

$$0 \to \mathbf{Z}/2 \to K^*/K^{*2} \to L^*/L^{*2} \xrightarrow{N} K^*/K^{*2}.$$

To prove this, use Hilbert's theorem 90 for the cyclic extension L/K. (As a matter of fact, this sequence is part of a well known infinite exact sequence in Galois cohomology.) Let us now take $K = \mathbf{C}(x)$ and $L = \mathbf{C}(x, \sqrt{x(x-1)(x+1)})$. From Lemma 1.2, the norm map induces a map

$Q(L) \to Q(K)$. Now we have $Q(K) = Q(\mathbf{C}(x)) = 0$, as already mentioned. From the above sequence we conclude that any element of $Q(L)$ comes from K^*/K^{*2} via the natural map $K^*/K^{*2} \to L^*/L^{*2}$ induced by $K \subset L$. We now need to determine which classes in K^*/K^{*2} have image in L^*/L^{*2} contained in $Q(L)$. Suppose that $\alpha \in K^*$ defines such a class. Consider the ring extension $\mathbf{C}[x, y]/(y^2 - x(x-1)(x+1))/\mathbf{C}[x]$. This is a finite extension of Dedekind rings and $\mathbf{C}[x, y]/(y^2 - x(x-1)(x+1))$ is the integral closure of $K = \mathbf{C}[x]$ in L (indeed, this $\mathbf{C}$-algebra is smooth over $\mathbf{C}$, hence is a regular ring), and one immediately checks that over any maximal ideal of $\mathbf{C}[x]$ corresponding to a point of the affine line different from $0, 1, -1$, there lie two distinct prime ideals of $\mathbf{C}[x, y]/(y^2 - x(x-1)(x+1))$. For any such point, taking valuations at that point and at the two points lying above it yields a commutative diagram

$$
\begin{array}{ccc}
\mathbf{C}(X, \sqrt{X(X-1)(X+1)})^* & \longrightarrow & \mathbf{Z} \oplus \mathbf{Z} \\
\uparrow & & \uparrow \\
\mathbf{C}(X)^* & \longrightarrow & \mathbf{Z}
\end{array}
$$

where the right hand side vertical map is $1 \to (1, 1)$. Let $P(x) \in \mathbf{C}(x)$, and suppose that its image in L has all its valuations even; then as an element of $\mathbf{C}(x)$, at all primes of the affine line different from $(x), (x-1), (x+1)$, $P(x)$ must have even valuation. Any element in K^*/K^{*2} may be represented by a square free polynomial in $\mathbf{C}[x]$. We conclude that α, up to a square in K^*, may be represented by a polynomial $x^a.(x-1)^b.(x+1)^c$, where a, b, c are 0 or 1. Now each of $x, x-1, x+1$ actually gives rise to an unramified element in L^*/L^{*2}, by the same argument that was used for x in Proposition 2.1. One easily checks that the classes of $x, x-1, x+1$ are linearly independent over $\mathbf{Z}/2$ in K^*/K^{*2}. They thus span a group $(\mathbf{Z}/2)^3 \subset K^*/K^{*2}$. On the other hand, the map $(\mathbf{Z}/2)^3 \to L^*/L^{*2}$ which sends a, b, c to $x^a.(x-1)^b.(x+1)^c \in L^*/L^{*2}$ has kernel exactly equal to the subgroup $\mathbf{Z}/2$ spanned by $(1, 1, 1)$, in view of the exact sequence

$$
0 \to \mathbf{Z}/2 \to K^*/K^{*2} \to L^*/L^{*2}.
$$

Thus, $Q(L) \simeq (\mathbf{Z}/2)^2$, with the two generators given by the class of x and of $x-1$ in L^*/L^{*2}. $\square$

EXERCISE 1.3.1. Let e_i, $i = 1, \ldots, 2g+1$ be distinct elements of $\mathbf{C}$, and let L be the function field $L = \mathbf{C}(x, \sqrt{\prod_{i=1}^{2g+1}(x - e_i)})$. Using the same arguments as above, show $Q(L) = (\mathbf{Z}/2)^{2g}$. In particular, two such function fields with different values of g are not $\mathbf{C}$-isomorphic.

REMARK 1.3.2. The learned reader will have long recognized that the functor $L \mapsto Q(L)$ which we have been using in the above example is a familiar invariant in algebraic geometry. Namely, if $\mathbf{C}(X)$ is the function field L

of a smooth, projective, connected variety $X/\mathbf{C}$, the group $Q(\mathbf{C}(X))$ is isomorphic to the 2-torsion subgroup of the Picard group $\mathrm{Pic}(X)$. For a curve of genus g, that group is well known to be isomorphic to $(\mathbf{Z}/2)^{2g}$. This is what the above naive computation gives us for the function field of the hyperelliptic curve

$$y^2 = \prod_{i=1,\dots,2g+1} (x - e_i),$$

where the e_i's are distinct.

Note that the birational invariant given by the torsion in $\mathrm{Pic}(X)$ is not a very subtle invariant. For instance, Serre [Se59] proved that if a smooth projective variety X over $\mathbf{C}$ is dominated by a projective space (i.e., is unirational), then $\mathrm{Pic}(X)$ has no torsion (he actually proved more, namely, that the fundamental group of X is trivial). Subtler functors than $A \mapsto A^*/A^{*2}$ are required to tell us that some unirational varieties are not rational.

§2. Unramified elements

2.1. Injectivity, codimension one purity, homotopy invariance: a general formalism.

Let $\mathscr{R}$ be the category of commutative rings with unit, and let $\mathscr{C} \subset \mathscr{R}$ be one of the following subcategories.

(1) The category $\mathscr{R}$ of all commutative rings with unit, morphisms being arbitrary ring homomorphisms.

(2) The category $\mathscr{R}_{fl}$ of all commutative rings with unit, morphisms being flat ring homomorphisms.

(3) The category $k - Alg$ of all k-algebras (not necessarily of finite type) over a fixed field k, morphisms being k-homomorphisms of k-algebras.

(4) The category $k - Alg_{fl}$ of k-algebras (not necessarily of finite type) over a fixed field k, morphisms being flat k-homomorphisms of k-algebras.

Let Ab be the category of abelian groups. We shall be interested in (covariant) functors from $\mathscr{C}$ to Ab. Although we want to keep this section at a formal level, it might help the reader if we already revealed which functors we have in mind. Let us here quote:

(i) the (Azumaya or cohomological) Brauer group [Au/Go60], [Gr68];

(ii) the Witt group [Kn77];

(iii) various étale cohomology groups with values in commutative group schemes [SGA4];

(iv) K-theory [Qu73] and K-theory with coefficients.

We may also mention more exotic functors on the category of k-algebras, such as:

(v) the functor $A \mapsto A^*/\mathrm{Nrd}((D \otimes_k A)^*)$, where D is a central simple algebra over the field k and Nrd denotes the associated reduced norm;

(vi) the functor $A \mapsto A^*/\Phi(A)$ where Φ is a Pfister form over k ($\mathrm{char}(k) \neq 2$) and $\Phi(A)$ denotes the subgroup of elements of A^* represented by Φ over A [CT78];

(vii) finally, functors with values in the category of pointed sets such as the functor $A \mapsto H^1(A, G)$ where G is a linear algebraic group over k and $H^1(A, G)$ is the set of isomorphism classes of principal homogeneous spaces over $\mathrm{Spec}(A)$ under G.

Interesting functors are also obtained by replacing the functor $A \mapsto F(A)$ by the functor $A \mapsto F(A \otimes_k B)$ where B is some fixed k-algebra.

DEFINITION 2.1.1. Let F be a functor from $\mathscr{C}$ to Ab. Given any fields K and L in $\mathscr{C}$, with $K \subset L$, we say that $\alpha \in F(L)$ is unramified over K if for each rank one discrete valuation ring A with $K \subset A$ and quotient field $\mathrm{qf}(A) = L$, the element α belongs to the image of $F(A) \to F(L)$. We shall denote by $F_{nr}(L/K) \subset F(L)$ the group of all such elements, and refer to it as the unramified subgroup of $F(L)$.

REMARK 2.1.2. Suppose that F is a functor to the category of pointed sets. Then there is a similar definition of the unramified subset $F_{nr}(L/K) \subset F(L)$. Most of the statements below have analogues in this context.

LEMMA 2.1.3. (a) *Given fields* $K \subset L$ *in* $\mathscr{C}$, *the natural map* $F(K) \to F(L)$ *induces a map* $F(K) \to F_{nr}(L/K)$.

(b) *Let* $E \subset K \subset L$ *be fields in* $\mathscr{C}$. *The natural map* $F(K) \to F(L)$ *induces a map* $F_{nr}(K/E) \to F_{nr}(L/E)$.

PROOF. The first statement is clear. If $B \subset L$ is a discrete valuation ring with $E \subset B$ and $L = \mathrm{qf}(B)$, then either $K \subset B$ or $A = B \cap K$ is a discrete valuation ring of K. In either case, the image in L of any element of $F_{nr}(K/E) \subset K$ clearly lies in B. (Note that (a) is just a special case of (b).) $\square$

Let F be a functor from $\mathscr{C}$ to Ab. We shall be interested in various properties of such a functor.

DEFINITION 2.1.4. (a) *Injectivity property for a regular local ring* A. Let A be a regular local ring in $\mathscr{C}$, with field of fractions K. Then the map $F(A) \to F(K)$ has trivial kernel.

(b) *Codimension one purity property for a regular local ring* A. Let A be a regular local ring in $\mathscr{C}$, with field of fractions K. Then

$$\mathrm{Im}(F(A) \to F(K)) = \bigcap_{p \text{ height one}} \mathrm{Im}(F(A_p) \to F(K)).$$

(c) *Specialization property for a regular local ring* A. Suppose that $\mathscr{C}$ is either $\mathscr{R}$ or $k - Alg$. Let A be a regular local ring in $\mathscr{C}$, let K be its quotient field and κ its residue field. Then the kernel of $F(A) \to F(K)$ lies in the kernel of $F(A) \to F(\kappa)$.

The specialization property is of course much weaker than the injectivity property.

LEMMA 2.1.5. *Let* F *be a covariant functor from the category* $\mathscr{C}$ *to* Ab.

(a) *If F satisfies the specialization property for all complete discrete valuation rings A, then it satisfies the specialization property for arbitrary regular local rings, in particular for arbitrary discrete valuation rings.*

(b) *If F satisfies the injectivity property for all complete discrete valuation rings, then its satisfies the specialization property for arbitrary regular local rings.*

(c) *Let k be a field, and let A be a k-algebra which is a regular local ring, with quotient field K and residue field κ; assume that the composite map $k \to A \to \kappa$ is an isomorphism. If F satisfies the specialization property, then the map $F(k) \to F(K)$ is injective.*

PROOF. Statement (c) is obvious, and statement (a) implies (b). For the proof of (a), which uses a well known induction argument on the dimension of a regular local ring A, we refer the reader to [CT78, Proposition 2.1], [CT80, Lemme 1.1], [CT/Sa79, 6.6.1]. □

DEFINITION 2.1.6. Let F be a covariant functor from the category $\mathscr{C}$ to Ab. We shall say that F satisfies *field homotopy invariance* if whenever K is a field in $\mathscr{C}$ and $K(t)$ is the field of rational functions in one variable over K, the induced map $F(K) \to F_{nr}(K(t)/K)$ is an isomorphism.

REMARK 2.1.7. Note that there is another kind of homotopy invariance of the functor F, which is of a global nature, and may be taken into consideration. We shall say that the functor F from $\mathscr{C}$ to Ab satisfies *ring homotopy invariance* if for any commutative ring A in $\mathscr{C}$, if $A[t]$ denotes the polynomial ring in one variable over A, the map $F(A) \to F(A[t])$ is an isomorphism. If $A[t_1, \ldots, t_n]$ denotes the polynomial ring in n variables over A, *ring homotopy invariance* immediately implies that the map $F(A) \to F(A[t_1, \ldots, t_n])$ is an isomorphism.

Let F be a functor from $k - Alg$ to Ab. To any integral variety X/k, with field of functions $k(X)$, let us associate the following subgroups of $F(k(X))$:

$$F_1(X) = \{\alpha \in F(k(X)) | \forall P \in X^1, \alpha \in \text{Im}\, F(\mathscr{O}_{X,P})\},$$
$$F_{\text{loc}}(X) = \{\alpha \in F(k(X)) | \forall P \in X, \alpha \in \text{Im}\, F(\mathscr{O}_{X,P})\},$$
$$F_{nr}(k(X)/k) = \{\alpha \in F(k(X)) | \forall k \subset A \subset k(X), A \text{ discrete valuation ring},$$
$$\text{qf}(A) = k(X), \alpha \in \text{Im}\, F(A)\}$$
$$F_{\text{val}}(k(X)/k) = \{\alpha \in F(k(X)) | \forall k \subset A \subset k(X), A \text{ valuation ring},$$
$$\text{qf}(A) = k(X), \alpha \in \text{Im}\, F(A)\}.$$

(In the last definition, A runs through all Krull valuation rings.)

PROPOSITION 2.1.8. *Let F be a functor from $k - Alg$ to Ab. Let X/k be an integral variety, with function field $k(X)$.*

(a) *We have $F_{\text{val}}(k(X)/k) \subset F_{nr}(k(X)/k)$ and $F_{\text{loc}}(X) \subset F_1(X)$.*
(b) *If X is normal, then we have $F_{nr}(k(X)/k) \subset F_1(X)$.*
(c) *If X/k is proper, then we have $F_{\text{loc}}(X) \subset F_{\text{val}}(k(X)/k)$.*

(d) *If X/k is smooth, and the functor F satisfies the (codimension one) purity property for regular local rings, then $F_{\mathrm{loc}}(X) = F_1(X)$.*

(e) *If X/k is smooth and proper, and F satisfies the (codimension one) purity property for regular local rings, then the four subgroups $F_1(X)$, $F_{\mathrm{loc}}(X)$, $F_{nr}(k(X)/k)$ and $F_{\mathrm{val}}(k(X)/k)$ of $F(k(X))$ coincide, and they are all k-birational invariants of smooth, proper, integral k-varieties.*

PROOF. Statement (a) is clear. As for (b), if X is normal, then local rings at points of codimension 1 are discrete valuation rings, hence we have $F_{nr}(k(X)/k) \subset F_1(X)$. Assume that X/k is proper. Then the inclusion $\mathrm{Spec}(k(X)) \to X$ of the generic point extends to a morphism $\mathrm{Spec}(A) \to X$. We thus get inclusions $\mathscr{O}_{X,P} \subset A \subset k(X)$. Now if $\alpha \in F(k(X))$ belongs to $F_{\mathrm{loc}}(X)$, it is in the image of $F(\mathscr{O}_{X,P})$; thus, α comes from $F(A)$. This proves (c). Let us assume that X/k is smooth and that F satisfies the (codimension one) purity property for regular local rings. Then X is normal, hence $F_{\mathrm{loc}}(X) \subset F_1(X)$ as noted above. Conversely, given $\alpha \in F_1(X)$ and $P \in X$, apply the codimension one purity property to $A = \mathscr{O}_{X,P}$ to get $\alpha \in \mathrm{Im}\, F(\mathscr{O}_{X,P})$. This proves (d). Statement (e) gathers the previous results. The k-birational invariance of $F_{nr}(k(X)/k)$ is clear, hence also that of the other groups. $\square$

PROPOSITION 2.1.9. *Let k be a field, and let F be a functor from the category $\mathscr{C}$ of commutative k-algebras to abelian groups. Assume that F satisfies the specialization property for discrete valuation rings and the codimension one purity property for regular local rings A of dimension 2. Assume that for all fields $K \supset k$, the natural map $F(K) \to F_{nr}(K(t)/K)$ is a bijection, where $K(t)$ denotes the function field in one variable over K. Then for all fields $K \supset k$ and all positive integers n, the natural map $F(K) \to F_{nr}(K(t_1, \ldots, t_n)/K)$ is a bijection.*

PROOF. First note that the assumption implies that for any field $K \supset k$, the map $F(K) \to F(K(t))$ is injective. Induction then shows that for any such field K, and any positive integer n, the map $F(K) \to F(K(t_1, \ldots, t_n))$ is injective (injectivity also follows from the specialization property: simply use a local ring at a K-rational point).

We shall prove the theorem by induction on n. The case $n = 1$ holds by assumption. Suppose we have proved the theorem for n. Consider the projection p of $\mathbf{P}_K^1 \times_K \mathbf{P}_K^n$ onto the second factor $\mathbf{P}_K^n$. Note that these varieties are smooth over k, hence their local rings are regular. On function fields, the map p induces an inclusion $E = K(t_1, \ldots, t_n) \subset L = K(t_1, \ldots, t_{n+1})$.

Let $A \in \mathbf{P}_K^1(K)$ be a fixed K-rational point. The map $x \to (A, x)$ defines a section σ of the projection p (i.e. $p \circ \sigma = \mathrm{id}_{\mathbf{P}_K^n}$). Let η denote the generic point of $\mathbf{P}_K^n$.

Let now $\alpha \in F(K(t_1, \ldots, t_{n+1}))$ be an element of $F_{nr}(K(t_1, \ldots, t_{n+1})/K)$. We may view α as an element of $F(L) = F(E(\mathbf{P}_E^1))$. It certainly belongs to $F_{nr}(E(\mathbf{P}^1_E)/E)$. From the hypothesis, we conclude that $\alpha \in F(L)$ is the image of a unique $\beta \in F(E)$ under the inclusion $p^*: F(E) \hookrightarrow F(L)$.

Let $x \in \mathbf{P}^n_K$ be an arbitrary codimension one point. Let $y = \sigma(x) \in \mathbf{P}^1_K \times_K \mathbf{P}^n_K$. This is a codimension 2 point on $\mathbf{P}^1_K \times_K \mathbf{P}^n_K$, which is a specialization of the codimension one point $\omega = \sigma(\eta)$. Let A be the (dimension one) local ring at x, B the (dimension two) local ring at y and C the (dimension one) local ring at ω. The local ring at η is the field $E = K(t_1, \ldots, t_n)$. We have inclusions $A \subset E, B \subset C \subset L$. We also have compatible maps $\sigma^*: F(B) \to F(A)$ and $\sigma^*: F(C) \to F(E)$.

Since F satisfies the codimension one purity assumption for regular local rings of dimension two, and α belongs to $F_{nr}(K(\mathbf{P}^1_K \times_K \mathbf{P}^n_K)/K)$, there exists an element $\gamma \in F(B)$ whose image in $F(L)$ is α. Now the image γ_1 of γ in $F(C)$ under $F(B) \to F(C)$ and the image β_1 of β in $F(C)$ under the map $F(E) \to F(C)$ both restrict to α in $F(L)$. The ring C is a discrete valuation ring, the natural map $C \to E$ from C to its residue field E being given by σ^*.

By the specialization property, we conclude that $\sigma^*(\beta_1) = \sigma^*(\gamma_1) \in F(E)$. But $\sigma^*(\beta_1) = \sigma^*(p^*(\beta)) = \beta$ and $\sigma^*(\gamma_1)$ is the restriction to $F(E)$ of $\sigma^*(\gamma) \in F(A)$. We therefore conclude that $\beta \in F(E)$ lies in the image of $F(A)$. Since A was the local ring of $\mathbf{P}^n_K$ at an arbitrary codimension one point, we conclude that β lies in $F_{nr}(E/K) = F_{nr}(K(t_1, \ldots, t_n)/K)$, and hence by induction comes from $F(K)$. □

Our reason for phrasing the proposition as we have done is that purity in codimension one, for most functors of interest, is a property which is not easy to check. As a matter of fact, for some functors of interest, such as the Witt group, it is not yet known to hold for local rings of smooth varieties of arbitrary dimension over a field. For regular local rings of dimension two, there are specific arguments—many of them relying on the basic fact that a reflexive module over a regular local ring of dimension two is free. A well known case, due to Auslander and Goldman [Au/Go60] is that of the (Azumaya) Brauer group. This was extended by Grothendieck to the cohomological Brauer group ([Gr68]. Another case is that of the set of isomorphism classes of principal homogeneous spaces under a reductive group (see [CT/Sa79, §6], which in turn yields a similar result for the Witt group (cf. [CT/Sa79, §2]).

Suppose we know the specialization property for our given functor F, but that we do not know the purity property in codimension one for regular local rings of dimension 2. In the literature, one finds two other methods to use unramified elements, or some variant, in showing that some function fields are not purely transcendental over their ground field.

One of them uses residue maps and their functorial behaviour. We shall not formalize this proof here. The reader will see it at work in the context of étale cohomology in [CT/Oj89] (see also Proposition 3.3.1, Theorem 4.1.5

and §4.2.4 and 4.2.5 below). It is also used in the Witt group context in [CT/Oj89] and [Oj90].

The other method ([CT78], [CT80]) does not use unramified elements directly. It uses the functor $F_{\text{loc}}(X)$ associated to a smooth integral variety X over a field k.

PROPOSITION 2.1.10 ([CT80], Proposition 1.2). *Let k be a field and let F be a functor from the category $\mathscr{C}$ of commutative k-algebras to abelian groups. If the functor F satisfies the specialization property for discrete valuation rings, hence for regular local rings, then associating to any integral smooth k-variety X the group $F_{\text{loc}}(X)$ defines a contravariant functor on the category of all smooth integral k-varieties, with morphisms arbitrary k-morphisms.*

PROOF. Let $f: X \to Y$ be a k-morphism of integral k-varieties, with Y smooth. Let ξ be the generic point of X, let $P = f(\xi) \in Y$. The quotient field of the local ring $\mathscr{O}_{Y,P}$ is $k(Y)$. Let κ_P be its residue field. Let $\alpha \in F_{\text{loc}}(Y)$. Then α comes from an element α_P of $F(\mathscr{O}_{Y,P})$. Let $\beta = f^*(\alpha_p) \in F(k(X))$. Since F satisfies the specialization property, the element β is well defined. Indeed, even though $\alpha_P \in F(\mathscr{O}_{Y,P})$ may not be uniquely defined, its image γ_P in $F(\kappa_P)$ is uniquely defined by the specialization property for the regular local ring $\mathscr{O}_{Y,P}$, and β is obtained as the image of α_P under the composite map $F(\mathscr{O}_{Y,P}) \to F(\kappa_P) \to F(k(X))$.

Let us show that β belongs to $F_{\text{loc}}(X)$. Let $x \in X$ be an arbitrary point. We then have the commutative diagram of local homomorphisms of local rings, where the horizontal maps are induced by f^*:

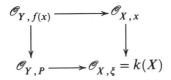

$$
\begin{array}{ccc}
\mathscr{O}_{Y,f(x)} & \longrightarrow & \mathscr{O}_{X,x} \\
\downarrow & & \downarrow \\
\mathscr{O}_{Y,P} & \longrightarrow & \mathscr{O}_{X,\xi} = k(X)
\end{array}
$$

Since α belongs to $F_{\text{loc}}(Y)$, there exists an element $\alpha_{f(x)} \in F(\mathscr{O}_{Y,f(x)})$ with image $\alpha \in F(k(Y))$. The image of that element in $F(\mathscr{O}_{Y,P})$ may differ from α_P, but their images in $F(k(Y))$ coincide. Arguing as above, we find that β is the image of $\alpha_{f(x)}$ in $F(k(Y))$ under the composite map $F(\mathscr{O}_{Y,f(x)}) \to F(\mathscr{O}_{Y,P}) \to F(k(X))$. From the above diagram we conclude that β is the image of $\alpha_{f(x)}$ under the composite map $F(\mathscr{O}_{Y,f(x)}) \to F(\mathscr{O}_{X,x}) \to F(k(X))$, hence that β comes from $F(\mathscr{O}_{X,x})$. Since x was arbitrary, we conclude that β belongs to $F_{\text{loc}}(X)$. We thus have a map $f^*: F_{\text{loc}}(Y) \to F_{\text{loc}}(X)$, under the sole assumption that X and Y are integral and that the specialization property holds for F and the local rings of Y.

It remains to show contravariance, i.e., given morphisms $X \xrightarrow{f} Y \xrightarrow{g} Z$ with Y and Z regular, we need to show that the two maps $(g \circ f)^*$ and $f^* \circ g^*$ from $F_{\text{loc}}(Z)$ to $F_{\text{loc}}(X)$ coincide. To check this, let $\xi \in X$, resp., $\eta \in Y$ be the generic points of X, resp., Y. Let $P = f(\xi) \in Y$,

$Q = g(P) \in Z$, $R = g(\eta) \in Z$. We then have the commutative diagram of local homomorphisms of local rings:

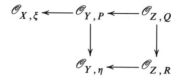

Given $\alpha \in F_{\mathrm{loc}}(Z) \subset F(k(Z))$, we may represent it as an element $\alpha_Q \in F(\mathcal{O}_{Z,Q})$, and this element restricts to an element $\alpha_R \in F(\mathcal{O}_{Z,R})$ with image α in $F(k(Z))$. Applying functoriality of F on rings to the above commutative diagram, we get

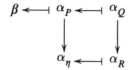

Now $(g \circ f)^*(\alpha) \in F_{\mathrm{loc}}(X)$, by definition, is nothing but $(g \circ f)^*(\alpha_Q) \in F(k(X))$, which by functoriality of F on rings coincides with $f^*(g^*(\alpha_Q)) = f^*(\alpha_P)$. On the other hand, $g^*(\alpha) \in F_{\mathrm{loc}}(Y)$, by definition, is $\alpha_\eta = g^*(\alpha_R) \in F(k(Y))$. Now to compute $f^*(g^*(\alpha))$, we must take a representative of $g^*(\alpha)$ in $F(\mathcal{O}_{Y,P})$ and take its image in $F(\mathcal{O}_{X,\xi}) = F(k(X))$ under f^*. But such a representative is given by α_P, which completes the proof of the equality

$$(g \circ f)^*(\alpha) = f^*(g^*(\alpha)) \in F_{\mathrm{loc}}(Z). \qquad \square$$

REMARK 2.1.11. As the reader will check, all that is needed in the definition of the functor F_{loc}, and in the above proof, is a (covariant) functor F on the category of local rings, where morphisms are local homomorphisms.

PROPOSITION 2.1.12. *Let k be a field, and let X/k be a smooth, proper, integral variety. Assume that the functor F satisfies the specialization property for discrete valuation rings containing k and that for all fields $K \supset k$, the projection $\mathbb{A}_K^1 = \mathrm{Spec}(K[t]) \to \mathrm{Spec}(K)$ induces a bijection $F(K) \xrightarrow{\sim} F_1(\mathbb{A}_K^1) = F_{\mathrm{loc}}(\mathbb{A}_K^1) \subset F(K(t))$. If X is k-birational to projective space, i.e., if the function field $k(X)$ is purely transcendental over k, then the natural map $F(k) \to F_{\mathrm{loc}}(X)$ is an isomorphism.*

PROOF. See [**CT80**, Théorème 1.5]. $\square$

REMARK 2.1.13. If X/k is smooth, Proposition 2.1.10 enables us to evaluate elements of $F_{\mathrm{loc}}(X)$ at k-points (rational points) of X. Let X/k be a smooth, proper, integral variety. If we find $\alpha \in F_{\mathrm{loc}}(X)$ and two k-points A and B such that $\alpha(A) \neq \alpha(B)$ in $F(k)$, then α does not come from $F(k)$, hence X is not k-birational to affine space by Proposition 2.1.12. The reader will find concrete examples in [**CT78**, §5] and [**CT80**, §2.5.2].

The two propositions above may be applied in various contexts ([**CT80**, §2]). However, the reader should keep in mind that application of this technique for a given proper variety X requires checking that a given element of $F(k(X))$ comes from $F(\mathscr{O}_{X,P})$ at every point $P \in X$. This may be much harder than checking such a condition on discrete valuation rings (and it might require working on an explicit smooth proper model...). Of course, if X is smooth and codimension one purity holds, these conditions are equivalent, but purity is also hard to establish.

§2.2. A survey of various functors. In this subsection, we shall briefly survey various functors which may be considered for the formalism above. For each of these functors, we shall mention the known results and open questions regarding the various properties mentioned above: specialization property, injectivity property, (codimension one) purity.

§2.2.1. Etale cohomology with coefficients $\mu_n^{\otimes j}$. This case will be discussed in full detail in §3 and §4. We let A be a ring, let n be a positive integer invertible on A, and we consider the cohomology groups $H_{\text{ét}}^i(A, \mu_n^{\otimes j}) = H_{\text{ét}}^i(\text{Spec}(A), \mu_n^{\otimes j})$ (see §3 for definitions). The injectivity property is known for discrete valuation rings (see §3.6 below), hence also the specialization property.

The injectivity property is known for local rings of smooth varieties over a field k ([**Bl/Og74**], see §3.8 below). The injectivity property for arbitrary regular local rings is known for $H_{\text{ét}}^1(A, \mu_n^{\otimes j})$ (it actually holds for any noetherian normal domain, cf. [**CT/Sa79**, Lemme 2.1]) and for $H_{\text{ét}}^2(A, \mu_n)$ (it is a consequence of the similar result for the Brauer group, see (3.2) and 2.2.2 below). For $i \geq 3$ and arbitrary regular local rings it is an open question, even when $\dim(A) = 2$.

Codimension 1 purity is known for local rings of smooth varieties over a field ([**Bl/Og74**], see §3.8 below; see also Theorem 5.2.7). For arbitrary regular local rings it is known for $H_{\text{ét}}^1(A, \mu_n^{\otimes j})$ (it actually holds for any noetherian normal domain, cf. [**CT/Sa79**, Lemme 2.1]). For regular local rings with $\dim(A) \leq 3$ it is known for $H_{\text{ét}}^2(A, \mu_n)$ (it is a consequence of the similar result for the Brauer group, see (3.2) and 2.2.2 below). For arbitrary regular local rings and $i \geq 3$ it is an open question, even when $\dim(A) = 2$.

The results valid for local rings of smooth varieties over a field also hold for local rings of schemes smooth over a discrete valuation ring (Gillet, unpublished).

§2.2.2. The Brauer group. Let $\text{Br}(A) = H_{\text{ét}}^2(A, \mathbb{G}_m)$ be the étale cohomological Brauer group. For an arbitrary regular domain A, with field of fractions K, the map $\text{Br}(A) \to \text{Br}(K)$ is injective (Auslander-Goldman [**Au/Go60**], Grothendieck [**Gr68**]) (whence $\text{Br}(A)$ is a torsion group). Thus, both injectivity and specialization hold for arbitrary regular local rings .

As for codimension one purity, it is known to hold for $\dim(A) = 2$ (Auslander-Goldman [Au/Go60], [Gr68]) and $\dim(A) \leq 3$ (Gabber [Ga81a]). If A is a local ring of a smooth variety over a field k, purity holds for the prime-to-p torsion of the (torsion group $\mathrm{Br}(A)$ (see §3.8). Gabber has very recently announced a proof of codimension one purity for arbitrary regular local rings.

§2.2.3. The Witt group. If A is a commutative ring with $2 \in A^*$, we let $W(A)$ be the Witt ring of A as defined in [Kn77]. If A is an arbitrary Dedekind domain, and K is its field of fractions, the map $W(A) \to W(K)$ is injective (see [Kn77]). Thus, the specialization property holds. If A is a (not necessarily local) regular domain and K is its quotient field, the map $W(A) \to W(K)$ is known to be injective if $\dim(A) = 2$ (Ojanguren [Oj82a], Pardon [Pa82b]), $\dim(A) = 3$ (Ojanguren [Oj82a] in the local case, Pardon [Pa82b]), $\dim(A) = 4$ in the local case (Pardon [Pa82b]). For dimension at least 4, and A global regular, counterexamples are known (Knus, see [Oj82]). The injectivity property is an open question for general regular local rings in dimension at least 5. If A is a local ring of a smooth variety over a field, the injectivity property is known (Ojanguren [Oj80], see §5.1 and §5.2 below).

Purity in codimension one for the Witt group is known for regular local rings of dimension two ([CT/Sa79]). It is unknown in higher dimension, even in the geometric case (local ring of a smooth variety over a field). It is quite likely that an L-theory version of Quillen's work [Qu73], in particular of his proof of Gersten's conjecture in the geometric case, could be developed (see [Pa82a]).

§2.2.4. The K-theory groups $K_n(A)$. The situation here is rather amazing: the injectivity property for arbitrary discrete valuation rings and $K_n(A)$ for $n \geq 3$ is not known. For K_1, injectivity is trivial. For K_2, it is known for regular local rings of dimension two (Van der Kallen [vdK76]). We refer the reader to Sherman's article [Sh89] for a recent survey.

Merkur'ev points out that specialization holds. It is enough to prove it for a discrete valuation ring A. Let E be the field of fractions of A, let κ be the residue field, and let $t \in A$ be a uniformizing parameter. For any positive integer n, let us consider the composite map $K_n(A) \to K_n(E) \to K_{n+1}(E) \to K_n \kappa$, where the first map is the obvious map, the second one is multiplication by $t \in E^* = K_1(E)$ and where the last map is the residue map. Then this composite map is none other than the reduction map $K_n(A) \to K_n(\kappa)$ (see [Gi86]). It is then clear that any element in the kernel of $K_n(A) \to K_n(E)$ lies in the kernel of $K_n(A) \to K_n(\kappa)$. The argument may be applied to other functors (the Witt group, K-theory with finite coefficients, étale cohomology).

For K-theory, both injectivity and codomension one purity are known for local rings of smooth varieties over a field, as follows from Quillen's proof of the Gersten conjecture ([Qu73]). A different proof of injectivity appears in [CT/Oj92], see §5.2 below.

§2.2.5. The K-theory groups with coefficients $K_n(A, \mathbf{Z}/m)$. The situation here is much better. The injectivity property for discrete valuation rings is known (Gillet [**Gi86**]). Thus, the specialization property certainly holds (alternatively, we could use the same argument as above).

In the geometric case, both injectivity and codimension one purity hold— one only needs to mimic Quillen's argument for K-theory.

§2.2.6. Isomorphism classes of principal homogeneous spaces. Given X a scheme and G/X a smooth affine group scheme, the étale Čech cohomology set $\check{H}^1_{\text{ét}}(X, G)$, denoted $H^1(X, G)$ further below, classifies the set of isomorphism classes of principal homogeneous spaces (torsors) over X under G (cf. [**Mi80**, Chapter III]). This is a pointed set, with distinguished element the class of the trivial torsor G/X.

If A is a regular local ring with quotient field K, and G/A is a reductive group scheme, one may ask whether the map $H^1(A, G) \to H^1(K, G)$ has trivial kernel (injectivity property—in a restricted sense), i.e., whether a torsor which is rationally trivial is trivial over the local ring A. One may ask whether an element of $H^1(K, G)$ which at each prime p of height one of A comes from an element of $H^1(A_p, G)$ actually comes from an element of $H^1(A, G)$.

The injectivity question was already raised by Serre and Grothendieck in the early days of étale cohomology (1958) (cf. [**CT79**]).

For arbitrary discrete valuation rings, injectivity was proved by Nisnevich [**Ni84**]. For split reductive groups, Nisnevich [**Ni89**] also proved injectivity over a regular local ring of dimension two. For local rings of smooth varieties over an infinite perfect field k, and for a reductive group G defined over k, injectivity was established in [**CT/Oj92**] (see §5 below). The case where k is infinite but not necessarily perfect has since been handled by Raghunathan [**Ra93**].

Codimension one purity holds over a regular local ring of dimension two ([**CT/Sa79**]). In higher dimension, it is an open question, even for A a local ring of a smooth k-variety and G a reductive k-group scheme. One special case is known, namely codimension one purity for $H^1(A, \mathbb{SL}(D))$ for A a local ring of a smooth variety over a field k and D/k a central simple algebra. Various proofs are available ([**CT/Pa/Sr89**] in the square free index case, [**Ro90**], [**CT/Oj92**]; another proof will be given in §5.3).

§2.2.7. One more general question. To conclude this section, let us mention a question related to the injectivity property. Let A be a regular local ring, let K be its field of fractions. Let G/A be a reductive group scheme and let X/A be an A-scheme which is a homogeneous space of G. If the set $X(K)$ of K-rational points is nonempty, does it follow that the set of A-points $X(A)$ is nonempty? If A is a discrete valuation ring, and X/A is proper, this is trivially so. But for $\dim(A) \geq 2$ and X/A proper, the answer is far from obvious. One known case is that of Severi-Brauer schemes [**Gr68**].

As a special example of the question, suppose $2 \in A^*$, let $a_i \in A^* (i = 1, \ldots, n)$. Suppose that the quadratic form $\sum_{i=1, \ldots, n} a_i X_i^2$ has a nontrivial zero with coordinates in K^n. Does it have a zero $(\alpha_1, \ldots, \alpha_n) \in A^n$ with at least one of the α_i's a unit? For more on this topic, see [CT79]. The case of a regular local ring of dimension two was handled by Ojanguren [Oj82b].

§3. Étale cohomology

The aim of this section is to recall some basic facts from étale cohomology, including cohomological purity, as proved in [SGA4], then to go on to the Gersten conjecture in the étale cohomological context, as stated and proved by Bloch and Ogus [Bl/Og74]. We shall try to motivate the Gersten conjecture (§3.5). In §3.8 we shall see that the injectivity and codimension one purity theorems for local rings of smooth varieties over a field are immediate consequences of the Gersten conjecture.

§3.1. A few basic properties of étale cohomology.
The reader is referred to [SGA4], to Deligne's introductory lectures [SGA4 1/2, pp. 4–75] or to Milne's book [Mi80] for the definition of étale cohomology. Given a scheme X and an étale sheaf $\mathscr{F}$ on X, the étale cohomology groups $H^i_{\text{ét}}(X, \mathscr{F})$ will often be denoted $H^i(X, \mathscr{F})$. Subscripts will be used when referring to other topologies, e.g., $H^i_{\text{Zar}}(X, \mathscr{F})$. In this section we shall only recall a few basic facts that will enable us to define and study unramified cohomology.

First recall that étale cohomology of an étale sheaf over the spectrum of a field k may be canonically identified with the Galois cohomology of the associated discrete module upon which the absolute Galois group of k acts continuously; exact sequences of étale cohomology correspond to exact sequences in Galois cohomology.

Let X be a scheme. The group scheme $\mathbb{G}_m = \mathbb{G}_{m,X}$ which to any X-scheme Y associated the group $\mathbb{G}_m(Y)$ of units of Y defines a sheaf on X for the flat topology, hence in particular for the étale topology and the Zariski topology.

There is a natural isomorphism

$$\text{Pic}(X) = H^1_{\text{Zar}}(X, \mathbb{G}_m) \simeq H^1_{\text{ét}}(X, \mathbb{G}_m)$$

(this is Grothendieck's version of Hilbert's theorem 90). Recall that given an integral noetherian scheme X one defines the group $\text{Div}(X)$ of Weil divisors as the free group on points of codimension 1, and the Chow group $CH^1(X)$ as the quotient of the group $\text{Div}(X)$ by the (Weil) divisors of rational functions on X. There is a natural map $\text{Pic}(X) = H^1_{\text{Zar}}(X, \mathbb{G}_m) \to CH^1(X)$ which is an isomorphism if X is locally factorial, for instance if X is regular.

In these notes, by Brauer group of X we shall mean Grothendieck's Brauer group, namely

$$\text{Br}(X) = H^2_{\text{ét}}(X, \mathbb{G}_m).$$

The torsion of this group is known to coincide with the Azumaya Brauer group in many cases (and O. Gabber has announced a proof that this is always the case). When X is regular, $\mathrm{Br}(X)$ is a torsion group, more precisely it is a subgroup of the Brauer group $\mathrm{Br}(k(X))$ of the function field of X.

Let n be a positive integer invertible on X. The X-group scheme of nth roots of unity defines a sheaf μ_n for the étale topology on X. Given any positive integer j, we define the étale sheaf $\mu_n^{\otimes j} = \mu_n \otimes \cdots \otimes \mu_n$ as the tensor product of j copies of μ_n. We let $\mu_n^{\otimes 0}$ be the constant sheaf $\mathbf{Z}/n$ on X. If j is a negative integer, we let $\mu_n^{\otimes j} = \mathrm{Hom}(\mu_n^{\otimes(-j)}, \mathbf{Z}/n)$, where the Hom is taken in the category of étale sheaves on X. For *any* integers j and k, we have $\mu_n^{\otimes j} \otimes \mu_n^{\otimes k} = \mu_n^{\otimes(j+k)}$.

The étale cohomology groups which will be of interest to us here will be the étale cohomology groups $H^i_{\text{ét}}(X, \mu_n^{\otimes j})$, along with $\mathrm{Pic}(X)$ and $\mathrm{Br}(X)$.

One basic property of étale cohomology with torsion coefficients (torsion prime to the characteristic) is *homotopy invariance*, namely for any scheme X, any integer n invertible on X, any nonnegative integer i and any integer j, the natural map

$$H^i_{\text{ét}}(X, \mu_n^{\otimes j}) \to H^i_{\text{ét}}(\mathbb{A}^m_X, \mu_n^{\otimes j})$$

induced by the projection of affine m-space $\mathbb{A}^m_X$ onto X is an isomorphism. The reader is referred to [**SGA4**, XV, Corollaire 2.2], or to [**Mi80**, VI, Corollary 4.20, p. 240] (for the definition of acyclicity, see [**Mi80**, p. 232]).

For X a scheme and n a positive integer invertible on X, there is a basic exact sequence of étale sheaves on X, the Kummer sequence

$$1 \to \mu_n \to \mathbb{G}_m \xrightarrow{n} \mathbb{G}_m \to 1,$$

where the maps $\mathbb{G}_m \to \mathbb{G}_m$ send x to x^n. This sequence gives rise to the short exact sequences

(3.1) $\qquad 0 \to H^0(X, \mathbb{G}_m)/H^0(X, \mathbb{G}_m)^n \to H^1_{\text{ét}}(X, \mu_n) \to {}_n\mathrm{Pic}(X) \to 0$

and

(3.2) $\qquad 0 \to \mathrm{Pic}(X)/n\,\mathrm{Pic}(X) \to H^2_{\text{ét}}(X, \mu_n) \to {}_n\mathrm{Br}(X) \to 0.$

Let X and n be as above, and let j be any integer. Let Y be a closed subscheme of X, and let U be the complement of Y in X. There is a long exact sequence of étale cohomology groups
(3.3)
$$\cdots \to H^i(X, \mu_n^{\otimes j}) \to H^i(U, \mu_n^{\otimes j}) \to H^{i+1}_Y(X, \mu_n^{\otimes j}) \to H^{i+1}(X, \mu_n^{\otimes j}) \to \cdots,$$

where $H^n_Y(X, .)$ denotes étale cohomology with support in the closed subset Y.

To any morphism $f \colon V \to X$ one associates the cohomology group $H^n_{f^{-1}(Y)}(V, \mu_n^{\otimes j})$. Letting f run through étale maps, we may sheafify this construction, thus giving rise to étale sheaves $\mathbf{H}^n_Y(\mu_n^{\otimes j})$ on X. These sheaves are actually concentrated on Y.

Another way to define them is as follows. To $\mathfrak{F}$, associate a new sheaf

$$\mathbf{H}^0_Y \mathfrak{F} = \mathrm{Ker}[\mathfrak{F} \to j_* j^* \mathfrak{F}]$$

and take derived functors.

There is a spectral sequence

(3.4) $$E^{pq}_2 = H^p_{\mathrm{\acute{e}t}}(Y, \mathbf{H}^q_Y \mathfrak{F}) \Rightarrow H^{p+q}_Y(X, \mathfrak{F}).$$

§3.2. The cohomological purity conjecture. Let X be a regular scheme, $Y \subset X$ a regular closed subscheme. Assume that $Y \subset X$ is everywhere of codimension $c > 0$. Let $n > 0$ be an integer invertible on X. There is a natural map given by the local fundamental class [SGA4 1/2, Cycle 2.2]:

$$(\mathbf{Z}/n)_Y \to \mathbf{H}^{2c}_Y(\mu^{\otimes c}_{n,X})$$

and similarly

$$\mu^{\otimes j-c}_{n,Y} \to \mathbf{H}^{2c}_Y(\mu^{\otimes j}_{n,X}).$$

In [SGA4, Chapter XVI and XIX], [SGA4 1/2, Cycle], and in [SGA5, I.3.1.4], there is a discussion of the (absolute).

COHOMOLOGICAL PURITY CONJECTURE 3.2.1. *For $i \neq 2c$ and any integer j, we have*

$$\mathbf{H}^i_Y((\mu^{\otimes j}_n)_X) = 0$$

and for $i = 2c$ and any integer j, the map

$$\mu^{\otimes j-c}_{n,Y} \to \mathbf{H}^{2c}_Y((\mu^{\otimes j}_n)_X)$$

is an isomorphism.

Assume that cohomological purity holds for $Y \subset X$ as above. Then the local-to-global spectral sequence degenerates and we get isomorphisms

$$H^i_Y(X, \mu^{\otimes j}_n) \simeq H^{i-2c}(Y, \mu^{\otimes j-c}_n).$$

Let U be the complement of Y in X. In view of spectral sequence (3.4), a consequence of purity would be an exact sequence, often referred to as the Gysin sequence

(3.5) $$\cdots \to H^i(X, \mu^{\otimes j}_n) \to H^i(U, \mu^{\otimes j}_n) \to H^{i+1-2c}(Y, \mu^{\otimes j-c}_n)$$
$$\to H^{i+1}(X, \mu^{\otimes j}_n) \to \cdots,$$

where the Gysin map $H^{i+1-2c}(Y, \mu^{\otimes j-c}_n) \to H^{i+1}(X, \mu^{\otimes j}_n)$ goes "the wrong way".

REMARK 3.2.2. There is a similar discussion in [Gr68, GBIII, §6]. Note however that the varieties Y_i there should at least be assumed to be normal.

REMARK 3.2.3. Very important progress on the absolute cohomological purity conjecture is due to Thomason [Th84] (see his Corollary 3.7).

§3.3. Cohomological purity for discrete valuation rings; the residue map. Let $X = \mathrm{Spec}(A)$, A a discrete valuation ring with residue field κ and quotient field K, let $Y = \mathrm{Spec}(\kappa) \subset X$ be the closed point and $\eta = \mathrm{Spec}(K) \subset X$ be the generic point. Let $n > 0$ be an integer invertible in A. The purity conjecture is known in that case and more generally for a Dedekind scheme. For the proof, see Grothendieck/Illusie, [SGA5, Chapitre I, §6].

The Gysin exact sequence (3.5) here reads

$$(3.6) \quad \cdots \to H^i_{\text{ét}}(A, \mu_n^{\otimes j}) \to H^i_{\text{ét}}(K, \mu_n^{\otimes j}) \xrightarrow{\partial_A} H^{i-1}_{\text{ét}}(\kappa, \mu_n^{\otimes j-1})$$
$$\to H^{i+1}_{\text{ét}}(A, \mu_n^{\otimes j}) \to \cdots.$$

In particular, we have

$$\mathrm{Ker}(\partial_A) = \mathrm{Im}[H^i(A, \mu_n^{\otimes j}) \to H^i(K, \mu_n^{\otimes j})].$$

The map ∂_A, which is called the *residue map*, has a description in terms of Galois cohomology (see the discussion in [SGA5, Chapitre I, pp. 50–52]). We may replace A by its completion and thus assume that A is a complete discrete valuation ring. Let $\overline{K}$ be a separable closure of K and let $K_{nr} \subset \overline{K}$ be the maximal unramified extension of K inside $\overline{K}$. Let $\mathfrak{G} = \mathrm{Gal}(\overline{K}/K)$ be the absolute Galois group, $I = \mathrm{Gal}(\overline{K}/K_{nr})$ be the inertia group and $G = \mathrm{Gal}(K_{nr}/K) = \mathrm{Gal}(\overline{\kappa}/\kappa)$. We then have the Hochschild-Serre spectral sequence for Galois cohomology

$$E_2^{pq} = H^p(G, H^q(I, \mu_n^{\otimes j})) \Rightarrow H^{p+q}(\mathfrak{G}, \mu_n^{\otimes j}).$$

Let p be the characteristic of κ. Now we have an exact sequence

$$1 \to I_p \to I \to \prod_{l \neq p} \mathbf{Z}_l(1) \to 1,$$

where $\mathbf{Z}_l(1)$ is the projective limit over m of the G-modules $\mu_{l^m} = \mu_{l^m}(\overline{K}) = \mu_{l^m}(K_{nr})$. (Here $I_p = 0$ if $p = 0$.) For this, see [Se68]. Also, we have $H^q(I, \mu_n^{\otimes j}) = 0$ for $q \geq 2$, and $H^1(I, \mu_n^{\otimes j}) = \mu_n^{\otimes j-1}$, as a G-module. The above spectral sequence thus gives rise to a long exact sequence

$$\cdots \to H^i(G, \mu_n^{\otimes j}) \to H^i(\mathfrak{G}, \mu_n^{\otimes j}) \to H^{i-1}(G, \mu_n^{\otimes j-1}) \to H^{i+1}(G, \mu_n^{\otimes j}) \to \cdots,$$

hence in particular to a map

$$\partial: H^i(K, \mu_n^{\otimes j}) \to H^{i-1}(\kappa, \mu_n^{\otimes j-1}),$$

and one may check that this map agrees (up to a sign) with the map ∂_A in (3.6).

Equipped with the previous description of the map ∂_A, one proves:

PROPOSITION 3.3.1. *Let $A \subset B$ be an inclusion of discrete valuation rings with associated inclusion $K \subset L$ of their fields of fractions (which need not be a finite extension of fields). Let κ_A and κ_B be their respective residue fields. Let n be an integer invertible in A, hence in B, and let $e = e_{B/A}$ be the* *ramification index of B over A, i.e., the valuation in B of a uniformizing parameter of A. The following diagram commutes:*

$$
\begin{array}{ccc}
H^i(K, \mu_n^{\otimes j}) & \xrightarrow{\ \partial_A\ } & H^{i-1}(\kappa_A, \mu_n^{\otimes j-1}) \\
\Big\downarrow{\scriptstyle \mathrm{Res}_{K,L}} & & \Big\downarrow{\scriptstyle \times e_{B/A}\cdot\,\mathrm{Res}\,\kappa_A,\kappa_B} \\
H^i(L, \mu_n^{\otimes j}) & \xrightarrow{\ \partial_B\ } & H^{i-1}(\kappa_B, \mu_n^{\otimes j-1})
\end{array}
$$

REMARK 3.3.2. Let A be a discrete valuation ring with quotient field K and residue field κ, with $\mathrm{char}(\kappa) = 0$. There also exist residue maps

$$\partial_A \colon \mathrm{Br}(K) \to H^1(\kappa, \mathbf{Q}/\mathbf{Z}).$$

Such maps may be defined in one of two ways.

The first one is by applying the definition above, since $\mathrm{Br}(K)$ is the union of all $H^2(K, \mu_n)$ and $H^1(\kappa, \mathbf{Q}/\mathbf{Z})$ is the union of all $H^1(\kappa, \mathbf{Z}/n)$.

The other one, which has the advantage of being defined under the sole assumption that the residue field κ is perfect, proceeds as follows. First of all, we may assume that A is complete (henselian would be enough). Let K_{nr} be the maximal unramified extension of K. One first shows $\mathrm{Br}(K_{nr}) = 0$, hence $\mathrm{Br}(K) = \mathrm{Ker}[\mathrm{Br}(K) \to \mathrm{Br}(K_{nr})] = H^2(G, K_{nr}^*)$, where $G = \mathrm{Gal}(\kappa_s/\kappa) = \mathrm{Gal}(K_{nr}/K)$. One now has the map $H^2(G, K_{nr}^*) \to H^2(G, \mathbf{Z})$ given by the valuation map. Finally, the group $H^2(G, \mathbf{Z})$ is identified with $H^1(G, \mathbf{Q}/\mathbf{Z}) = H^1(\kappa, \mathbf{Q}/\mathbf{Z})$.

Serre (unpublished) has checked that these two residue maps are actually the opposite of each other.

There is another way to define the residue map

$$\partial_A \colon \mathrm{Br}(K) \to H^1(\kappa, \mathbf{Q}/\mathbf{Z})$$

when the residue field is perfect. This appears in Grothendieck's paper [**GB III**, §2]. Comparing these various residue maps might be cumbersome. In many cases, the only thing of interest is that the kernel of each residue map $\partial_A \colon \mathrm{Br}(K) \to H^1(\kappa, \mathbf{Q}/\mathbf{Z})$ coincides with the image of $\mathrm{Br}(A) \to \mathrm{Br}(K)$.

To complete the picture, it should be noted that ring theorists have a way of their own to define the residue map $\mathrm{Br}(K) \to H^1(\kappa, \mathbf{Q}/\mathbf{Z})$ (K being complete and κ perfect). For this, see [**Dr/Kn80**].

§3.4. Cohomological purity for smooth varieties over a field.

For $Y \subset X$ smooth varieties over a field, purity is established by Artin in [**SGA4**, Chapitre XVI] (see 3.6, 3.7, 3.8, 3.9, 3.10).

THEOREM 3.4.1. *Let k be a field, let $Y \subset X$ be smooth k-varieties with Y closed in X of pure codimension c. Let $n > 0$ be prime to the characteristic of k. Then cohomological purity holds for $\mu_n^{\otimes j}$. In particular, we have a long* *exact sequence*

$$\cdots \to H^i(X, \mu_n^{\otimes j}) \to H^i(X - Y, \mu_n^{\otimes j})$$
$$\to H^{i+1-2c}(Y, \mu_n^{\otimes j-c}) \to H^{i+1}(X, \mu_n^{\otimes j}) \to \cdots.$$

This has the following corollary:

COROLLARY 3.4.2. *Let X/k be smooth and irreducible over a field k. Let $F \subset X$ be any closed subvariety (not necessarily smooth), $\mathrm{codim}_X F \geq c$. Then the restriction maps*

$$H_{\text{ét}}^i(X, \mu_n^{\otimes j}) \to H_{\text{ét}}^i(X - F, \mu_n^{\otimes j})$$

are injective for $i < 2c$ and isomorphisms for $i < 2c - 1$.

PROOF. Suppose first that k is a perfect field. Fix i. Let us prove that the restriction map is surjective for $c > (i + 1)/2$. The proof is by descending induction on the codimension of F in X, subject to the condition $c > (i + 1)/2$. Assume it has been proved for $c + 1$. There exists a closed subset $F_1 \subset F$ of codimension at least $c + 1$ in X such that $F - F_1$ is regular, hence smooth over the perfect field k, and of pure codimension c in $X - F_1$ (the closed set F_1 is built out of the components of F of dimension strictly smaller than the dimension of F and of the singular locus of F). From $i + 1 - 2c < 0$ and the theorem, we conclude that the restriction map $H_{\text{ét}}^i(X - F_1, \mu_n^{\otimes j}) \to H_{\text{ét}}^i(X - F, \mu_n^{\otimes j})$ is surjective. The induction assumption now implies that the restriction map $H_{\text{ét}}^i(X, \mu_n^{\otimes j}) \to H_{\text{ét}}^i(X - F, \mu_n^{\otimes j})$ is surjective, hence $H_{\text{ét}}^i(X, \mu_n^{\otimes j}) \to H_{\text{ét}}^i(X - F, \mu_n^{\otimes j})$ is surjective. That the restriction map is injective for $c > i/2$ is proved in a similar manner. The proof will be left to the reader. For the induction to work we need to be over a perfect field, to ensure that regular schemes are smooth. But since étale cohomology is invariant under purely inseparable extensions ([SGA4, VIII 1.1], [Mi80, Remark 3.17, p. 77]), the corollary holds over all fields. □

Let X be a smooth integral k-variety. For $c = 1$, Corollary 3.4.2 says that for a nonempty open set $U \subset X$, the restriction maps on sections $H_{\text{ét}}^0(X, \mu_n^{\otimes j}) \to H_{\text{ét}}^0(U, \mu_n^{\otimes j})$ are isomorphisms, which is nearly obvious, and that the maps $H_{\text{ét}}^1(X, \mu_n^{\otimes j}) \to H_{\text{ét}}^1(U, \mu_n^{\otimes j})$ are injective.

For $c = 2$, Corollary 3.4.2 says that if F is a closed subset of codimension at least 2, the restriction maps $H_{\text{ét}}^i(X, \mu_n^{\otimes j}) \to H_{\text{ét}}^i(X - F, \mu_n^{\otimes j})$ are injective for $i \leq 3$ and bijective for $i \leq 2$. Using purity for discrete valuation rings, commutativity of étale cohomology with direct limits ([Mi80, III. 3.17, p. 119]), and the Mayer-Vietoris sequence ([Mi80, III. 2.24, p. 110]), from the result just proved one may deduce the existence of exact sequences

$$(3.7) \qquad 0 \to H^1(X, \mu_n^{\otimes j}) \to H^1(k(X), \mu_n^{\otimes j}) \to \bigoplus_{x \in X^{(1)}} H^0(k(x), \mu_n^{\otimes j-1})$$

and

$$(3.8) \qquad H^2(X, \mu_n^{\otimes j}) \to H^2(k(X), \mu_n^{\otimes j}) \to \bigoplus_{x \in X^{(1)}} H^1(k(x), \mu_n^{\otimes j-1}).$$

(A similar argument, with more details, will appear in the proof of Theorem 3.8.2 below.)

From the Kummer sequence we get a functorial surjection $H^2(X, \mu_n) \to {}_n\mathrm{Br}(X)$ (see (3.2)). The map $H^2_{\text{ét}}(X, \mu_n) \to H^2_{\text{ét}}(X - F, \mu_n)$ is bijective by Corollary 3.4.2. Hence if X/k is a smooth integral variety over a field k of characteristic zero, and F is a closed subset of codimension at least 2, the restriction map $\mathrm{Br}(X) \to \mathrm{Br}(X - F)$ is surjective. It is therefore an isomorphism, since over any regular integral scheme X, for any nonempty open set U, the restriction map $\mathrm{Br}(X) \to \mathrm{Br}(U)$ is injective [**GB II**, 1.10]. All in all, for X a smooth variety over a field k and $\mathrm{char}(k) = 0$, there is an exact sequence

$$(3.9) \qquad 0 \to \mathrm{Br}(X) \to \mathrm{Br}(k(X)) \to \bigoplus_{x \in X^{(1)}} H^1(k(x), \mathbf{Q}/\mathbf{Z}).$$

If $\mathrm{char}(k) = p \neq 0$, then for any prime $l \neq p$, we have a similar exact sequence for the l-primary torsion subgroups of the groups above.

For applications of purity to the Brauer group, see [**Fo89**] and [**Fo92**].

§**3.5. The Gersten conjecture.** In general, for $\mathrm{codim}(F) \geq 2$, the restriction map $H^3_{\text{ét}}(X, \mu_n^{\otimes j}) \to H^3_{\text{ét}}(X - F, \mu_n^{\otimes j})$ need not be surjective. Here is the simplest example. Let $k = \mathbf{C}$, let X be a smooth, irreducible, affine surface, and let P be a closed point of X. Since X is affine over an algebraically closed field, one basic theorem of étale cohomology [**Mi80**, VI.7.2, p. 253] ensures that $H^i_{\text{ét}}(X, \mu_n^{\otimes j}) = 0$ for $i > 2 = \dim(X)$. The Gysin sequence (3.5) for the pair (X, P) and $j = 2$ simply reads

$$0 = H^3(X, \mu_n^{\otimes 2}) \to H^3(X - P, \mu_n^{\otimes 2}) \to H^0(P, \mathbf{Z}/n) \to H^4(X, \mu_n^{\otimes 2}) = 0$$

and since $H^0(P, \mathbf{Z}/n) = \mathbf{Z}/n$, clearly the map $H^3(X, \mu_n^{\otimes 2}) \to H^3(X - P, \mu_n^{\otimes 2})$ is not surjective. The same example could be given with $X = \mathrm{Spec}(A)$ and A the local ring of X at P.

Let k be an arbitrary field, let n be prime to the characteristic of k. Let X/k be a smooth irreducible surface over k, let P be a k-point, and let $Y \subset X$ be a smooth irreducible curve going through P. We then have the

following diagram

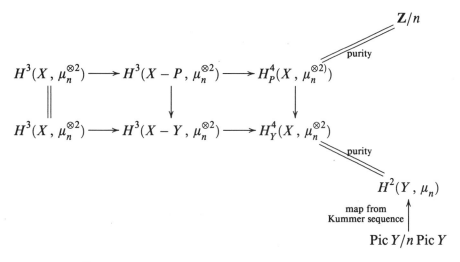

Let $\alpha \in H^2_{\text{ét}}(Y, \mu_n)$ be the image of $1 \in \mathbf{Z}/n$ under the composite map

$$\mathbf{Z}/n \simeq H^4_P(X, \mu_n^{\otimes 2}) \to H^4_Y(X, \mu_n^{\otimes 2}) \simeq H^2_{\text{ét}}(Y, \mu_n).$$

The point P defines a divisor on the curve Y. Let $[P] \in \operatorname{Pic}(Y)$ be its class. Checking through the definitions [**SGA4 1/2**, Cycle], one sees that the image of $[P]$ under the map $\operatorname{Pic}(Y) \to H^2_{\text{ét}}(Y, \mu_n)$ deduced from the Kummer sequence is none other than α.

On the smooth irreducible curve Y, we may find a rational function $f \in k(Y)^*$ with divisor $\operatorname{div}_Y(f) = P + \sum_{i=1,\dots,n} m_i P_i$, where the P_i are closed points of Y, distinct from P, and the m_i are integers. If we now define X_1 to be X with the closed points P_i deleted, and Y_1 to be Y with these closed points deleted, we find that the map

$$H^4_P(X, \mu_n^{\otimes 2}) = H^4_P(X_1, \mu_n^{\otimes 2}) \to H^4_{Y_1}(X_1, \mu_n^{\otimes 2})$$

is zero—the first equality follows from the excision property for cohomology with support [**Mi80**, III. 1.27, p. 92] or [**CT/Oj92**, Proposition 4.4].

In particular, this proves that any element of $H^3(X - P, \mu_n^{\otimes 2})$, once restricted to $H^3(X_1 - Y, \mu_n^{\otimes 2})$, comes from $H^3(X_1, \mu_n^{\otimes 2})$.

Passing over to the cohomology of the local ring of X at P, one gets the following perhaps more striking statement. Let $X = \operatorname{Spec}(A)$ be the local ring of a k-surface at a smooth k-point P. Let $Y = \operatorname{Spec}(A/f) \subset X$ be defined by a regular parameter $f \in A$. An element $\gamma \in H^3(X - P, \mu_n^{\otimes 2})$ need not lift to an element in $H^3(X, \mu_n^{\otimes 2})$. However, its image in $H^3(X - Y, \mu_n^{\otimes 2})$ does lift to an element of $H^3(X, \mu_n^{\otimes 2})$.

What the Gersten conjecture postulates is that what we have just observed is a general phenomenon.

Let X be a regular noetherian scheme of finite Krull dimension. Following Bloch and Ogus [Bl/Og74], we shall say that the Gersten conjecture holds for étale cohomology over X if the following key property holds:

LOCAL ACYCLICITY (Gersten's conjecture for étale cohomology). *Let $n > 0$ be an integer invertible on X and let j be an integer. Let S be a finite set of points in an affine open set of X. Let $Y \subset X$ be a closed subset of codimension at least $p + 1$ in X. Then there exists a closed subset $Z \subset X$, with $Y \subset Z$ and $\mathrm{codim}(Z) = p$ and an open set $U \subset X$ containing S such that the composite map*

$$H_Y^i(X, \mu_n^{\otimes j}) \to H_Z^i(X, \mu_n^{\otimes j}) \to H_{Z \cap U}^i(U, \mu_n^{\otimes j})$$

is zero.

§3.6. The Gersten conjecture for étale cohomology: discrete valuation rings.

Let $X = \mathrm{Spec}(A)$, A a discrete valuation ring with residue field κ and quotient field K, let $Y = \mathrm{Spec}(\kappa) \subset X$ be the closed point and $\eta = \mathrm{Spec}(K) \subset X$ be the generic point. Let $n > 0$ be an integer invertible in A.

The Gersten conjecture in that case simply asserts that all maps $H_Y^i(X, \mu_n^{\otimes j}) \to H_{\text{ét}}^i(X, \mu_n^{\otimes j})$ are zero, in other words, in view of purity, that the maps $H_{\text{ét}}^{i-1}(\kappa, \mu_n^{\otimes(j-1)}) \to H_{\text{ét}}^{i+1}(A, \mu_n^{\otimes j})$ in the long exact sequence (3.6) are all zero, which in turn is equivalent to saying that this long exact sequence breaks up into short exact sequences

$$(3.10) \qquad 0 \to H_{\text{ét}}^i(A, \mu_n^{\otimes j}) \to H_{\text{ét}}^i(K, \mu_n^{\otimes j}) \xrightarrow{\partial_A} H_{\text{ét}}^{i-1}(\kappa, \mu_n^{\otimes(j-1)}) \to 0.$$

This conjecture is known. Although it does not seem to be written explicitly in the literature, the reader will have no difficulty to write down a proof by copying the proof given by Gillet [Gi86] for the analogous statement for K-theory with coefficients. The two key ingredients are the *cup-product* and the *norm map* (also called transfer, or corestriction) on cohomology for finite flat extensions of schemes.

On the discrete valuation ring A, the Kummer sequence gives rise to isomorphisms $H^2(A, \mu_n) \simeq_n \mathrm{Br}(A)$ and $H^2(K, \mu_n) \simeq_n \mathrm{Br}(K)$. Provided we restrict attention to torsion prime to the residue characteristic (denoted by a dash), we thus have an exact sequence

$$(3.11) \qquad 0 \to \mathrm{Br}(A)' \to \mathrm{Br}(K)' \to H'(\kappa, \mathbf{Q}/\mathbf{Z})' \to 0.$$

This exact sequence was first discussed by Auslander and Brumer [Au/Br68].

§3.7. The Gersten conjecture for smooth varieties over a field: results of Bloch and Ogus and some recent developments.

In 1974, inspired by the work of Gersten and Quillen [Qu73] in algebraic K-theory, Bloch and Ogus stated the analogue of the Gersten conjecture—initially stated by Gersten in the K-theory context—for étale cohomology of smooth varieties over a field. They proved the local acyclicity result, as stated at the end of §3.5, for varieties X smooth over an infinite field [Bl/Og74, §5].

They actually worked in the broader set-up of a Poincaré duality theory with support. The proofs in [Bl/Og74] are by no means easy to follow, if only because they rely on the rather formidable machinery of Grothendieck's duality theory, and on étale homology theory.

As far as the geometry is concerned, their proof relied on Quillen's (semilocal) presentation of hypersurfaces of smooth varieties over an infinite field [Qu73]. Let us recall here what this presentation is. Given $X = \mathrm{Spec}(A)$ a smooth, integral, affine variety of dimension d over a field k, a finite set $S \subset X$ of points of X and $Y = \mathrm{Spec}(A/f) \subset X$ a hypersurface in X, after suitable shrinking of X around S, Quillen produces a projection $X \to \mathbb{A}_k^{d-1}$ such that the induced map $Y \to \mathbb{A}_k^{d-1}$ is finite and étale at the points of $S \cap Y$. As in Quillen's proof, Bloch and Ogus then consider the fibre product of $X \to \mathbb{A}_k^{d-1}$ and $Y \to \mathbb{A}_k^{d-1}$ over $\mathbb{A}_k^{d-1}$.

Since 1974, there have been various efforts to obtain a better understanding of the Gersten conjecture in the étale cohomological context and in other contexts. Let us here mention the paper of Gabber [Ga81b] and various talks by Gillet around 1985. Gillet advocated an approach that would only use the basic properties of étale cohomology with support (functoriality, excision, purity), but would not rely on Grothendieck's duality theory nor on étale homology. This approach was taken anew by Srinivas in a lecture at the Tata Institute in 1992. Gillet and Srinivas both used Quillen's presentation of hypersurfaces of smooth varieties over an infinite field.

On the other hand, a different geometric local presentation of subvarieties of smooth varieties over an infinite field was introduced around 1980.

Motivated by a quadratic version of the Gersten conjecture in K-theory, Ojanguren ([Oj80], [Kn91]) proved the following geometric lemma. Given $X = \mathrm{Spec}(A)$ a smooth, integral, affine variety of dimension d over an infinite field k, a closed point $P \in X$ and $Y = \mathrm{Spec}(A/f) \subset X$ a hypersurface in X, after suitable shrinking of X around P, one may find an étale map $p : \mathrm{Spec}(A) \to \mathbb{A}_k^d$ and a nonzero element $g \in k[t_1, \ldots, t_d]$ such that the induced map $k[t_1, \ldots, t_d]/g \to A/f$ is an isomorphism after localization at $Q = f(P)$. That approach was again used in [CT/Oj92] to get some analogues of the Gersten conjecture in a noncommutative context (principal homogeneous spaces under linear algebraic groups, over a base which is smooth over a field).

But let us go back to the early eighties. Independently of Ojanguren, in his I.H.E.S. preprint [Ga81b], Gabber proved a preparation lemma (op. cit., Lemma 3.1) which covered Ojanguren's result. This preparation lemma was used again by Gros and Suwa [Gr/Su88, §2]. In his talk at Santa Barbara in July 1992, Hoobler sketched a proof of Gersten's conjecture in the geometric case (the Bloch-Ogus result) based on this presentation of geometric semilocal rings on the one hand, on cohomology with support (and such basic property as homotopy invariance) on the other hand. This approach seems to be the nicest one to date, and it gives some additional properties (universal exactness, see already [**Ga81b**]) which might not have been so easy to prove along the original lines of [**Bl/Og74**]. The details of this approach are presently being written down [**CT/Ho/Ka94**]. Some simple cases of this approach will be discussed in §5 below.

An amusing aspect of this historical sketch is that not until June 1993 did I realize that Ojanguren's presentation lemma and Gabber's presentation were so close to each other.

To close this discussion of the Gersten conjecture, let us point out that in his 1985 talks, Gillet had announced a proof of the Gersten conjecture for étale cohomology (local acyclicity lemma) for schemes smooth over a Dedekind ring.

§3.8. Injectivity property and codimension one purity property.

THEOREM 3.8.1 (injectivity property). *Let* $i > 0$, $n > 0$, *and* j *be integers. Let* k *be a field of characteristic prime to* n. *If* A *is a semilocal ring of a smooth integral* k-*variety* X, *and* $K = k(X)$ *is the quotient field of* A, *then the natural map*

$$H^i(A, \mu_n^{\otimes j}) \to H^i(K, \mu_n^{\otimes j})$$

is injective.

PROOF. Let us first assume that k is infinite. We may assume that X is affine. Let S be the finite set of points of X with associated semilocal ring A. Let $\alpha \in H^i(A, \mu_n^{\otimes j})$ be in the kernel of the above map. Using the commutativity of étale cohomology with direct limits of commutative rings with affine flat transition homomorphisms, after shrinking X we may assume that α is represented by a class $\beta \in H^i(X, \mu_n^{\otimes j})$ which vanishes in $H^i(U, \mu_n^{\otimes j})$ for $U \subset X$ a suitable nonempty open set. Let Y be the closed set which is the complement of U in X. We have the exact sequence of cohomology with support

$$H_Y^i(X, \mu_n^{\otimes j}) \to H^i(X, \mu_n^{\otimes j}) \to H^i(U, \mu_n^{\otimes j}),$$

hence β comes from $\gamma \in H_Y^i(X, \mu_n^{\otimes j})$. Now the local acyclicity theorem ensures that there exists a closed set Z containing Y and a nonempty open set $V \subset X$ containing S such that the composite map

$$H_Y^i(X, \mu_n^{\otimes j}) \to H_Z^i(X, \mu_n^{\otimes j}) \to H_{Z \cap V}^i(V, \mu_n^{\otimes j})$$

is zero. By functoriality, this implies that the image of γ, hence also of β in $H^i(V, \mu_n^{\otimes j})$, vanishes. Thus $\alpha = 0$. The proof is therefore complete when k is infinite. If k is finite, one may find two infinite extensions of k of coprime pro-order. If $\alpha \in H^i(A, \mu_n^{\otimes j})$ is as above, then by the commuting property of étale cohomology just mentioned, one finds two finite extensions k_1/k and k_2/k of coprime degrees, such that α vanishes in $H^i(A \otimes_k k_r, \mu_n^{\otimes j})$

for $r = 1, 2$. A standard argument using traces (transfers) then shows $\alpha = 0 \in H^i(A, \mu_n^{\otimes j})$. $\square$

THEOREM 3.8.2 (codimension one purity theorem). *Let $i > 0$, $n > 0$, and j be integers. Let k be a field of characteristic prime to n. Let A be a semilocal ring of a smooth integral k-variety X, with quotient field $K = k(X)$, and let α be an element of $H^i(K, \mu_n^{\otimes j})$. If for each height one prime p of A, α belongs to the image of $H^i(A_p, \mu_n^{\otimes j})$, then α comes from a (unique) element of $H^i(A, \mu_n^{\otimes j})$.*

PROOF. Unicity of the lift in $H^i(A, \mu_n^{\otimes j})$ follows from the previous theorem. Let us first assume that k is infinite. Let $S \subset X$ be the finite set of points corresponding to the maximal ideals of A.

We first claim that there exists an open set U of X which contains the generic points of all the codimension one irreducible closed subvarieties of X going through a point of S and is such that the class α comes from $\beta \in H^i(U, \mu_n^{\otimes j})$. Since étale cohomology commutes with filtering projective limits of schemes with flat affine transition morphisms, we may find an open set U of X such that $\alpha \in H^i(K, \mu_n^{\otimes j})$ comes from $\alpha_U \in H^i(U, \mu_n^{\otimes j})$. Assume that U does not contain the generic point P of a codimension one irreducible closed subvariety of X going through a point of S. By assumption α comes from some class $\alpha_P \in H^i(A_P, \mu_n^{\otimes j})$, which we may extend to a class in $H^i(V, \mu_n^{\otimes j})$ for V some open set of X containing P. We would like α_U and α_V to agree on the overlap $V \cap U$. But (P being fixed) $\lim_{P \in V}(U \cap V) = \operatorname{Spec} k(X)$ and α_U and α_V agree when restricted to $\operatorname{Spec} k(X)$, hence they agree on some $U \cap V$ with V small enough. From the Mayer-Vietoris sequence in étale cohomology [**Mi80**, III. 2.24, p. 110], we conclude that α comes from $H^i(U \cup V, \mu_n^{\otimes j})$.

Shrinking X around S, we may further assume that the closed set $F = X - U$ is of codimension at least 2 in X. In the case $i \leq 2$, cohomological purity ensures that the restriction map $H^i(X, \mu_n^{\otimes j}) \to H^i(U, \mu_n^{\otimes j})$ is surjective (§3.4), and the theorem follows. Suppose $i \geq 3$. By the local acyclicity theorem (§3.6 and §3.7) there exists a closed set F_1 containing F, with $\operatorname{codim}(F_1) \geq 1$, and an open set $V \subset X$ containing S such that the composite vertical map on the right hand side of the following diagram is zero:

$$
\begin{array}{ccccc}
H^i(X, \mu_n^{\otimes j}) & \longrightarrow & H^i(X - F, \mu_n^{\otimes j}) & \longrightarrow & H_F^i(X, \mu_n^{\otimes j}) \\
\downarrow & & \downarrow & & \downarrow \\
H^i(X, \mu_n^{\otimes j}) & \longrightarrow & H^i(X - F_1, \mu_n^{\otimes j}) & \longrightarrow & H_{F_1}^i(X, \mu_n^{\otimes j}) \\
\downarrow & & \downarrow & & \downarrow \\
H^i(V, \mu_n^{\otimes j}) & \longrightarrow & H^i(V - F_1, \mu_n^{\otimes j}) & \longrightarrow & H_{F_1 \cap V}^i(V, \mu_n^{\otimes j})
\end{array}
$$

Now the image of β in $H^i(V - F_1, \mu_n^{\otimes j})$ comes from $H^i(V, \mu_n^{\otimes j})$, and the theorem is proved when k is infinite. For k, the result follows as in 3.8.1. $\square$

There are Brauer group versions of the above results, which we gather in one result.

THEOREM 3.8.3. *Let k be a field of characteristic zero. Let A be a semilocal ring of a smooth integral k-variety X with quotient field $K = k(X)$, and let α be an element of $\mathrm{Br}(K)$. If for each prime p of height one of A, the element α belongs to the image of $\mathrm{Br}(A_p)$, then α comes from a (unique) element of $\mathrm{Br}(A)$. There is an exact sequence*

$$0 \to \mathrm{Br}(A) \to \mathrm{Br}(K) \to \bigoplus_{p \in A^{(1)}} H^1(\kappa_p, \mathbf{Q}/\mathbf{Z}),$$

where the last map is the direct sum of the residue maps $\partial_{A_p} : \mathrm{Br}(K) \to H^1(\kappa_p, \mathbf{Q}/\mathbf{Z})$.

PROOF. It is enough to note that for any positive n (here n prime to the characteristic of k would be enough), and A as above, in particular local, the Kummer exact sequence gives rise to compatible isomorphisms $H^2(A, \mu_n) \simeq {}_n\mathrm{Br}(A)$ and $H^2(K, \mu_n) \simeq {}_n\mathrm{Br}(K)$ (see (3.2)). The first result now follows from the previous results, and the exact sequence then follows from an application of (3.6) to each discrete valuation ring A_p. $\square$

REMARK 3.8.4. Except for the injectivity of the Brauer group of A into the Brauer group of K—which actually holds for any regular integral domain—the previous theorem is nothing but the semilocal version of sequence (3.9). Indeed, at the level of H^2, only the purity theorem comes into play. It is only from H^3 onwards that one must appeal to the local acyclicity theorem (Gersten's conjecture).

§4. Unramified cohomology

§4.1. Equivalent definitions of unramified cohomology, homotopy invariance.

THEOREM 4.1.1. *Let k be a field, X/k a smooth integral k-variety and $n > 0$ be an integer prime to $\mathrm{char}(k)$. Let $k(X)$ be the function field of X. Let i and j be integers. The following subgroups of $H^i(k(X), \mu_n^{\otimes j})$ coincide:*

(a) *the group of elements $\alpha \in H^i(k(X), \mu_n^{\otimes j})$ which, at any point P of codimension 1 in X with local ring $\mathcal{O}_{X,P}$ and residue field κ_P, come from a class in $H^i(\mathcal{O}_{X,P}, \mu_n^{\otimes j})$, or equivalently which satisfy $\partial_{\mathcal{O}_{X,P}}(\alpha) = 0 \in H^{i-1}(\kappa_P, \mu_n^{\otimes j-1})$;*

(b) *the group of elements of $H^i(k(X), \mu_n^{\otimes j})$ which at any point $P \in X$ come from a class in $H^i(\mathcal{O}_{X,P}, \mu_n^{\otimes j})$;*

(c) *the group $H^0(X, \mathcal{H}^i(\mu_n^{\otimes j}))$ of global sections of the Zariski sheaf $\mathcal{H}^i(\mu_n^{\otimes j})$, which is the sheaf associated to the Zariski presheaf $U \mapsto H^i_{\text{ét}}(U, \mu_n^{\otimes j})$;*

(d) *if* X/k *is complete, the group* $H_{nr}^i(k(X)/k, \mu_n^{\otimes j})$ *consisting of elements* α *of* $H^i(k(X), \mu_n^{\otimes j})$ *such that for any discrete valuation ring* $A \subset k(X)$ *with* $k \subset A$ *and with field of fractions* $k(X)$, *the element* α *comes from* $H^i(A, \mu_n^{\otimes j})$, *or equivalently the residue* $\partial_A(\alpha) = 0 \in H^{i=1}(\kappa_A, \mu_n^{\otimes(j-1)})$;

(e) *if* X/k *is complete, the group of elements* $\alpha \in H^i(k(X), \mu_n^{\otimes j})$ *such that for any valuation ring* (*not necessarily discrete*) A *of* $k(X)$ *containing* k, α *comes from* $H^i(A, \mu_n^{\otimes j})$.

PROOF. The proof follows from the codimension one purity property for local rings of X (Theorem 3.8.2) and from the general argument given in Proposition 2.1.8. (Proposition 2.1.8 assumed the codimension one purity property for all regular local rings, but only used it for those of geometric origin.) The characterisation of "unramified classes" by means of residues at discrete valuation rings comes from cohomological purity for such rings (§3.3).

REMARK 4.1.2. In the notation of §2, the group appearing in (b), is none other than the group $F_{\mathrm{loc}}(X)$ associated to the functor $F(A) = H^i(A, \mu_n^{\otimes j})$. According to the above theorem, the group in (b) coincides with the group $H^0(X, \mathscr{H}^i(\mu_n^{\otimes j}))$ in (c). The functor $X \mapsto H^0(X, \mathscr{H}^i(\mu_n^{\otimes j}))$ is a natural contravariant functor on the category of all k-varieties. Restricting ourselves to the category of smooth integral k-varieties, we thus get another proof of Proposition 2.1.10 for the functor $F(A) = H^i(A, \mu_n^{\otimes j})$. Note however that the proof of Proposition 2.1.10 only used the specialization property, whereas the present one uses the codimension one purity theorem, hence the Gersten conjecture.

REMARK 4.1.3. The k-birational invariance of the groups $H^0(X, \mathscr{H}^i(\mu_n^{\otimes j}))$ on smooth, proper, integral varieties, was also observed (in zero characteristic) by Barbieri-Viale [BV92a]. M. Rost tells me that the k-birational invariance of the group of elements in Theorem 4.1.1(a) can be proved in a "field-theoretic" fashion, without using the Gersten conjecture (see [Ro93]).

PROPOSITION 4.1.4. *Let* k *be a field and* $n > 0$ *be an integer prime to* char.(k). *Let* i, j *and* $m > 0$ *be integers. Let* $K = k(t)$ *be the rational field in one variable over* k. *Then the natural map* $H^i(k, \mu_n^{\otimes j}) \to H^i(k(t), \mu_n^{\otimes j})$ *induces a bijection* $H^i(k, \mu_n^{\otimes j}) \simeq H_{nr}^i(k(t)/k, \mu_n^{\otimes j})$.

PROOF. First assume that k is perfect. For the affine line $\mathbb{A}_k^1$, and for a closed, reduced, proper subset $F \subset \mathbb{A}_k^1$, we may write the long exact sequence of cohomology with support and use purity (§3.3 or §3.4) to translate it as

$$\cdots \to H^i(\mathbb{A}_k^1, \mu_n^{\otimes j}) \to H^i(\mathbb{A}_k^1 - F, \mu_n^{\otimes j}) \to \bigoplus_{P \in F} H^{i-1}(k(P), \mu_n^{\otimes j-1}) \to \cdots.$$

Now homotopy invariance for étale cohomology (see §3.1) guarantees that the pull-back map $H^i(k, \mu_n^{\otimes j}) \to H^i(\mathbb{A}_k^1, \mu_n^{\otimes j})$ induced by the structural morphism is an isomorphism. For any F, the map $H^i(k, \mu_n^{\otimes j}) \to H^i(\mathbb{A}_k^1 - F, \mu_n^{\otimes j})$ is injective, as may be seen by specializing to a k-rational point if k is infinite, or by using a 0-cycle of degree one and a norm argument if k is finite (as a matter of fact, $H^i(k, \mu_n^{\otimes j}) = 0$ for k finite and $i > 1$). We thus get short exact sequences

$$0 \to H^i(k, \mu_n^{\otimes j}) \to H^i(\mathbb{A}_k^1 - F, \mu_n^{\otimes j}) \to \bigoplus_{P \in F} H^{i-1}(k(P), \mu_n^{\otimes j-1}) \to 0.$$

We may now let F be bigger and bigger, and we ultimately get the exact sequence

$$(4.1) \quad 0 \to H^i(k, \mu_n^{\otimes j}) \to H^i(k(t), \mu_n^{\otimes j}) \to \bigoplus_{P \in \mathbb{A}_k^{1(1)}} H^{i-1}(k(P), \mu_n^{\otimes j-1}) \to 0$$

from which the proposition follows. (If k is not perfect, simply observe that étale cohomology does not change by purely inseparable extensions.) □

THEOREM 4.1.5. *Let k be a field and $n > 0$ be an integer prime to* char(k). *Let E be a function field over k. Let $i, j,$ and $m > 0$ be integers. Let $K = E(t_1, \ldots, t_m)$ be the rational field in m variables over E. Then the natural map $H^i(E, \mu_n^{\otimes j}) \to H^i(K, \mu_n^{\otimes j})$ induces a bijection $H_{nr}^i(E/k, \mu_n^{\otimes j}) \simeq H_{nr}^i(K/k, \mu_n^{\otimes j})$. In particular, the natural map $H^i(k, \mu_n^{\otimes j}) \to H^i(k(t_1, \ldots, t_m), \mu_n^{\otimes j})$ induces a bijection*

$$H^i(k, \mu_n^{\otimes j}) \xrightarrow{\sim} H_{nr}^i(k(t_1, \ldots, t_m)/k, \mu_n^{\otimes j}).$$

PROOF. Let t be a variable, and let E be a function field over k. It is enough to show that the map $H^i(E, \mu_n^{\otimes j}) \to H^i(E(t), \mu_n^{\otimes j})$ induces an isomorphism

$$H_{nr}^i(E/k, \mu_n^{\otimes j}) \xrightarrow{\sim} H_{nr}^i(E(t)/k, \mu_n^{\otimes j}).$$

First note that the map $H^i(E, \mu_n^{\otimes j}) \to H^i(E(t), \mu_n^{\otimes j})$ is injective (any class $\alpha \in H^i(E, \mu_n^{\otimes j})$ which vanishes in $H^i(E(t), \mu_n^{\otimes j})$ actually vanishes in $H^i(U, \mu_n^{\otimes j})$ for some open set U of the affine line over E; if E is infinite, evaluation at an E-point shows that $\alpha = 0$, if E is finite, see above). We have a natural embedding (Lemma 2.1.3)

$$H_{nr}^i(E/k, \mu_n^{\otimes j}) \subset H_{nr}^i(E(t)/k, \mu_n^{\otimes j}).$$

Take $\beta \in H_{nr}^i(E(t)/k, \mu_n^{\otimes j})$. Restricting attention to discrete valuation ring of $E(t)$ which contain E, and using Proposition 4.1.4, one sees that β comes from a class γ in $H^i(E, \mu_n^{\otimes j})$. Let $A \subset E$ be a discrete valuation ring with $k \subset A$ and $E = \mathrm{qf}(A)$. Let $\pi \in A$ be a uniformizing parameter of A, and let $B \subset E(t)$ be the discrete valuation ring which is the local ring of $A[t]$ at the ideal of height one defined by π in $A[t]$. If κ_A is the residue field of A, the residue field κ_B of B is none other than the rational function field $\kappa_A(t)$. The discrete valuation ring B is unramified over A, in other words $e_{B/A} = 1$. By the functorial behaviour of the residue map (Proposition 3.3.1), we have a commutative diagram:

$$
\begin{array}{ccc}
H^i(E, \mu_n^{\otimes j}) & \xrightarrow{\ \partial_A\ } & H^{i-1}(\kappa_A, \mu_n^{\otimes j-1}) \\
\downarrow{\scriptstyle \mathrm{Res}_{E, E(t)}} & & \downarrow{\scriptstyle \mathrm{Res}_{\kappa_A, \kappa_A(t)}} \\
H^i(E(t), \mu_n^{\otimes j}) & \xrightarrow{\ \partial_B\ } & H^{i-1}(\kappa_A(t), \mu_n^{\otimes j-1})
\end{array}
$$

Now the map $\mathrm{Res}_{\kappa_A, \kappa_B} : H^{i-1}(\kappa_A, \mu_n^{\otimes j-1}) \to H^{i-1}(\kappa_B, \mu_n^{\otimes j-1})$ is *injective* (same arguments as above), hence $\partial_B(\beta) = 0$ implies $\partial_A(\gamma) = 0$. Since A was arbitrary, we conclude that γ belongs to $H^i_{nr}(E/k, \mu_n^{\otimes j})$. $\square$

REMARK 4.1.6. In the context of the Witt group, a similar proof may be given (see [CT/Oj89] and [Oj90, §7]). This is also the case in the context of the Brauer group (see [Sa85], [CT/S93]). Another proof could be given along the lines of Proposition 2.1.9; however such a proof relies on the (known, but difficult) codimension one purity theorem 3.8.2 for a two-dimensional regular local ring of a smooth variety over a field.

§4.2. **Computing unramified cohomology.** In this section, we list cases where unramified cohomology has been computed, or at least where methods have been devised to detect nontrivial elements in unramified cohomology groups.

Let us first discuss unramified H^1.

PROPOSITION 4.2.1. *Let X be a smooth, complete, connected variety over a field k. Let n be a positive integer prime to $\mathrm{char}(k)$.*

(a) *For any $j \in \mathbf{Z}$, the natural map $H^1(X, \mu_n^{\otimes j}) \to H^1(k(X), \mu_n^{\otimes j})$ induces an isomorphism between $H^1(X, \mu_n^{\otimes j})$ and $H^1_{nr}(k(X)/k, \mu_n^{\otimes j})$.*

(b) *If k is algebraically closed, this map induces an isomorphism of finite groups $_n\mathrm{Pic}(X) \simeq H^1_{nr}(k(X)/k, \mu_n)$.*

(c) *If k is algebraically closed and $\mathrm{char}(k) = 0$, then there is a (noncanonical) isomorphism $H^1_{nr}(k(X)/k, \mu_n^{\otimes j}) \simeq (\mathbf{Z}/n)^{2q} \oplus_n\mathrm{NS}(X)$, where q denotes the dimension of the Picard variety of X, also equal to the dimension of the coherent cohomology group $H^1(X, \mathscr{O}_X)$, and where $\mathrm{NS}(X)$ denotes the Néron-Severi group of X, which is a finitely generated abelian group.*

PROOF. After Theorem 4.1.1, Statement (a) is just a reinterpretation of exact sequence (3.7) (as a matter of fact, the special case of Purity required for (a) is easy to prove and holds under quite general assumptions).

Statement (b) then follows from exact sequence (3.1), since for X/k as above, $H^0(X, \mathbb{G}_m) = k^*$ and $k^*/k^{*n} = 1$.

For X/k as in (c), the Néron-Severi group $\mathrm{NS}(X)$ of classes of divisors modulo algebraic equivalence is well known to be a finitely generated group. The kernel of the natural surjective map $\mathrm{Pic}(X) \to \mathrm{NS}(X)$ is the group $A(k)$ of k-points of an abelian variety A, the Picard variety of X. As such, it is a divisible group. Thus, as abelian groups, and in a non-canonical way, we have $\mathrm{Pic}(X) \simeq A(k) \oplus \mathrm{NS}(X)$. Since $\mathrm{char}(k) = 0$, the dimension of the Picard variety is equal to $q = \dim H^1(X, \mathscr{O}_X)$. By the theory of abelian varieties, this implies that $_nA(k) \simeq (\mathbf{Z}/n)^{2q}$ for all $n > 0$ (here again, use is made of the $\mathrm{char}(k) = 0$ hypothesis). $\square$

REMARK 4.2.2. Let k be a field of characteristic prime to the positive integer n. Let A be a discrete valuation ring with $k \subset A$, and let K be the quotient field of A. By means of the Kummer sequence one may identify the natural map $A^*/A^{*n} \to K^*/K^{*n}$ with the map $H^1(A, \mu_n) \to H^1(K, \mu_n)$. Thus for any smooth, complete, integral variety X/k, the group $H^1_{nr}(k(X)/k, \mu_n)$ coincides with the group of unramified elements in $k(X)$ associated to the functor $R \to R^*/R^{*n}$, in the sense of Definition 2.1. For any A as above, there is an obvious exact sequence

$$1 \to A^*/A^{*n} \to K^*/K^{*n} \to \mathbf{Z}/n \to 0.$$

We may now interpret the computations of §1 as a computation of unramified H^1 with coefficients $\mathbf{Z}/2$ for hyperelliptic curves over the complex field, that is as a computation of the 2-torsion subgroup of the Picard group of such curves.

Let us now discuss unramified H^2. The following proposition is just a reformulation of some results of Grothendieck [Gr68].

PROPOSITION 4.2.3. Let X be a smooth, projective, connected variety over a field k.

(a) Let n be a positive integer prime to $\mathrm{char}(k)$. The isomorphism $H^2(k(X), \mu_n) \simeq {}_n\mathrm{Br}(k(X))$ coming from the Kummer exact sequence induces an isomorphism between $H^2_{nr}(k(X)/k, \mu_n)$ and $_n\mathrm{Br}(X)$.

(b) If k is algebraically closed and n is a power of a prime number $l \neq \mathrm{char}(k)$, then $H^2_{nr}(k(X)/k, \mu_n) \simeq (\mathbf{Z}/n)^{B_2-\rho} \oplus {}_nH^3(X, \mathbf{Z}_l)$, where $B_2 - \rho$ is the "number of transcendental cycles", i.e., the difference between the rank of the étale cohomology group $H^2(X, \mathbf{Q}_l)$ and the rank of $\mathrm{NS}(X) \otimes \mathbf{Q}_l$, and where $H^3(X, \mathbf{Z}_l)$ denotes the third étale cohomology group with $\mathbf{Z}_l$-coefficients.

PROOF. After Theorem 4.1.1, statement (a) is just a reformulation of the discussion at the end of §3.4 (see the exact sequences (3.7) and (3.8)). As for (b), Grothendieck explains in [Gr68] how starting from exact sequence (3.2), for any prime $l \neq \mathrm{char}(k)$, one gets the exact sequence

$$0 \to \mathrm{NS}(X) \otimes \mathbf{Q}_l/\mathbf{Z}_l \to H^2(X, \mathbf{Q}_l/\mathbf{Z}_l) \to \mathrm{Br}(X)\{l\} \to 0,$$

where $\mathrm{Br}(X)\{l\}$ denotes the l-primary subgroup of the torsion group $\mathrm{Br}(X)$ (twisting by the roots of unity has been ignored). Statement (b) easily follows. $\square$

Let us assume that k is algebraically closed of characteristic zero. If the variety X is unirational, then it may be shown that $B_2 - \rho = 0$. Thus the only possible nontrivial part in the Brauer group (i.e., in $H^2_{nr}(k(X), \mu_n)$ for some n) comes from the torsion subgroup of $H^3(X, \mathbf{Z}_l)$ for some prime l. It may be shown by comparision with classical cohomology that this torsion subgroup is zero for almost all l.

4.2.4. Literature on unramified H^2. The invariant appearing in Proposition 4.2.3 has been much studied. The first computation in this direction is due to Artin and Mumford [**Ar/Mu72**]. They produced one of the first examples of unirational varieties that are not rational. Their example is a threefold, a conic bundle over the complex plane. Their proof used an explicit smooth projective variety X and a direct computation of the torsion subgroup in the third integral cohomology group of this variety.

That there was something to be gained in apprehending the Brauer group as an unramified Brauer group was revealed by Saltman [**Sa84**], when he exhibited fields $k(V)^G$ of invariants for a faithful linear action of a finite group G on a vector space V, over an algebraically closed field k of characteristic zero, that are not rational because the unramified Brauer group is nonzero. In this case, one would have been at a loss to construct a smooth projective model.

The unramified point of view was further developed by Saltman ([**Sa85**], [**Sa88**], [**Sa90**]) and Bogomolov ([**Bo87**], [**Bo89**]). Their work is discussed in the survey [**CT/Sa88**].

Bogomolov [**Bo87**] (see also [**CT/Sa88**]) gave a formula for $\mathrm{Br}_{nr}(k(V)^G)$ for G finite and V/k as above, namely he showed that this group may be identified with the subgroup of the integral cohomology group $H^2(G, \mathbf{Q}/\mathbf{Z}) \simeq H^3(G, \mathbf{Z})$ consisting of classes whose restrictions to all abelian subgroups of G vanish. This condition is equivalent to the vanishing of restrictions to all abelian subgroups with at most two generators.

Bogomolov's formula in turn is subsumed in Saltman's formula [**Sa90**] for the unramified Brauer group of twisted multiplicative field invariants.

Let G be a connected, reductive group over an algebraically closed field k of characteristic zero. Let $G \subset GL_n$ be an injective homomorphism. One may then consider the quotient variety $X = GL_n/G$, which is an affine variety since G is reductive. It is an open question whether the (obviously unirational) variety X is a rational variety. Saltman ([**Sa85**], [**Sa88**]) and Bogomolov [**Bo89**] have studied this problem. For $G = PGL_r$, Saltman [**Sa85**] (see [**CT/Sa87**] for a different proof) and Bogomolov for G arbitrary ([**Bo89**], see also [**CT/Sa88**]) showed that the unramified Brauer group of $k(GL_n/G)$ is always zero.

Taking the point of view of unramified cohomology, one may handle the Artin-Mumford example quoted above, as well as other examples, in a manner totally parallel to that of §1. Indeed, in §1, the function field $L = \mathbf{C}(X)$ is a quadratic field extension of the rational function field in one variable $K = \mathbf{C}(t)$; that is, it is the function field of a 0-dimensional conic over $\mathbf{C}(t)$.

A key point in the proof we gave in §1 is the exact sequence

$$1 \to \mathbf{Z}/2 \to K^*/K^{*2} \to L^*/L^{*2},$$

that is,

$$1 \to \mathbf{Z}/2 \to H^1(K, \mathbf{Z}/2) \to H^1(L, \mathbf{Z}/2),$$

associated to the quadratic field extension L/K, which gives control on the kernel of the map $K^*/K^{*2} \to L^*/L^{*2}$. Let $\alpha \in H^1(K, \mathbf{Z}/2)$ be the generator of the above kernel. Note that α is a ramified element in K^*/K^{*2}. Nontrivial unramified elements in L^*/L^{*2} are then produced in the following way. Let $\beta \neq 1$, α be a ramified element in K^*/K^{*2}. Suppose that all ramification of β is "contained" in the ramification of α. Since α dies in L^*/L^{*2}, so does its ramification, hence so does the ramification of β when going over to L. However, because β does not lie in the kernel of $K^*/K^{*2} \to L^*/L^{*2}$, β itself does not die when going up to L: this produces a nontrivial unramified element in L^*/L^{*2}.

Now let L be the function field of a 1-dimensional conic over the rational function field in two variables $K = \mathbf{C}(t_1, t_2)$. If L is the function field of a smooth projective conic over a field K, it is a special case, already known to Witt [Wi35, p. 465], of a result of Amitsur that the kernel of the map $H^2(K, \mathbf{Z}/2) \to H^2(L, \mathbf{Z}/2)$ is either 0 or $\mathbf{Z}/2$, depending on whether the conic has a K-rational point or not. In the second case, the nontrivial element in $H^2(K, \mathbf{Z}/2) \simeq {}_2\mathrm{Br}(K)$ is the class of the quaternion algebra α associated to the conic. Now this is an exact analogue of the situation above. To produce an unramified element in $H^2(L, \mathbf{Z}/2)$, one should therefore look for an element $\beta \in H^2(K, \mathbf{Z}/2)$ which is neither 0 nor α, but whose ramification (as an element of $H^2(K, \mathbf{Z}/2)$) is "somehow" dominated by the ramification of α, so that when going over to L, all ramification of $\beta_L \in H^2(L, \mathbf{Z}/2)$ vanishes—although β itself does not vanish. This is precisely what happens in the Artin-Mumford counterexample. More examples of the kind may be produced along these lines. For more details, see [CT/Oj89].

4.2.5 Beyond H^2. In the following discussion, unless otherwise specified, we assume that the ground field k is algebraically closed of characteristic zero. Since algebraic geometry often produces unirational varieties whose rationality is an open question (such as $X = GL_n/G$ for G a reductive subgroup of GL_n, see above), it would be very useful to be able to compute the higher unramified cohomology groups $H^i_{nr}(k(X), \mathbf{Z}/n)$ for such varieties. Also, for some of these varieties the lower unramified cohomology groups might be trivial. This is the case for unramified H^2 of $k(X)$ for $X = GL_n/G$ with G as above and connected (Bogomolov, see comments above), and it may be the case for some finite groups $G \subset GL_n$. However, for higher unramified cohomology groups, there is no ready-made formula available, such as those appearing in Propositions 4.2.1 and 4.2.3 (see however the exact sequence (4.2) below). As a matter of fact, one does not even know whether the groups $H^i_{nr}(k(X), \mathbf{Z}/n)$ for $i \geq 3$ are finite. For some special varieties for which finiteness of $H^3_{nr}(k(X), \mathbf{Z}/n)$ is known, see §4.3.

The first task when dealing with higher unramified cohomology has been to produce examples where it is nontrivial. In [CT/Oj89], Ojanguren and I produced an example of a unirational variety (of dimension 6) for which none of the previously used birational invariants could be used, but for which unramified H^3 could detect nonrationality. The construction was inspired by the Artin-Mumford example, as reexamined above. As a substitute for the Witt/Amitsur result, we used a result of Arason [Ar75] computing the kernel of $H^3(K, \mathbf{Z}/2) \to H^3(F, \mathbf{Z}/2)$ when F is the function field of a quadric defined by a 3-fold Pfister form $\langle 1, -a \rangle \otimes \langle 1, -b \rangle \otimes \langle 1, -c \rangle (a, b, c \in K^*)$. The kernel is spanned by the cup-product $(a) \cup (b) \cup (c)$, where $(a), (b), (c) \in H^1(K, \mathbf{Z}/2) \simeq K^*/K^{*2}$ are the cohomology classes associated to the respective classes of a, b, c in K^*/K^{*2}.

(Similar computations may be done at the level of the Witt group. The analogue of Arason's result in that case is an earlier and simpler result of Arason and Pifster. This was briefly noticed in [CT/Oj89]. For a pleasant approach to this purely quadratic form theoretic point of view, see Ojanguren's book [Oj90].)

Arason's theorem computes the kernel of $H^3(K, \mathbf{Z}/2) \to H^3(F, \mathbf{Z}/2)$ for F the function field of the affine quadric

$$X^2 - aY^2 - bZ^2 + abT^2 = c$$

$(a, b, c \in K^*)$. Now such an equation also reads:

$$\mathrm{Nrd}_{D/k}(\Xi) = c,$$

where D denotes the quaternion algebra (a, b) over K,

$$\Xi = X + iY + jZ + kT$$

(with the usual notation for quaternions) and Nrd denotes the reduced norm. Using hard techniques of algebraic K-theory, Suslin [Su91] has generalized the latter result. Let D/K be a central simple algebra of prime index p over a field K $(\mathrm{char}(K) \neq p)$. Let X/K be the affine $(p^2 - 1)$-dimensional affine variety given by the equation

$$\mathrm{Nrd}_{D/k}(\Xi) = c$$

for some $c \in K^*$. Then (Suslin, loc. cit.) the kernel of the map $H^3(K, \mu_p^{\otimes 2})$ $\to H^3(K(X), \mu_p^{\otimes 2})$ is spanned by the class $(D) \cup (c)$ where $(c) \in H^1(K, \mu_p)$ is the class of c in K^*/K^{*p} (under the Kummer identification) and $(D) \in H^2(K, \mu_p) \simeq_p \mathrm{Br}(K)$ is the class of the central simple algebra D in the Brauer group of K.

On the basis of this result, and with some inspiration from Bogomolov's extension [**Bo87**] of Saltman's results [**Sa84**], for arbitrary prime p, E. Peyre [**Pe93**] managed to produce many examples of unirational varieties X over an algebraically closed field k for which the unramified cohomology group $H^3(k(X), \mathbf{Z}/p)$ does not vanish, hence which are not rational, even though the whole unramified Brauer group of $k(X)$ vanishes.

Let us come back to the case $p = 2$. In this case, Amitsur's H^2 result and Arason's H^3 result have been extended by Jacob and Rost [**Ja/Ro89**] to H^4. Let K be a field, $\mathrm{char}(K) \neq 2$, let a, b, c, d be elements of K^*, and let X be the smooth projective quadric associated to the 4-fold Pfister form $\langle 1, -a \rangle \otimes \langle 1, -b \rangle \otimes \langle 1, -c \rangle \otimes \langle 1, -d \rangle$ $(a, b, c, d \in K^*)$. Then, with notation as above, the kernel of the map $H^4(K, \mathbf{Z}/2) \to H^4(K(X), \mathbf{Z}/2)$ is spanned by the class of the cup-product $(a) \cup (b) \cup (c) \cup (d)$. Here again, Peyre was able to use this result to produce unirational varieties whose nonrationality is detected by unramified H^4 (with coefficients $\mathbf{Z}/2$).

When the ground field is not algebraically closed, unramified cohomology may still provide some information (such is already the case for the Brauer group). For instance, it may detect whether some varieties over k, even though they are rational over the algebraic closure of the ground field, are not rational over the ground field.

I shall here mention two general cases where some higher cohomology groups for varieties over a non-algebraically closed field have been computed.

Suppose that k is the field $\mathbf{R}$ of real numbers. Let $X/\mathbf{R}$ be a smooth variety over $\mathbf{R}$. Then for any integer $n > \dim(X)$, the group $H^0(X, \mathscr{H}^n(\mathbf{Z}/2))$ is isomorphic to $(\mathbf{Z}/2)^s$, where s denotes the number of real components of the topological space $X(\mathbf{R})$ ([**CT/Pa90**]; that result has since been proved for arbitrary separated varieties). In particular, if $X/\mathbf{R}$ is a smooth, proper, integral variety over the reals, for any $n > \dim(X)$, the unramified cohomology group $H^n_{nr}(\mathbf{R}(X)/\mathbf{R}, \mathbf{Z}/2)$ is isomorphic to $(\mathbf{Z}/2)^s$, where s denotes the number of connected components of $X(\mathbf{R})$ (well known not to depend on the particular smooth complete model). These results are proved in a joint paper with Parimala [**CT/Pa90**].

Suppose that $k = \mathbf{F}$ is a finite field of characteristic p. Let $X/\mathbf{F}$ be a smooth, projective, integral variety of dimension d. Let n be a positive integer prime to p. According to a conjecture of Kato [**Ka86**], the unramified cohomology group $H^{d+1}_{nr}(\mathbf{F}(X)/\mathbf{F}, \mu_n^{\otimes d})$ should be trivial. For a curve, this is just a reformulation of the classical result $\mathrm{Br}(X) = 0$. For surfaces, this was proved by Sansuc, Soulé, and the author in 1983, and independently by K. Kato (see [**Ka86**]). For threefolds, the limit result $H^4_{nr}(\mathbf{F}(X)/\mathbf{F}, \mathbf{Q}_l/\mathbf{Z}_l(3)) = 0$ (here $l \neq p$ is a prime and $\mathbf{Q}_l/\mathbf{Z}_l(3)$ denotes the direct limit of all $\mu_{l^m}^{\otimes 3}$ for m tending to infinity) has recently been given two independent proofs (S. Saito [**Sa92**], the author [**CT92**]).

There are many open problems in this area. Here are three of them.

(i) Let k be field, char$(k) = 0$. Let G be a finite group, and let V be a faithful finite dimensional representation of G over k. Let $k(V)$ be the quotient field of the symmetric algebra on V, and let $K = k(V)^G$ be the field of G-invariants.

For any integer i and any positive integer n, give a formula for $H^i_{nr}(K/k, \mu_n^{\otimes i-1})$.

(One reason for expecting a nicer formula for the $(i-1)$th twist is that one expects—and one knows for $i \leq 3$—that for n dividing m the natural map from $H^i_{nr}(K/k, \mu_n^{\otimes i-1})$ to $H^i_{nr}(K/k, \mu_m^{\otimes i-1})$ is an injection.)

It is worth observing that, thanks to the "no-name lemma" (see [CT/Sa88]), for fixed i, j, and n, the group $H^i_{nr}(k(V)^G/k, \mu_n^{\otimes j})$ does not depend on the particular faithful representation V of G. For k algebraically closed and $i = 2$, such a formula, purely in terms of the cohomology of G and of its subgroups, was given by Bogomolov [Bo87] (see above) (see also [Sa90] and [CT/Sa88]).

(ii) Let k be a (non-algebraically closed!) field and let T/k be an algebraic k-torus (i.e., an algebraic group over k, which over an algebraic closure of k becomes isomorphic to a product of multiplicative groups $\mathbf{G}_m$). Let $k(T)$ be the function field of T.

For any integer i and any positive integer n prime to the characteristic of k, give a formula for the group $H^i_{nr}(k(T)/k, \mu_n^{\otimes i-1})$.

Here the k-torus T is entirely determined by its character group $\hat{T}$, which is a lattice with an action of the absolute Galois group $\mathscr{G}$ of k. For $i = 2$, there is such a formula, purely in terms of the cohomology of the $\mathscr{G}$-module $\hat{T}$ ([CT/Sa87]). Namely, if $G = \mathrm{Gal}(K/k)$ is the Galois group of a finite splitting extension of the torus T, the group $\mathrm{Br}_{nr}(k(X)/k)$ coincides with the group of elements in $H^2(G, \hat{T})$ which vanish by restriction to all cyclic subgroups of G. The group $H^2_{nr}(k(T)/k, \mu_n)$ coincides with the n-torsion subgroup of $\mathrm{Br}_{nr}(k(X)/k)$.

(iii) Let k be an algebraically closed field, $G \subset GL_n$ be a connected reductive subgroup of GL_n and $X = GL_n/G$.

Compute the unramified cohomology groups $H^i_{nr}(k(X), \mathbf{Z}/n)$.

For $i = 3$ and $G = PGL_r$, this is the problem addressed by Saltman [Sa93] in these Proceedings.

§4.3. Finiteness results for unramified H^3 and for $CH^2(X)/n$.

Given a smooth variety X over a separably closed field k, an integer $n > 0$ prime to the characteristic of k, i, and j integers, the cohomology groups $H^i(X, \mu_n^{\otimes j})$ are finite ([SGA4, XVI, 5.2]; [Mi80, VI. 5.5] restricted to separably closed fields). In characteristic zero, this is known to hold for arbitrary varieties [SGA4, XVI, 5.1].

There is a natural map $H^i(X, \mu_n^{\otimes j}) \to H^0(X, \mathscr{H}^i(\mu_n^{\otimes j}))$. If X is smooth, and $i \leq 2$, this map is surjective by Purity (see §3.4), hence the groups

$H^0(X, \mathcal{H}^i(\mu_n^{\otimes j}))$ are finite for $i \leq 2$. However, for $i \geq 3$, the above map need not be surjective—however the surjectivity for smooth, projective, varieties is an open question. [1]

One may wonder whether for X/k smooth the groups $H^0(X, \mathcal{H}^i(\mu_n^{\otimes j}))$ are finite. In this section, for $i = 3$, we shall give sufficient conditions on X for this to be the case. We shall also be interested in non-algebraically closed ground fields. First, we recall various tools which will be used in the proof of each of our finiteness theorems.

Quite generally, for any variety X over a field k, and $n > 0$ prime to $\operatorname{char}(k)$, there is a spectral sequence

$$E_2^{pq} = H_{Zar}^p(X, \mathcal{H}^q(\mu_n^{\otimes j})) \Rightarrow H_{\text{ét}}^{p+q}(X, \mu_n^{\otimes j}).$$

The terms E_2^{pq} are zero for $p > d = \dim(X)$. If X is smooth, the terms E_2^{pq} are zero for $p > q$. This is a consequence of the Gersten conjecture for étale cohomology, as proved by Bloch and Ogus [Bl/Og74]. In particular, for X/k smooth (not necessarily proper), there is a short exact sequence

$$H^3(X, \mu_n^{\otimes j}) \to H^0(X, \mathcal{H}^3(\mu_n^{\otimes j})) \to H^2(X, \mathcal{H}^2(\mu_n^{\otimes j})) \to H^4(X, \mu_n^{\otimes j})$$

which for $j = 2$ may be rewritten

$$(4.2) \qquad H^3(X, \mu_n^{\otimes 2}) \to H^0(X, \mathcal{H}^3(\mu_n^{\otimes 2})) \to CH^2(X)/n \to H^4(X, \mu_n^{\otimes 2}).$$

Recall that $CH^i(X)$, resp. $CH_i(X)$, denotes the ith Chow group of codimension i cycles, resp., dimension i cycles on X, modulo rational equivalence (see [Fu84] for the theory of Chow groups). That $H^2(X, \mathcal{H}^2(\mu_n^{\otimes 2}))$ may be identified with $CH^2(X)$ on a smooth variety (over an arbitrary field k of characteristic prime to n) also follows from the work of Bloch/Ogus (see [Bl/Og74], [Bl80], [Bl81], and [CT93]).

Let us recall the localization sequence. Given X/k an arbitrary variety, $U \subset X$ an open set, and F the complement of U in X, for any integer i, there is an exact sequence, the localization sequence (see [Fu84, I, Proposition 1.8, p. 21]):

$$(4.3) \qquad\qquad CH_i(F) \to CH_i(X) \to CH_i(U) \to 0.$$

Given X and Y two k-varieties and $f\colon X \to Y$ a finite and flat map of constant rank d, given any integer i, the composite map $p_* \circ p^*$ of the flat pull-back p^* with the proper push-forward map p_* is multiplication by d on $CH_i(X)$ ([Fu84, I, Example 1.7.4, p. 20]).

We shall make use of the following fact, which is a special case of Theorem 4.1.1. If X/k is smooth and integral, there is a natural embedding

$$(4.4) \qquad\qquad H_{nr}^i(k(X)/k, \mu_n^{\otimes j}) \hookrightarrow H^0(X, \mathcal{H}^i(\mu_n^{\otimes j})).$$

---

[1] Surjectivity may already fail for X a smooth projective surface over a p-adic field and $i = 3$ (Parimala and Suresh, September 1993).

Finally, we shall also use the following consequence, foreseen by Bloch [Bl81], of the Merkur'ev-Suslin theorem [Me/Su82]. If X/k is a smooth variety over the field k, for any integer $m > 0$ prime to $\mathrm{char}(k)$, the m-torsion subgroup ${}_m CH^2(X)$ is a subquotient of the étale cohomology group $H^3(X, \mu_m^{\otimes 2})$ (see [CT93, 3.3.2]).

Now we are prepared to state and prove some finiteness theorems. The first explicit appearance of such theorems is in Parimala's paper [Pa88] (with application to finiteness theorems for the Witt group of unirational threefolds over the reals). Over $\mathbf{C}$, the first theorem below was stated and proved independently by L. Barbieri-Viale [BV92]. The theorem applies to unirational varieties; it also applies to threefolds which are conic bundles over arbitrary surfaces. Concerning these last varieties, it may actually be shown that for such a threefold X over an algebraically closed field k of characteristic prime to n, all the groups $H_{nr}^3(k(X)/k, \mu_n^{\otimes j})$ vanish (see [Pa89, appendix]).

THEOREM 4.3.1. *Let k be an algebraically closed field of characteristic zero, and let X be a smooth integral variety of dimension d over k. Assume that there exists a dominant rational map of degree N from $\mathbb{A}^{d-2} \times S$ to X for some integral surface S. Then the groups $CH^2(X)/n$ and $H_{nr}^3(k(X)/k, \mathbb{Z}/n)$ are finite, and this last group is zero if n is prime to N.*

PROOF. The assumption may also be phrased as follows: there is a function field K/k of transcendence degree 2, such that the rational function field $K(t_1, \dots, t_{d-2})$ is a field extension of degree N of the function field $k(X)$.

We may and will assume that S is smooth and affine. First consider the maps

$$H_{nr}^i(k(X)/k, \mathbb{Z}/n) \to H_{nr}^i(k(S)(t_1, \dots, t_{d-2})/k, \mathbb{Z}/n) \overset{\mathrm{Cores}}{\to} H^i(k(X)/k, \mathbb{Z}/n),$$

where the composition is multiplication by N (as a matter of fact, the image of the last map lies in $H_{nr}^i(k(X)/k, \mathbb{Z}/n)$, but this is not needed for the proof). The central term is isomorphic to $H_{nr}^i(k(S)/k, \mathbb{Z}/n)$ by Theorem 4.1.5, hence is a subgroup of $H^i(k(S)/k, \mathbb{Z}/n)$ which is zero for $i \geq 3$ [Se65]. This shows that N kills $H_{nr}^3(k(X)/k, \mathbb{Z}/n)$, hence that this group is zero for all integers n prime to N.

We may find smooth open sets $V \subset \mathbb{A}^{d-2} \times S$ and $U \subset X$ and a finite and flat morphism $p: V \to U$ of constant rank N such that the composite morphism $V \to U \to X$ defines the given rational map from $\mathbb{A}^{d-2} \times S$ to X.

Projection $f: \mathbb{A}^{d-2} \times S \to S$ induces an isomorphism

$$f^*: CH_0(S) \to CH_{d-2}(\mathbb{A}^{d-2} \times S)$$

[Fu84, Theorem 3.3, p. 64]. The restriction map $CH_{d-2}(\mathbb{A}^{d-2} \times S) \to CH_{d-2}(V)$ is trivially surjective. Now since S is a connected affine variety, the group $CH_0(S)$ is divisible. Indeed, to prove this, it is enough to do it for an affine connected curve. By normalization, this last case reduces to that of a smooth connected affine curve C. Let J be the jacobian of the smooth projective completion of C. The divisibility of $CH_0(C)$ follows from that of the group $J(k)$. Thus for our surface S we have $CH_0(S)/n = 0$, from which we deduce $CH_{d-2}(\mathbf{A}^{d-2} \times S)/n = 0$, hence $CH_{d-2}(V)/n = 0$ for all $n > 0$.

Since the open set $U \subset X$ is smooth, as recalled above we have an embedding

$$H^3_{nr}(k(X)/k, \mathbb{Z}/n) \hookrightarrow H^0(U, \mathscr{H}^3(\mathbb{Z}/n)).$$

Thus to show that $H^3_{nr}(k(X)/k, \mathbb{Z}/n)$ is finite, it suffices to show that $H^0(U, \mathscr{H}^3(\mathbb{Z}/n))$ is finite. Since, as recalled at the beginning of this section, the group $H^3(U, \mu_n^{\otimes 2})$ is finite (because k is algebraically closed), by (4.2), finiteness of $H^0(U, \mathscr{H}^3(\mathbb{Z}/n))$ is equivalent to finiteness of $CH^2(U)/nCH^2(U)$ (note that $\mathbb{Z}/n \simeq \mu_n^{\otimes 2}$ since k is algebraically closed).

The finite and flat morphism p gives rise to morphisms $p^* \colon CH_{d-2}(U) \to CH_{d-2}(V)$ and $p_* \colon CH_{d-2}(V) \to CH_{d-2}(U)$. The composite map

$$p_* \circ p^* \colon CH_{d-2}(U) \to CH_{d-2}(V) \to CH_{d-2}(U)$$

is multiplication by N.

The group ${}_m CH^2(U)$ is a subquotient of the group $H^3(U, \mu_m^{\otimes 2})$, as recalled above (the Merkur'ev-Suslin result is used here), and this last group is finite since k is algebraically closed.

Letting $A = CH_{d-2}(U)$ and $B = CH_{d-2}(V)$, and bearing in mind the vanishing, hence finiteness of $CH_{d-2}(V)/n = 0$ for all $n > 0$, the finiteness of $CH^2(U)/nCH^2(U)$ now follows from the purely formal lemma:

LEMMA 4.3.2. *Let* $f \colon A \to B$ *and* $g \colon B \to A$ *be homomorphisms of abelian groups. Assume that* $g \circ f$ *is multiplication by the positive integer* $N > 0$. *Assume that* ${}_N A$ *and* B/NB *are finite. Let* $n \geq 0$. *Then, if* B/nB *is finite, so is* A/nA.

PROOF. Let $b_i \in B$, $i \in I$ be representatives of the finite set $I = B/NB$. For each $i \in I$ such that $g(b_i)$ belongs to NA, fix an element $a_i \in A$ with $g(b_i) = Na_i$. For any $a \in A$ we may find $b \in B$ and $i \in I$ such that $f(a) = Nb + b_i$. Now $Na = g \circ f(a) = Ng(b) + g(b_i)$. Hence $a - g(b) - a_i$ belongs to the N-torsion subgroup ${}_N A$, which is finite. We conclude that the quotient $C = A/g(B)$ is a finite abelian group. For any positive integer n, we have an exact sequence

$$B/nB \to A/nA \to C/nC \to 0,$$

hence A/nA is finite if B/nB is finite. $\square$

It remains to prove the finiteness of $CH^2(X)/n$. The argument given above shows that there exists a nonempty open set of $U \subset X$ such that

$CH^2(U)/n$ is finite. Let $F \subset X$ be the complement of U in X. We have the localization sequence

$$CH_{d-2}(F) \to CH_{d-2}(X) \to CH_{d-2}(U) \to 0.$$

Since $\dim(F) \leq d - 1$, the finiteness of $CH_{d-2}(X)/n$ follows from the lemma:

LEMMA 4.3.3. *Let Z/k be a variety of dimension $\leq d$ over the algebraically closed field k. Then for any $n > 0$, the group $CH_{d-1}(Z)/n$ is finite.*

PROOF. If $\dim(Z) < d - 1$, then $CH_{d-1}(Z) = 0$, and if $\dim(Z) = d-1$, then $CH_{d-1}(Z)$ is finitely generated (one generator for each irreducible component of Z). Let us assume $\dim(Z) = d$. We may also assume that all components of Z have dimension d. Let Z_{sing} be the singular locus of Z and let $U \subset Z$ be the complement. We have the localization sequence

$$CH_{d-1}(Z_{\text{sing}}) \to CH_{d-1}(Z) \to CH_{d-1}(U) \to 0.$$

By the argument above, the abelian group $CH_{d-1}(Z_{\text{sing}})$ is finitely generated. We have $CH_{d-1}(U) = CH^1(U) \simeq \text{Pic}(U)$, hence for any $n > 0$, we have $CH_{d-1}(U)/n \simeq \text{Pic}(U)/n \hookrightarrow H^2(U, \mu_n)$, and this last group is finite since k is algebraically closed. □

REMARK 4.3.4. As indicated above, over $\mathbb{C}$, essentially the same theorem was proved by L. Barbieri-Viale [BV92b]. His proof, which is shorter, uses a result of Bloch-Srinivas [Bl/Sr83], namely the fact that for a smooth variety $X/\mathbb{C}$, the Zariski sheaf $\mathscr{H}^3(\mathbf{Z})$ associated to the Zariski presheaf $U \mapsto H^3_{\text{Betti}}(U(\mathbb{C}), \mathbf{Z})$ has no torsion (that result in its turn is a consequence of the Merkur'ev-Suslin theorem). Assume that $X/\mathbb{C}$ is smooth and projective. Let $NS^2(X)$ be the quotient of $CH^2(X)$ by the subgroup $A^2(X)$ of classes algebraically equivalent to zero. From the geometric assumption in Theorem 4.3.1, one deduces that the group $H^0(X, \mathscr{H}^3(\mathbf{Z}))$ is torsion, hence that it is zero by the Bloch/Srinivas result. Now from [Bl/Og74] we have an exact sequence

$$H^0(X, \mathscr{H}^3(\mathbf{Z})) \to NS^2(X) \to H^4_{\text{Betti}}(X, \mathbf{Z})$$

and we conclude that the group $NS^2(X)$ is a finitely generated group. Since the group $A^2(X)$ is divisible, for any positive integer n, the induced map $CH^2(X)/n \to NS^2(X)/n$ is an isomorphism, hence $CH^2(X)/n$ is finite.

REMARK 4.3.5. The same result can also be obtained under any of the following weaker (and related) hypotheses (cf. [Bl/Sr83]):

(i) There exists a morphism f from a smooth projective surface $S/\mathbb{C}$ to the smooth projective variety $X/\mathbb{C}$ such that the induced map $f_*: CH_0(S) \to CH_0(X)$ is surjective.

(ii) Let $\Delta \subset X \times X$ be the diagonal. There exists a positive integer m such that the cycle $m\Delta$ is rationally equivalent to a sum of codimension two cycles $Z_1 + Z_2$, the support of Z_1 being included in $S \times X$ and the support of Z_2 in $X \times Y$, where $S \subset X$ is a surface and $Y \subset X$ is of codimension 1.

Indeed the correspondence formalism ([**Bl/Sr83**]) and the torsion-freeness of $\mathscr{H}^3(\mathbf{Z})$ (see above) yield $H^0(X, \mathscr{H}^3(\mathbf{Z})) = 0$ and we may conclude just as above.

The more complicated proof of Theorem 4.3.1 given above is however better suited for extensions of this finiteness theorem to varieties over non algebraically closed fields, as we shall see in the balance of §4.3.

For $S = \mathbb{A}^2$, and $d = 3$, statement (b) in the following theorem is due to Parimala [**Pa88**].

THEOREM 4.3.6. *Let k be a real closed field, and let X be a smooth, geometrically integral variety of dimension d over k. Assume that there exists a dominant rational map of degree N from $\mathbb{A}^{d-2} \times_{\overline{k}} S$ to $\overline{X}$, for some integral surface $S/\overline{k}$. Let $n > 0$ be an integer. Then*

(a) *The group $H^3_{nr}(k(X)/k, \mu_n^{\otimes 2})$ is finite, and it is zero for all integers n prime to $2N$.*

(b) *If X/k is smooth, the group $CH^2(X)/n$ is finite.*

PROOF. One may assume that $S/\overline{k}$ is smooth and affine and find a finite and flat map of constant degree $2N$ from a nonempty open set V of $\mathbb{A}^{d-2} \times_{\overline{k}} S$ to an open set U of X. The proof of the theorem is then essentially identical to the proof of Theorem 4.3.1, in view of the following facts:

(i) For any variety U/k with k real closed, the étale cohomology groups $H^i(U, \mu_n^{\otimes j})$ are finite. This follows from the Hochschild-Serre spectral sequence for the extension $\overline{k}/k$ together with the finiteness of étale cohomology of varieties over an algebraically closed field and finiteness of the Galois cohomology of finite $\mathrm{Gal}(\overline{k}/k)$-modules.

(ii) For any smooth variety X over a real closed field k, and any positive integer n, the group $_nCH^2(X)$ is finite. Indeed, it is a subquotient of $H^3(X, \mu_n^{\otimes 2})$ which we have just seen is finite.

(iii) $H^3(k, \mu_n^{\otimes 2}) = 0$ for n odd.

(iv) Lemma 4.3.3 still holds, with the same proof, when k is real closed. $\square$

THEOREM 4.3.7. *Let k be a p-adic field (finite extension of $\mathbb{Q}_p$), let X a smooth geometrically integral variety of dimension d over k.*

(a) *Assume that $\overline{X} = X \times_k \overline{k}$ is rationally dominated by the product of an integral curve and an affine space of dimension $d - 1$; then for any integer $n > 0$, the groups $CH^2(X)/n$ and $H^3_{nr}(k(X)/k, \mu_n^{\otimes 2})$ are finite.*

(b) *If n is prime to p, then the same finiteness results hold if $\overline{X}$ is rationally dominated by the product of an integral surface and an affine space of dimension $d - 2$.*

PROOF. The proof of the first statement is essentially identical to the proof of Theorems 4.3.1 and 4.3.6. One chooses a finite extension K/k of fields, a smooth integral curve C/K, resp., surface S/K, and a finite, flat morphism of constant degree N from a nonempty open set of $C \times_K \mathbb{A}_K^{d-1}$ to an open set of X. One then combines the following facts:

(i) For any variety U/k with k p-adic, the étale cohomology groups $H^i(U, \mu_n^{\otimes j})$ are finite. This follows from the Hochschild-Serre spectral sequence for the extension $\overline{k}/k$ together with the finiteness of étale cohomology of varieties over an algebraically closed field and finiteness of the Galois cohomology of finite $\mathrm{Gal}(\overline{k}/k)$-modules [Se65]. This implies that for any smooth variety U over k p-adic, and any positive integer n, the group $_nCH^2(U)$ is finite.

(ii) For the curve C, we obviously have $CH^2(C \times_K \mathbb{A}_K^{d-1}) = CH^2(C) = 0$. For the smooth surface S, if n is prime to p, the group $CH^2(S)/n$ is finite. It is enough to prove this for a smooth projective surface S over the p-adic field K. In that case, Saito and Sujatha [Sa/Su93] have proved that $H^0(X, \mathscr{H}^3(S, \mu_n^{\otimes 2}))$ is finite. The result then follows from exact sequence (4.2) together with the finiteness results for étale cohomology in (i) above.

(iii) Lemma 4.3.3 still holds, with the same proof, when k is a p-adic field (indeed, with notation as in that Lemma, $H^2(U, \mu_n)$ is finite according to (i) above). $\square$

The same line of investigation as above, combined with results of [CT/Ra91], [Sal93], and [Sa91], enables one to prove theorems which may be regarded as weak Mordell-Weil theorems for codimension two cycles. Whether such theorems hold for arbitrary smooth varieties is a big open problem.

THEOREM 4.3.8. *Let X/k be a smooth geometrically integral variety of dimension d over a number field k. If $\overline{X} = X \times_k \overline{k}$ is rationally dominated by the product of a smooth integral curve C and an affine space of dimension $d - 1$, then $CH^2(X)/n$ is a finite group.*

PROOF. Given an m-dimensional variety Z over a number field k, the group $CH_{m-1}(Z)$ is a finitely generated abelian group: this is a known consequence of the Mordell-Weil theorem and of the Néron-Severi theorem. This holds more generally if k is finitely generated over $\mathbf{Q}$. Let n be a positive integer. So Lemma 4.3.3 still holds over such ground fields. Using the localization sequence, we see that if Y is smooth and k-birational to X, then $CH^2(Y)/n$ is finite if and only if $CH^2(X)/n$ is finite. Similarly, we see that $_nCH^2(Y)$ is finite if and only if $_nCH^2(X)$ is finite.

Let Y/k be a smooth projective model of X. Since $\overline{Y}$ is dominated by the product of a curve C and a projective space, we conclude that the coherent cohomology group $H^2(Y, \mathscr{O}_Y)$ vanishes. Here is one way to see this. Let Ω be a universal domain containing k. Since the Chow group of zero-cycles of smooth projective varieties is a birational invariant, the hypothesis implies that the Chow group $CH_0(Y_\Omega)$ of 0-dimensional cycles is representable. An argument of Roitman (see [**Ja90**, p. 157]) then implies that all $H^i(Y, \mathcal{O}_Y)$ vanish for $i \geq 2$. According to Raskind, the author, and Salberger ([**CT/Ra91**], [**Sal93**]), this implies that for any integer $n > 0$, the group $_nCH^2(Y)$ is finite. Hence $_nCH^2(X)$ and $_nCH^2(U)$ for any open set U of X are also finite groups.

We may now proceed as in the proof of Theorem 4.3.7. We find a finite extension K/k of fields, a smooth integral curve C/K and a finite, flat morphism of constant degree N from a nonempty open set V of $C \times_K \mathbb{A}_K^{d-1}$ to an open set U of X. We have $CH^2(C \times_K \mathbb{A}_K^{d-1}) = CH^2(C) = 0$, hence $CH^2(V) = 0$. Since $_nCH^2(U)$ is finite, Lemma 4.3.2 now implies that $CH^2(U)/n$ is finite for any $n > 0$, hence also $CH^2(X)/n$. $\square$

THEOREM 4.3.9. *Let X/k be a smooth integral variety of dimension d over a field k finitely generated over $\mathbb{Q}$. If $\overline{X} = X \times_k \overline{k}$ is rationally dominated by d-dimensional projective space, and if X has a rational k-point, then $CH^2(X)/n$ is a finite group.*

PROOF. The proof differs from the proof of the previous theorem only in two points. The assumption that $\overline{X} = X \times_k \overline{k}$ is rationally dominated by d-dimensional projective space is used to ensure that both $H^2(Y, \mathcal{O}_Y)$ and $H^1(Y, \mathcal{O}_Y)$ vanish for Y a smooth projective model of X (use [**Ja90**] as above). Together with the assumption that X has a k-rational point, one may then use a result of Saito ([**Sa91**], see also [**CT93**, §7, Theorem 7.6]) ensuring that the torsion subgroup of $CH^2(Y)$ is finite. $\square$

Versions of the following theorem, which combines many of the previous results, were brought to my attention by N. Suwa and L. Barbieri-Viale (independently).

THEOREM 4.3.10. *Let X/k be a smooth, projective, geometrically integral variety of dimension d over a field $k \subset \mathbb{C}$. Under any of the two sets of hypotheses:*

(i) *the field k is a number field and $CH_0(X_\mathbb{C})$ is represented by a curve;*

(ii) *the field k is finitely generated over $\mathbb{Q}$, there is a k-rational point on X, and the degree map $CH_0(X_\mathbb{C}) \to \mathbb{Z}$ is an isomorphism;*

the group $CH^2(X)$ is a finitely generated abelian group.

REMARKS. (a) The geometric assumption in (i) is that there exist a curve $C/\mathbb{C}$ and a $\mathbb{C}$-morphism from C to $X_\mathbb{C}$ such that the induced map $CH_0(C_\mathbb{C}) \to CH_0(X_\mathbb{C})$ is surjective. Concrete examples are provided by varieties dominated by the product of a curve and an affine space. A special case is that of quadric bundles (of relative dimension at least one) over a curve.

(b) The geometric assumption in (ii) is satisfied by unirational varieties, by quadric bundles (of relative dimension at least one) over rational varieties, and also by Fano varieties (Miyaoka, Campana [Ca92]).

PROOF OF THE THEOREM. Let $\overline{k} \subset \mathbf{C}$ be a fixed algebraic closure of k, let $G = \mathrm{Gal}(\overline{k}/k)$, let $\overline{X} = X \times_k \overline{k}$ and $X_{\mathbf{C}} = X \times_k \mathbf{C}$. There is a natural filtration on the Chow group $CH^2(\overline{X})$:

$$CH^2(\overline{X})_{\mathrm{alg}} \subset CH^2(\overline{X})_{\mathrm{hom}} \subset CH^2(\overline{X}).$$

The smallest subgroup is that of cycles algebraically equivalent to zero, the middle subgroup is that of cycles homologically equivalent to zero, that is those cycles which are in the kernel of the composite map

$$CH^2(X \times_k \overline{k}) \to CH^2(X_{\mathbf{C}}) \to H^4_{\mathrm{Betti}}(X(\mathbf{C}), \mathbf{Z}).$$

A standard specialization argument shows that the map

$$CH^2(X \times_k \overline{k}) \to CH^2(X_{\mathbf{C}})$$

is injective. On the other hand, the group $H^4_{\mathrm{Betti}}(X(\mathbf{C}), \mathbf{Z})$ is a finitely generated abelian group. Thus the quotient $CH^2(\overline{X})/CH^2(\overline{X})_{\mathrm{hom}}$ is a finitely generated group. By the same specialization argument, the quotient $CH^2(\overline{X})_{\mathrm{hom}}/CH^2(\overline{X})_{\mathrm{alg}}$ is a subgroup of the classical Griffiths group $CH^2(X_{\mathbf{C}})_{\mathrm{hom}}/CH^2(X_{\mathbf{C}})_{\mathrm{alg}}$.

Under any of the two assumptions in the theorem, and even under the weaker assumption that $CH_0(X_{\mathbf{C}})$ is representable by a surface, the group $CH^2(X_{\mathbf{C}})_{\mathrm{hom}}/CH^2(X_{\mathbf{C}})_{\mathrm{alg}}$ vanishes: this is a result of Bloch and Srinivas [Bl/Sr83], whose proof relies on the Merkur′ev-Suslin theorem.

Thus the group $CH^2(\overline{X})/CH^2(\overline{X})_{\mathrm{hom}}$ is an abelian group of finite type.

According to standard usage, let us write $A^2(\overline{X}) = CH^2(\overline{X})_{\mathrm{hom}}$. Murre [Mu83], relying on work of H. Saito [Sa79] and Merkur′ev-Suslin [Me/Su82], has shown that there exists a universal regular map $\rho: A^2(\overline{X}) \to A(\overline{k})$, where A is a certain abelian variety defined over $\overline{k}$. Under the assumption that $CH_0(X_{\mathbf{C}})$ is represented by a curve, Bloch and Srinivas (loc. cit.) show that the map ρ is an isomorphism. There exists an abelian variety B over $\overline{k}$ and a cycle $Z \in CH^2(B \times \overline{X})$ which induces maps $B(\overline{k}) \to A^2(\overline{X}) \to A(\overline{k})$, the composite map being induced by a $\overline{k}$-morphism of abelian varieties $\varphi: B \to A$ (see [Sa79, Proposition 1.2 (ii)]). Now the varieties A, B, the morphism φ and the cycle Z are all defined over a finite field extension L of k. Let $H = \mathrm{Gal}(\overline{k}/L)$. It then follows that the isomorphism $\rho: A^2(\overline{X}) \to A(\overline{k})$ is H-equivariant. Thus the group $(A^2(\overline{X}))^G \subset (A^2(\overline{X}))^H$ is a subgroup of $A(L)$, and for k, hence L, finitely generated over $\mathbf{Q}$, this last abelian group is finitely generated by the Mordell-Weil-Néron theorem. Hence $(A^2(\overline{X}))^G$ is a finitely generated abelian group. To conclude the proof, it only remains to show that under the assumptions of (i) and (ii), the group

$$\mathrm{Ker}[CH^2(X) \to CH^2(\overline{X})]$$

is a finite group. In case (i), observe that the geometric assumption on 0-cycles implies the vanishing of $H^2(X, \mathscr{O}_X)$ (see [**Ja90**, p. 157]). The required finiteness now follows from [**CT/Ra91**] and [**Sal93**]. In case (ii), the geometric assumption on 0-cycles implies the vanishing of both $H^2(X, \mathscr{O}_X)$ and $H^1(X, \mathscr{O}_X)$. The required finiteness result is then due to S. Saito ([**Sa91**], [**CT93**]). □

§**4.4. Rigidity for unramified cohomology.** The following theorem is an adaptation to étale cohomology of Suslin's celebrated rigidity theorem for K-theory with coefficients ([**Su83**], [**Su88**], see also Lecomte's paper [**Le86**] for interesting variants in the Chow group context):

THEOREM 4.4.1. *Let* $k \subset K$ *be separably closed fields, let* X/k *be a smooth, integral, proper k-variety. Let* $k(X)$ *be the function field of* X *and* $K(X)$ *the function field of* $X_K = X \times_k K$. *Let* $n > 0$ *be an integer prime to* $\mathrm{char}(k)$. *Let* $i > 0$ *and* j *be integers. Then the natural map* $H^i(k(X), \mu_n^{\otimes j}) \to H^i(K(X), \mu_n^{\otimes j})$ *induces an isomorphism*

$$H^i_{nr}(k(X)/k, \mu_n^{\otimes j}) \simeq H^i_{nr}(K(X)/K, \mu_n^{\otimes j}),$$

i.e., an isomorphism

$$H^0(X, \mathscr{H}^i(\mu_n^{\otimes j})) \simeq H^0(X_K, \mathscr{H}^i(\mu_n^{\otimes j})).$$

In other words, unramified cohomology is rigid: it does not change by extension of separably closed field.

PROOF. (1) Since étale cohomology does not change under inseparable extensions, we may assume that k and K are algebraically closed.

(2) The functor $V \mapsto H^0(V, \mathscr{H}^i(\mu_n^{\otimes j}))$ is functorial contravariant on the category of all k-schemes. Indeed, étale cohomology is functorial contravariant under arbitrary morphisms. If $f: V \to W$ is any k-morphism of k-varieties, we have an induced map of sheaves on W: $\mathscr{H}^i_W(\mu_n^{\otimes j}) \to f_*(\mathscr{H}^i_V(\mu_n^{\otimes j}))$, hence a map

$$f^*: H^0(W, \mathscr{H}^i_W(\mu_n^{\otimes j})) \to H^0(V, \mathscr{H}^i_V(\mu_n^{\otimes j})),$$

and one easily checks that this map is functorial, i.e., respects composition of morphisms.

(3) Given a finite field extension $K \subset L$ with $k \subset K$, an excellent discrete valuation ring A with $k \subset A$, with residue field κ and with field of fractions K, let B be the integral closure of A in L, which we assume to be of finite type over A. Let q_α be the finitely many maximal ideals of B, and let κ_α be the corresponding residue field. There is a commutative diagram

$$
\begin{array}{ccc}
H^i(L, \mu_n^{\otimes j}) & \xrightarrow{(\partial_\alpha)} & \bigoplus_\alpha H^{i-1}(\kappa_\alpha, \mu_n^{\otimes j-1}) \\
\downarrow{\scriptstyle \mathrm{Cores}_{L,K}} & & \downarrow{\scriptstyle \sum_\alpha \mathrm{Cores}_{\kappa_\alpha, \kappa}} \\
H^i(K, \mu_n^{\otimes j}) & \xrightarrow{\partial_A} & H^{i-1}(\kappa, \mu_n^{\otimes j-1})
\end{array}
$$

This is most easily proved by means of the Galois cohomological description of the residue map.

If now $f: V \to W$ is a finite flat morphism of smooth integral k-varieties, we deduce that there is an induced norm map

$$f_*: H^0(V, \mathscr{H}_V^i(\mu_n^{\otimes j})) \to H^0(W, \mathscr{H}_W^i(\mu_n^{\otimes j})).$$

In particular, given a smooth, proper, integral k-variety X and $f: V \to W$ a finite flat morphism of smooth integral k-curves, there is an induced norm map

$$f_*: H^0(X \times V, \mathscr{H}_V^i(\mu_n^{\otimes j})) \to H^0(X \times W, \mathscr{H}_W^i(\mu_n^{\otimes j})).$$

Suppose that V is a smooth affine curve and that $W = \mathbf{A}_k^1$. Given any point $x \in \mathbf{A}^1(k)$, we have a commutative diagram

$$
\begin{array}{ccc}
H^0(X \times C, \mathscr{H}^i) & \longrightarrow & H^0(X, \mathscr{H}^i) \\
\text{Cores} \downarrow & & \text{id} \downarrow \\
H^0(X \times \mathbf{A}_k^1, \mathscr{H}^i) & \longrightarrow & H^0(X, \mathscr{H}^i)
\end{array}
$$

Here the top row is the sum $\sum_{i \in I} e_i \rho_i$, where I denotes the set of points $y_i \in C(k)$ above x, where e_i denotes the multiplicity of y_i in the fibre $f^{-1}(x) \subset C$, and where ρ_i denotes evaluation at y_i. The right vertical arrow is the identity. The left vertical map is the norm map described above. The bottom horizontal map is evaluation at x.

In order to show that this diagram actually commutes, one may restrict $\alpha \in H^0(X \times C, \mathscr{H}^i)$ to the semilocal ring S of $X \times C$ defined by the points above the generic point of $(X \times x) \subset X \times C$, whose local ring on $X \times C$ will be denoted by R. We now use Gersten's conjecture for semilocal rings of smooth varieties over a field (see §3.7 and §3.8). In particular, for such a semilocal ring S, we have $H^i(S, \mu_n^{\otimes j}) = H^0(\mathrm{Spec}(S), \mathscr{H}^i)$. The commutativity to be proved now boils down to a functoriality for the evaluation of the norm map on étale cohomology, for which we refer to [SGA4, Exposé XVII, Théorème 6.2.3, p. 422].

(4) We are now ready to use the machinery described by Suslin in his ICM86 address ([Su88], §2) (and which is an adaptation by Gabber, Gillet, and Thomason of an older argument of Suslin). We will not repeat the arguments here, but we shall give all the ingredients which make the machinery work in the case in point. We consider the functor V which to a k-scheme Y associates the torsion group $V(Y) = H^0(X \times Y, \mathscr{H}^i)$.

(5) Arguing as in Theorem 4.1.5 above, one proves that the map $H^0(X, \mathscr{H}^i) \to H^0(X \times \mathbf{A}_k^1, \mathscr{H}^i)$ is an isomorphism for all smooth k-varieties X. (This is the homotopy invariance for the functor V.)

(6) Using the various functorialities described above, and the divisibility of generalized jacobians, together with the fact that the unramified cohomology

groups are torsion groups, one finds that for any two points $x, y \in C(k)$, where C is a smooth connected affine curve, and any $\alpha \in H^0(X \times C, \mathscr{H}^i)$, the evaluations of α at x and y give the same element in $H^0(X, \mathscr{H}^i)$.

(7) In order to complete the proof of the theorem, one needs to show that any class in $H^0(X_K, \mathscr{H}^i)$ comes from some class in $H^0(X_A, \mathscr{H}^i)$ for a suitable smooth k-algebra A, $A \subset K$ (cf. [**Su88**, Corollary 2.3.3]).

We first note that K is the direct limit of all (finitely generated) smooth k-algebras $A_\alpha \subset K$. Given a class $\xi \in H^0(X_K, \mathscr{H}^i)$, one easily produces a finite covering $X_K = \bigcup_{r \in R} U_r$, finite coverings $U_r \cap U_s = \bigcup_{t \in I_{r,s}} W_t$ for each $r, s \in R$, elements $\xi_r \in H^i(U_r, \mu_n^{\otimes j})$ such that each ξ_r restricts to $\xi \in H^0(X_K, \mathscr{H}^i) \subset H^i(K(X), \mu_n^{\otimes j})$ and such that ξ_r and ξ_s have equal restrictions in $H^i(U_r \cap U_s, \mu_n^{\otimes j})$. Now the open sets U_r, W_t are defined by finitely many equations. Hence they are already defined at the level of some A_α, i.e. $X_{A_\alpha} = \bigcup_{r \in R} U_{\alpha,r}$ with $U_{\alpha,r} \times_{A_\alpha} K = U_r$, and similarly $W_t = W_{\alpha,t} \times_{A_\alpha} K = W_t$. Because cohomology commutes with filtering limits, after replacing A_α by a bigger smooth k-algebra inside K, which we will still call A_α, we may ensure that ξ_r comes from $\xi_{\alpha,r} \in H^i(U_{\alpha,r}, \mu_n^{\otimes j})$, and that the restrictions of $\xi_{\alpha,r}$ and $\xi_{\alpha,s}$ to each $W_{\alpha,t}$, $t \in I_{r,s}$ coincide. The elements ξ_r thus give rise to an element $\xi_\alpha \in H^0(X_{A_\alpha}, \mathscr{H}^i)$ whose restriction to $H^0(X_K, \mathscr{H}^i)$ is ξ.

REMARK 4.4.2. As pointed out to me by Jannsen, analogous arguments should give a similar rigidity theorem for all groups $H^r_{\mathrm{Zar}}(X, \mathscr{H}^i(\mu_n^{\otimes j}))$.

§5. Back to the Gersten conjecture

As already mentioned in §2.2.3 and §3.7, in 1980, Ojanguren [**Oj80**] proved that the Witt group of a local ring of a smooth variety over a field injects into the Witt group of its field of fractions. In 1989, Ojanguren and I [**CT/Oj92**] axiomatized Ojanguren's method. We were thus able to prove injectivity, in the sense of §2.1, for various functors. In this section I will describe the method (§5.1) and give a few more injectivity results (§5.2). We shall actually prove injectivity results "with parameters", i.e., for functors on the category of k-algebras, of the shape $A \mapsto F(Z \times_k A)$, where Z/k is a fixed k-variety. Based on one such injectivity result for the Chow groups, in §5.3 we shall give a new proof of a codimension one purity theorem due to M. Rost.

§5.1. A general formalism. Let k be a field and F be a covariant functor $A \mapsto F(A)$ from the category of noetherian k-algebras (not necessarily of finite type), with morphisms the flat homomorphisms of rings, to the category of pointed sets, i.e., sets equipped with a distinguished element. The distinguished element in $F(A)$ shall be denoted $\mathbf{1}_A$, and often simply $\mathbf{1}$. Given $A \to B$, the kernel of $F(A) \to F(B)$ is the set of elements of $F(A)$ whose image is $\mathbf{1}_B$. Consider the following properties.

A1. F commutes with filtering direct limits of rings (with flat transition homomorphisms).

A2. *Weak homotopy.* For all fields L containing k, and for all $n \geq 0$, the map

$$F(L[t_1, \ldots, t_n]) \to F(L(t_1, \ldots, t_n))$$

has trivial kernel (i.e., kernel reduced to **1**.)

A3. *Patching.* Given an étale inclusion of integral k-algebras $A \to B$ and given a nonzero element $f \in A$ such that the induced map $A/f \to B/f$ is an isomorphism, the induced map on kernels

$$\mathrm{Ker}[F(A) \to F(A_f)] \to \mathrm{Ker}[F(B) \to F(B_f)]$$

is onto.

THEOREM 5.1.1 ([**CT/Oj92**]). *Let k be an infinite field. Assume that F satisfies* **A1**, **A2**, *and* **A3**. *If $L \supset k$ and A is a local ring of a smooth L-variety, with fraction field K, then*

$$\ker[F(A) \to F(K_A)] = \mathbf{1}.$$

PROOF. (1) One starts from the presentation for principal hypersurfaces in a smooth variety over an infinite field already mentioned in §3.7. This presentation, which differs from Quillen's, will henceforth be referred to as Ojanguren's presentation. Let us repeat the description here.

Let $\mathrm{Spec}(B)/\mathrm{Spec}(k)$ be smooth and integral of dimension d, let $f \in B$ be nonzero and let P be a closed point on the zero set of f. Then up to shrinking $\mathrm{Spec}(B)$ around P, there is an étale map $\mathrm{Spec}(B) \to \mathrm{Spec}(R)$, where $\mathrm{Spec}(R)$ is an open set in affine space $\mathbb{A}_k^d$ and a $g \in R$ such that $g \mapsto f$ and $R/g \overset{\approx}{\to} A/f$. That is: "*A germ of a hypersurface on a smooth variety is analytically isomorphic to a germ of a hypersurface in affine space of the same dimension*".

(2) One then uses a patching argument to reduce the problem to the case of a local ring of an affine space.

Start with A a local ring of a smooth variety X/k at a closed point P (one easily reduces to the case of a closed point). Let K be the quotient field of A. Assume $\alpha \in F(A)$ is sent to $\mathbf{1} \in F(K)$. By Axiom **A1**, α comes from $\alpha_B \in F(B)$ for some B of finite type and smooth over k, and $\alpha \mapsto \mathbf{1} \in F(B_g)$ for some $g \in B$. By Ojanguren's presentation, shrinking $\mathrm{Spec}(B)$ somewhat around the closed point P, we find an étale map $\mathrm{Spec}(B) \to \mathrm{Spec}(R)$, where $\mathrm{Spec}(R)$ is an open set in affine space $\mathbb{A}_k^d$ and a $g \in R$ such that $g \mapsto f$ and $R/g \overset{\approx}{\to} A/f$. By Axiom **A3**, α_B lifts to $\alpha_R \in F(R)$ such that $\alpha_R \mapsto \mathbf{1} \in F(R_f)$. This reduces us to the case of a local ring of a closed point in affine space $\mathbb{A}_k^n$.

(3) Let A be the local ring at a closed point M of $\mathbb{A}_k^n$, let K be the quotient field of A and let $\alpha \in F(A)$ have image $\mathbf{1} \in F(K)$. Rather than giving the complete argument, for which we refer to [**CT/Oj92**], we do the

case $n = 1$ and sketch the case $n = 2$. Let $\mathfrak{m}$ be the maximal ideal of $k[t_1, \ldots, t_n]$ corresponding to M.

CASE $n = 1$. Let $k[t]$ be the polynomial ring in one variable. By axiom **A1**, there exists an $f \in k[t]$, $f \notin \mathfrak{m}$ such that α comes from some $\alpha_1 \in F(k[t]_f)$. By **A1** again, there exists a $g \in k[t]$ such that α_1 has image $\mathbf{1} \in F(k[t]_{fg})$. We may change $g(t)$ so that it becomes coprime with $f(t)$. The map $k[t] \to k[t]_f$ then induces an isomorphism $k[t]/g \simeq (k[t]_f)/g$. (This uses one-dimensionality: if the two elements f and g in the ring $k[t]$ have no common divisor, then they span the whole of $k[t]$.) From the patching axiom **A3** we conclude that there exists $\alpha_2 \in F(k[t])$ with image α_1 in $F(k[t]_f)$, and with image $\mathbf{1}$ in $F(k[t]_{fg})$, hence also in $F(k(t))$. Now the weak homotopy axiom **A2** implies $\alpha_2 = \mathbf{1} \in F(k[t])$, hence $\alpha_1 = \mathbf{1}$ and $\alpha = \mathbf{1} \in F(A)$.

CASE $n = 2$ (detailed sketch). By axiom **A1**, α comes from a class $\alpha_1 \in F(k[x, y]_f)$ for some $f \notin \mathfrak{m}$. Then α_1 maps to $\mathbf{1} \in F(k[x, y])_{fg}$) for some $g \in k[x, y]$. We want to patch α to $\mathbf{1}$ as before. Since $k[x, y]$ is a unique factorization domain, we may still change g so that f and g have no common divisor. But now the closed set defined by $f = g = 0$ may be nonempty and the map $k[x, y]/g \to (k[x, y]_f)/g$ need not be an isomorphism!

To overcome this problem, one makes a general position argument—here we assume k is infinite and perfect. One thus reduces to the case where

(a) the projection of M on $\mathbb{A}_k^1 = \operatorname{Spec}(k[x])$ is the point O defined by $x = 0$,

(b) f and g are monic in the second variable y,

(c) $f(0, y)$ and $g(0, y)$ are coprime, i.e., $f = 0$ and $g = 0$ do not meet above O as shown in Figure 1.

Over some interval around $O \in \mathbb{A}_k^1$, say on $\operatorname{Spec}(k[X]_{h(X)}[Y])$, $f = 0$ and $g = 0$ do not meet and we can use the patching axiom **A3** to produce $\beta \in \operatorname{Spec}(k[X]_{h(X)}[Y])$ which lifts $\alpha \in F(A)$. Now β becomes $\mathbf{1} \in F(k(X, Y))$ and so by the homotopy axiom **A2** (with the ground field $k(X)$) it becomes $\mathbf{1}$ already in $F(k(X)[Y])$. Thus by axiom **A1** there exists a polynomial $r(X)$ such that β becomes $\mathbf{1}$ in $F(k[X]_{h(X)r(X)}[Y])$. We may change r to ensure that h and r are coprime. Now the open sets $r \neq 0$ and $h \neq 0$ cover the entire line $\mathbb{A}_k^1 = \operatorname{Spec}(k[x])$, hence their inverse images cover the whole plane $\mathbb{A}_k^2$, and we may use axiom **A3** to produce a class $\gamma \in F(k[x, y])$ with image $\beta \in \operatorname{Spec}(k[X]_{h(X)}[Y])$ and with trivial image in $F(k(x, y))$. From the weak homotopy axiom, we conclude $\gamma = \mathbf{1} \in F(k[x, y])$, hence also $\beta = \mathbf{1}$, hence finally $\alpha = \mathbf{1}$.

This completes the proof in the two-dimensional case (k perfect). The higher dimension, non-perfect case can be done by a more intricate version of the same argument [**CT/Oj92**]. □

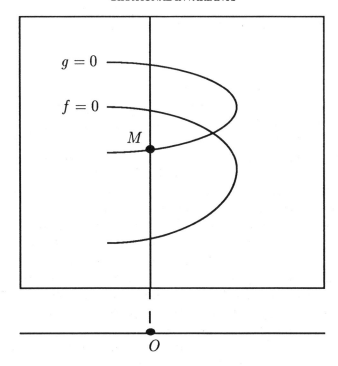

Figure 1

§5.2. The injectivity property.

Let $W(R)$ denote the Witt group of a ring R with $2 \in R^*$. When $B = k$, the following result is Ojanguren's initial result [Oj80].

THEOREM 5.2.1. *Let k be a field, and let A be a local ring of a smooth k-variety. Let K be the quotient field of A. Assume $2 \in A^*$. Let B be a k-algebra of finite type. Then the map $W(B \otimes_k A) \to W(B \otimes_k K)$ on Witt groups is an injection.*

PROOF. Axiom **A1** is easy. Axiom **A2** here is just a special case of a result of Karoubi ([Ka73], [Ka75], [Oj84]): for any commutative ring R with $2 \in R^*$, the map $W(R) \to W(R[t])$ is an isomorphism. To show that axiom **A3** (patching) holds, one reduces to a patching property for quadratic spaces. That property is Theorem 1 of [Oj80]. See also [CT/Oj92, Proposition 2.6]. □

THEOREM 5.2.2. *Let k be an infinite field. Let Z/k be an arbitrary variety. If A is a local ring of a smooth variety over k, and K is its quotient field, then for any integer $n \geq 0$:*
  (1) *the maps on G-theory groups*

$$G_n(Z \times_k A) \to G_n(Z \times_k K)$$

are injective;

(2) *if Z/k is smooth, the maps on K-theory groups*

$$K_n(Z \times_k A) \to K_n(Z \times_k K)$$

are injective.

PROOF. By $Z \times_k A$ we actually mean $Z \times_{\mathrm{Spec}(k)} \mathrm{Spec}(A)$. The groups $G_n(Y)$, also denoted $K_n'(Y)$, are the groups associated by Quillen to the exact category of coherent modules on a scheme Y and the groups $K_n(Y)$ are the groups associated to the exact category of locally free modules on Y. For each n there is a natural map $K_n(Y) \to G_n(Y)$ and this map is an isomorphism if Y is a regular scheme [**Qu73**, §4, Corollary 2, p. 26]. The second statement therefore follows from the first one.

We only have to check axioms **A1**, **A2**, and **A3** for the covariant functor from the category of k-algebras (with morphisms the flat k-homomorphisms) to abelian groups given by $F(A) = G_n(Z \times_k A)$. For axiom **A1**, see [**Qu73**, §2 (9), p. 20] and [**Qu73**, §7, Prop. 2.2, p. 41]. Much more than axiom **A2** holds in the present context. Namely, for A noetherian, the maps $G_n(Z \times_k A) \to G_n(Z \times_k A[t_1, \ldots, t_m])$ are isomorphisms [**Qu73**, §6, Theorem 8, p. 38]. Thus if $L \supset k$ is a field and α is in the kernel of $G_n(Z \times_k L[t_1, \ldots, t_m]) \to G_n(Z \times_k L(t_1, \ldots, t_m))$, then α comes from an element $\beta \in G_n(Z \times_k L)$ which vanishes in $G_n(Z \times_k L(t_1, \ldots, t_m))$, hence by axiom **A1** vanishes in $G_n(Z \times_k L[t_1, \ldots, t_m]_f)$ for some nonzero $f \in L[t_1, \ldots, t_m]$. Now, since k and L are infinite, we may specialize to an L-rational point of $\mathrm{Spec}(L[t_1, \ldots, t_m]_f)$ and we conclude $\alpha = 0$. As for the patching axiom **A3**, it follows from the localization sequence for G-theory ([**Qu73**, §7, Proposition 3.2 and Remark 3.4, p. 44]; see [**CT/Oj92**, p. 112], where the whole argument is developed in the special case $Z = \mathrm{Spec}(k)$). $\square$

REMARK 5.2.3. When $Z = \mathrm{Spec}(k)$, the above result is just a special case of the Gersten conjecture for K-theory of smooth k-varieties, as proved by Quillen ([**Qu73**]). It is likely that Quillen's approach could also yield the more general result given above.

REMARK 5.2.4. For A as above, the injection $G_n(A) \to G_n(K)$ is part of the long exact sequence

$$0 \to G_n(A) \to G_n(K) \to \bigoplus_{x \in A^{(1)}} G_{n-1}(k(x)) \to \cdots,$$

the Gersten sequence. In my Santa Barbara lectures, I mentioned that the theorem above makes it likely that for an arbitrary variety Z/k the complex

$$0 \to G_n(Z \times_k A) \to G_n(Z \times_k K) \to \bigoplus_{x \in A^{(1)}} G_{n-1}(Z \times_k k(x)) \to \cdots$$

is exact. This has since been checked (see [**CT/Ho/Ka94**]).

THEOREM 5.2.5. *Let k be an infinite field. Let Z/k be an arbitrary variety. If A is a local ring of a smooth variety over k, and K is its quotient field,* *then for any integer* $n \geq 0$ *the map*

$$CH^n(Z \times_k A) \to CH^n(Z \times_k K)$$

is injective. In particular, the group $CH^n(Z \times_k A)$ *is zero for* $n > \dim Z$.

PROOF. Rather than invoking the general formalism, let us go through a variant of it. First of all, Chow groups are generally defined only for varieties over a field [**Fu84**]. Let us therefore make the statement more precise. Let R be a k-algebra of finite type which is an integral domain. Let $A = R_p$ be the local ring of R at a prime ideal p. We *define* $CH^n(Z \times_k A)$ as the direct limit of the $CH^n(Z \times_k R_f)$ for all $f \notin p$. The transition maps are given by the flat pull-back map [**Fu84**, I.1.7]. Similarly, $CH^n(Z \times_k K)$ is defined as the direct limit of the $CH^n(Z \times_k R_f)$ for all nonzero f's.

Let α be in the kernel of $CH^n(Z \times_k A) \to CH^n(Z \times_k K)$. Changing R into R_f for suitable $f \notin p$, we may represent α by an element $\beta \in CH^n(Z \times_k R)$. There exists a nonzero $g \in R$ such that β becomes zero when restricted to $CH^n(Z \times_k R_g)$. By Ojanguren's presentation of hypersurfaces in smooth varieties, replacing R by R_f for some suitable $f \notin p$, we may assume that there exists an étale map $\mathrm{Spec}(R) \to \mathrm{Spec}(S)$ where $S = k[t_1, \ldots, t_d]_r$ and an element $h \in S$ such that the image of h in R differs from g by a unit, and that moreover the induced map $S/h \to R/f$ is an isomorphism. We then have the commutative diagram of localization sequences for the Chow group [**Fu84**, I. 1.8]

$$
\begin{array}{ccccccc}
CH^{n-1}(Z \times_k (S/h)) & \longrightarrow & CH^n(Z \times_k S) & \longrightarrow & CH^n(Z \times_k S_h) & \longrightarrow & 0 \\
\downarrow & & \downarrow & & \downarrow & & \\
CH^{n-1}(Z \times_k (R/g)) & \longrightarrow & CH^n(Z \times_k R) & \longrightarrow & CH^n(Z \times_k R_g) & \longrightarrow & 0
\end{array}
$$

In this diagram, the map $CH^{n-1}(Z \times_k (S/h)) \to CH^{n-1}(Z \times_k (R/g))$ is an isomorphism. A straightforward diagram chase then shows that β comes from an element $\gamma \in CH^n(Z \times_k S)$ with image 0 in $CH^n(Z \times_k S_h)$. Now the restriction map $CH^n(Z \times_k k[t_1, \ldots, t_n]) \to CH^n(Z \times_k S)$ is (trivially) surjective. The flat pull-back map $CH^n(Z) \to CH^n(Z \times_k k[t_1, \ldots, t_d])$ is surjective [**Fu84**, I. 1.9]. Thus γ comes from an element $\delta \in CH^n(Z)$ with trivial image in $CH^n(Z \times_k S_h)$. Since k is infinite, we may *specialize* at some k-point of the open set $\mathrm{Spec}(S_h)$ of $\mathbb{A}_k^d$ to conclude that $\delta = 0$, hence $\gamma = 0$, hence $\beta = 0$, hence finally $\alpha = 0$. Note that the inclusion of a k-point in $\mathrm{Spec}(S_h)$ is not a flat map and that some care should be exercised here, since Chow groups are not a priori functorial contravariant with respect to arbitrary morphisms. However, it is a regular embedding, hence so is the embedding $Z \subset Z \times_k S_h$, and we may use the pull-back maps as defined in [**Fu84**, Chapter VI], and use the functoriality [**Fu84**, VI, Proposition 6.5, p. 110]. $\square$

REMARK 5.2.6. A rather convoluted proof of this theorem in a very special case was given in [CT/Pa/Sr89].

THEOREM 5.2.7. *Let k be an infinite field. Let Z/k be an arbitrary variety. If A is a local ring of a smooth variety over k, and K is its quotient field, then for any integer $n \geq 0$, any positive integer m prime to $\mathrm{char}(k)$ and any integer j, the maps on étale cohomology groups*

$$H^n(Z \times_k A, \mu_m^{\otimes j}) \to H^n(Z \times_k K, \mu_m^{\otimes j})$$

are injective.

PROOF. Axiom **A1** is proved in [**SGA4**, t.2, exp. VII, Corollaire 5.9, p. 362]. For any scheme X with m invertible on X, the pull-back maps

$$H^n(X, \mu_m^{\otimes j}) \to H^n(X[t], \mu_m^{\otimes j})$$

are isomorphisms ([**SGA4**, XV, Corollaire 2.2]; see also [**Mi80**, VI. 4.20, p. 240]), hence so are the maps

$$H^n(Z \times_k A, \mu_m^{\otimes j}) \to H^n(Z \times_k A[t_1, \ldots, t_r], \mu_m^{\otimes j})$$

for any positive integer r. Axiom **A2** follows just as above from a specialization argument. As for axiom **3** it follows from the excision property in étale cohomology [**Mi80**, III. 1.27, p. 92], completed in [**CT/Oj92**, p. 114/115]. □

REMARK 5.2.8. When $\mathrm{char}(k) = 0$ and Z/k is smooth, a similar statement holds with étale cohomology with coefficients in the sheaf $\mathbf{G}_m$ (see [**CT/Oj92**, loc. cit.] in the special case $Z = \mathrm{Spec}(k)$).

REMARK 5.2.9. Just as with G-theory, it is natural to think that the injection in the theorem above is just a special case of a statement that would say: the Bloch-Ogus-Gersten exact sequence for A, with coefficients $\mu_n^{\otimes j}$ remains exact when crossed with an arbitrary k-variety Z. This is proved in [**CT/Ho/Ka94**].

In all the examples above, the functors F we were considering were from the category of k-algebras to the category of *abelian groups*. The motivation for phrasing the axioms in terms of pointed sets, rather than groups, comes from the study of principal homogeneous spaces under linear algebraic groups. Let k be a field, $\mathrm{char}(k) = 0$, let G be a linear algebraic group over k. If X is a k-scheme, a principal homogeneous space (or torsor) over X under G is a k-scheme Y equipped with a faithfully flat k-morphism $p: Y \to X$ (the structural morphism) and with an action $G \times_k Y \to Y$, $(g, y) \mapsto g.y$ which respects the projection (namely $p(gy) = p(y)$). Also, the group G must act faithfully and transitively in the fibres of the projection. In other words the map $G \times_k Y \to Y \times_X Y$ given by $(g, y) \mapsto (gy, y)$ is an isomorphism. Under the assumptions above, one may show that the set of isomorphism classes of torsors over X under G is classified by the Čech cohomology set $\check{H}^1_{\text{ét}}(X, G)$, henceforth simply denoted $H^1(X, G)$. This is a pointed set, the class **1** corresponding to the isomorphism class of the trivial torsor $Y = X \times_k G$ equipped with the projection onto the first factor X. A torsor is in the trivial class if and only if the projection $p: Y \to X$ has a section. (For more details on principal homogeneous spaces, the reader may consult [**Mi80**, III, §4] and the literature quoted there.)

In [**CT/Oj92**], Ojanguren and I study the functor from k-algebras to pointed sets given by $A \mapsto H^1(A, G)$. For simplicity, let us assume that $\text{char}(k) = 0$ and that G is a reductive k-group. Axioms **A1** and **A3** (patching) may be checked in this context. However, axiom **A2** (the weak homotopy axiom) does not hold in general. It holds when one goes from L to $L[t]$ (one variable) as proved by Raghunathan and Ramanathan [**Ra/Ra84**]. However, in the general case it fails as soon as one goes from L to $L[t_1, t_2]$, as demonstrated by various counterexamples due to Parimala, Ojanguren, Sridharan, Raghunathan (see [**Ra89**]). However, in [**Ra89**], Raghunathan essentially shows that axiom **A2** holds if all the k-simple components of the derived group of G are k-isotropic (e.g., k is algebraically closed). So the formalism above applies. Using an idea of Raghunathan, Ojanguren and I could produce a variant of the formalism above and prove the general

THEOREM 5.2.10 ([**CT/Oj92**]). *Let k be a field, $\text{char}(k) = 0$, and let G/k be a linear algebraic group over k. Let A be a local ring of a smooth variety over k, and let K be its quotient field. The map $H^1(A, G) \to H^1(K, G)$ has trivial kernel. In other words, if a principal homogeneous space $p: Y \to X$ under G over a smooth integral k-variety X is rationally trivial, i.e., has a section over a nonempty open set, then it is everywhere locally trivial, i.e., for any point $x \in X$ there exists a neighbourhood U of x such that $p_U: Y \times_X U \to U$ has a section.*

Variants over a field of arbitrary characteristic are available ([**CT/Oj92**] for k infinite and perfect, [**Ra93**] for k infinite). The case of a finite ground field has resisted many assaults.

A PARTICULAR EXAMPLE. Let k be as above, let D/k be a central simple algebra, and let $G = \mathbb{SL}(D)$ be the special linear group on D. We have an exact sequence of algebraic groups

$$1 \to \mathbb{SL}(D) \to \mathbb{GL}(D) \to \mathbb{G}_{m,k} \to 1$$

given by the reduced norm map $\mathbb{GL}(D) \to \mathbb{G}_{m,k}$. A variant of Hilbert's theorem 90 says that for any local k-algebra A, the set $H^1(A, \mathbb{GL}(D))$ is reduced to **1**. One therefore gets a bijection

$$A^* / \text{Nrd}((D \otimes_k A)^*) \simeq H^1(A, \mathbb{SL}(D)).$$

The previous theorem therefore implies that if A is a local ring of a smooth k-variety and K is its field of fractions, then the map

$$A^* / \text{Nrd}((D \otimes_k A)^*) \to K^* / \text{Nrd}((D \otimes_k K)^*)$$

is injective. In words, if an element of A^* is a reduced norm from $D \otimes_k K$, then it is also a reduced norm from $D \otimes_k A$. As a very special case, one may take D to be the usual quaternion algebras, and for A as above, we obtain: if $a \in A^*$ is a sum of 4 squares in K, then it is also a sum of 4 squares in A.

One may wonder whether a similar property holds for the reduced norm of an arbitrary Azumaya algebra over a regular local ring A. This is known when $\dim(A) = 2$ ([**Oj82b**], [**CT/Oj92**]) and for A a local ring of a smooth variety of dimension 3 over an infinite field k [**CT/Oj92**].

§5.3. Codimension one purity for some functors. Let k be a field. For any k-variety Z we define the *norm subgroup* $N_Z(k) \subset k^*$ to be the subgroup spanned by the norm subgroups $N_{K/k}(K^*)$, where K/k runs through all finite field extensions of k such that the set $Z(K) = \mathrm{Hom}_{\mathrm{Spec}(k)}(\mathrm{Spec}(K), Z)$ of K-rational points is not empty. Equivalently, $N_Z(k) \subset k^*$ is the subgroup spanned by the norm subgroups $N_{k(P)/k}(k(P)^*) \subset k^*$, for P running through the closed points of the k-variety Z—here $k(P)$ denotes the residue field at P. Of course, if $Z(k) \neq \varnothing$, then $N_Z(k) = k^*$. If $L \supset k$ is any field extension, we let $N_Z(L) = N_{Z \times_k L}(L)$.

Here are two concrete examples. If D/k is a central simple algebra and Z/k is the associated Severi-Brauer variety, then $N_Z(k) \subset k^*$ coincides with the subgroup $\mathrm{Nrd}(D^*) \subset k^*$. If q is a nondegenerate quadratic form in $n \geq 3$ variables and q represents 1 over k, and if Z is the projective quadric defined by q, then $N_Z(k)$ is the subgroup of k^* spanned by the nonzero values of q on k^n. If q is a Pfister form, then $N_Z(k)$ is simply the group of (nonzero) values of the Pfister form.

Given any integral variety X/k with function field $k(X)$, we then define two subgroups of $k(X)^*$:

$$D_Z^1(X) = \{f \in k(X)^* | \forall M \in X^1 \, f = u_M g_M, \, u_M \in \mathscr{O}_{X,M}^*, \, g_M \in N_Z(k(X))\},$$

$$D_Z(X) = \{f \in k(X)^* | \forall M \in X \, f = u_M g_M, \, u_M \in \mathscr{O}_{X,M}^*, \, g_M \in N_Z(k(X))\}.$$

We trivally have $D_Z(X) \subset D_Z^1(X)$.

The following theorem is due to Markus Rost [**Ro90**]. Rost's proof relied on Quillen's presentation of hypersurfaces in smooth varieties. The proof given below starts with a reduction also due to Rost, but to complete the proof, I use the formalism described above, applied to the Chow group functor.

THEOREM 5.3.1. *Assume that Z/k is a proper equidimensional variety. If X/k is a smooth integral k-variety, then $D_Z(X) = D_Z^1(X)$.*

PROOF. Let $p: Z \to \mathrm{Spec}(k)$ be the structural morphism. Let d be the dimension of Z. Let R be the local ring of X at a point $M \in X$. Consider the commutative diagram

$$
\begin{array}{ccc}
Z_R & \supset & Z_K \\
{\scriptstyle p_R}\downarrow & & \downarrow{\scriptstyle p_K} \\
\operatorname{Spec} R & \supset & \operatorname{Spec} K
\end{array}
$$

where the vertical maps are proper. From [**Fu84**, I, Proposition 1.4], we then get the commutative diagram

$$
\begin{array}{ccccc}
\bigoplus\limits_{P\in Z_R^{(d)}} k(P)^* & \longrightarrow & \bigoplus\limits_{P\in Z_R^{(d+1)}} \mathbb{Z} & \longrightarrow CH^{d+1}(Z_R) \longrightarrow 0 \\
{\scriptstyle p_*}\downarrow & & \downarrow{\scriptstyle p_*} & \\
R^* \longrightarrow K^* & \longrightarrow & \bigoplus\limits_{p\in R^{(1)}} \mathbb{Z} &
\end{array}
$$

where the middle horizontal maps are the divisor maps. The top row is exact by the definition of Chow groups. The bottom row is exact since an element of K^* with trivial divisor is a unit in the regular ring R (a very special case of Gersten's conjecture!). We have

$$
p_* \left(\bigoplus_{P\in Z_R^{(d)}} k(P)^* \right) = p_* \left(\bigoplus_{P\in Z_K^{(d)}} k(P)^* \right) = N_Z(K).
$$

On the other hand, if $f \in K^*$ belongs to $D_Z^1(X)$, by the analogous and compatible diagram over each local ring R_p for p prime of height one, we conclude that $\operatorname{div}(f) = p_*(z)$ for some $z \in \bigoplus_{P\in Z_R^{(d+1)}} \mathbb{Z}$.

By Theorem 5.2.5 (whose proof uses Ojanguren's presentation and the formalism described above) we know that $CH^{d+1}(Z_R) = 0$. From the commutative diagram we conclude that there exists $g \in p_*(\bigoplus_{P\in Z_R^{(d)}} k(P)^*) = N_Z(K)$ such that $\operatorname{div}(f/g) = 0$, hence by the exactness of the sequence below, $f = ug$ with $u \in R^*$. $\square$

COROLLARY 5.3.2. *Let* Z/k *be a proper* k-*variety. Let* X *and* Y *be smooth, proper, integral* k-*varieties. If* X *is* k-*birational to* Y, *then the group* $D_Z^1(X)$ *is isomorphic to* $D_Z^1(Y)$, *and the quotient* $D_Z^1(X)/N_Z(k(X))$ *is isomorphic to* $D_Z^1(Y)/N_Z(k(Y))$. *If* $X = \mathbb{P}_k^d$, *then the natural map* $k^*/N_Z(k) \to D_Z(X)/N_Z(k(X))$ *is an isomorphism.*

PROOF. Only the last statement has not been proved. But if in the diagram above one lets $R = k[t_1, \ldots, t_d]$ the proof follows, since $k^* = R^*$. $\square$

REMARK 5.3.3. Let $k = \mathbf{R}$ and let Z be the smooth projective quadric over k defined by the Pfister form $\langle 1, 1\rangle^{\otimes d}$. If $X/\mathbf{R}$ is a smooth, projective, geometrically integral variety of dimension d, the group $D_Z(X)/N_Z(k(X))$ may be identified with the group $(\mathbb{Z}/2)^s$ where s is the number of connected components of the topological space $X(\mathbf{R})$ of real points of X [CT78]. This number is a birational invariant. Rost's theorem is a far reaching generalization of this fact.

Let now D be a central simple algebra over a field k. Let Z/k be the associated Severi-Brauer variety. Let $G = \mathbb{SL}(D)$ be the special linear group on D. Let X be a smooth integral k-variety, and let $E/k(X)$ be a principal homogeneous space under G over the generic point of X. Assume that E may be extended to a principal homogeneous space in some neighbourhood of each codimension 1 point on X. Then Theorem 5.3.1 implies that everywhere locally on X the principal homogeneous space $E/k(X)$ may be extended (up to isomorphism) to a principal homogeneous space. This result, first proved in a special case in [CT/Pa/Sr89] has also been proved in a more K-theoretical way in [CT/Oj92, Theorem 5.3] (the proof there uses Quillen's approach).

Whether the same codimension 1 purity statement holds true for principal homogeneous spaces under an arbitrary linear algebraic group G/k over a smooth k-variety X is an open question. Only the case $\dim(X) = 2$ is known [CT/Sa79]; the basic ingredient of the proof being that reflexive modules over two-dimensional regular local rings are free.

Acknowledgments

Section 5 dwells on a formalism due to M. Ojanguren in the Witt group context [Oj80] and further developed in our joint paper [CT/Oj92]. I had helpful discussions with my Indian colleagues, in particular Parimala and Sujatha, both during the preparation of the lectures and during the writing up of the notes. Some of these discussions were made possible by support from the Indo-French Centre for the Promotion of Advanced Research (IFCPAR). I thank Steve Landsburg for producing an on-the-spot preliminary write-up of the lectures, and I thank Wayne Raskind for going through the final version of this text.

Finally, I thank Bill Jacob and Alex Rosenberg for having given me an opportunity to reflect on some basic concepts, and for the superb setting of the conference.

REFERENCES

[Ar75] J. Kr. Arason, *Cohomologische Invarianten quadratischer Formen*, J. Algebra **36** (1975), 448–491.

[Au/Br68] M. Auslander and A. Brumer, *Brauer groups of discrete valuation rings*, Nederl. Akad. Wetensch. Proc. Ser. A. **71** (1968), 286–296.

[Au/Go60] M. Auslander and O. Goldman, *The Brauer group of a commutative ring*, Trans. Amer. Math. Soc. **97** (1960), 367–409.

[Ar/Mu72] M. Artin and D. Mumford, *Some elementary examples of unirational varieties which are not rational*, Proc. London Math. Soc. **25** (1972), 75–95.

[BV92a] L. Barbieri-Viale, *Des invariants birationnels associés aux théories cohomologiques*, C. R. Acad. Sci. Paris **315** Série I (1992), 1259–1262.

[BV92b] _____, *Cicli di codimensione 2 su varietà unirazionali complesse*, no. 214, Diparti-
mento di Matematica, Università di Genova, 1992 (To appear in the Proceedings of the
Strasbourg *K*-theory Conference, Astérisque.).

[Bl80] S. Bloch, *Lectures on algebraic cycles*, Duke Univ. Math. Ser. 4, Duke Univ. Press,
Durham, NC, 1980.

[Bl81] _____, *Torsion algebraic cycles, K_2 and Brauer groups of function fields*, Groupe de
Brauer, Séminaire, Les Plans-sur-Bex, Suisse 1980 (M. Kervaire et M. Ojanguren, éd.),
Lecture Notes in Math., vol. 844, Springer-Verlag, Berlin and New York, pp. 75–102.

[Bl/Og74] S. Bloch and A. Ogus, *Gersten's conjecture and the homology of schemes*, Ann. Sci.
École Norm. Sup. 7 (1974), 181–202.

[Bl/Sr83] S. Bloch and V. Srinivas, *Remarks on correspondences and algebraic cycles*, Amer. J.
Math. 105 (1983), 1235–1253.

[Bo87] F. A. Bogomolov, *The Brauer group of quotient spaces by linear group actions*, Izv. Akad.
Nauk SSSR Ser. Mat. 51 (1987), 485–516; English transl. in Math. USSR Izv. 30 (1988).

[Bo89] _____, *Brauer groups of fields of invariants of algebraic groups*, Mat. Sb. 180 (1989),
279–293; English transl. in Math. USSR-Sb. 66 (1990).

[Ca92] F. Campana, *Connexité rationnelle des variétés de Fano*, Ann. Sci. École Norm. Sup. 25
(1992), 539–545.

[CT78] J.-L. Colliot-Thélène, *Formes quadratiques multiplicatives et variétés algébriques*, Bull.
Soc. Math. France 106 (1978), 113–151.

[CT79] _____, *Formes quadratiques sur les anneaux semi-locaux réguliers*, Colloque sur les
formes quadratiques, Montpellier 1977, Bull. Soc. Math. France, Mémoire 59, Soc. Math.,
France, Paris, 1979, pp. 13–31.

[CT80] _____, *Formes quadratiques multiplicatives et variétés algébriques: deux compléments*,
Bull. Soc. Math. France 108 (1980), 213–227.

[CT92] _____, *On the reciprocity sequence in higher class field theory of function fields*, Algebraic
K-Theory and Algebraic Topology (Lake Louise, 1991) (P. G. Goerss and J. F. Jardine
eds.), NATO Adv. Sci. Inst. Ser. C, vol. 407, Kluwer, Dordrecht, 1993, pp. 35–55.

[CT93] _____, *Cycles algébriques de torsion et K-théorie algébrique*, Arithmetical Algebraic Geometry (E. Ballico, ed.), C.I.M.E., 1991, Lecture Notes in Math., vol. 1553, Springer-Verlag,
Berlin and New York, 1993, pp. 1–49.

[CT/Ho/Ka94] J.-L. Colliot-Thélène, R. Hoobler, and B. Kahn, *Equivariant refinements of Gersten's conjecture*, work in progress.

[CT/Oj89] J.-L. Colliot-Thélène et M. Ojanguren, *Variétés unirationnelles non rationnelles: au-
delà de l'exemple d'Artin et Mumford*, Invent. Math. 97 (1989), 141–158.

[CT/Oj92] _____, *Espaces principaux homogènes localement triviaux*, Publ. Math. Inst. Hautes
Études Sci. 75 (1992), 97–122.

[CT/Pa90] J.-L. Colliot-Thélène and R. Parimala, *Real components of algebraic varieties and
étale cohomology*, Invent. Math. 101 (1990), 81–99.

[CT/Pa/Sr89] J.-L. Colliot-Thélène, R. Parimala et R. Sridharan, *Un théorème de pureté locale*,
C. R. Acad. Sci. Paris 309 Série I (1989), 857–862.

[CT/Ra91] J.-L. Colliot-Thélène et W. Raskind, *Groupe de Chow de codimension deux des
variétés définies sur un corps de nombres: un théorème de finitude pour la torsion*, Invent. Math. 105 (1991), 221–245.

[CT/Sa79] J.-L. Colliot-Thélène et J.-J. Sansuc, *Fibrés quadratiques et composantes connexes
réelles*, Math. Ann. 244 (1979), 105–134.

[CT/Sa87] _____, *Principal homogeneous spaces under flasque tori: applications*, J. Algebra 106
(1987), 148–205.

[CT/Sa88] _____, *The rationality problem for fields of invariants under linear algebraic groups
(with special regards to the Brauer group)*, notes from the 1988 ELAM conference, San-
tiago de Chile.

[Dr/Kn80] P. Draxl und M. Kneser, *SK_1 von Schiefkörpern*, Seminar Bielefeld-Göttingen 1976
Lecture Notes in Math. vol. 778, Springer-Verlag, Berlin and New York, 1980.

[Fo89] T. J. Ford, *On the Brauer group of $k[x_1, \ldots, x_n, 1/f]$*, J. Algebra 122 (1989), 410–424.

[Fo92] _____, *On the Brauer group of a localization*, J. Algebra 147 (1992), 365–378.

[Fu84] W. Fulton, *Intersection Theory*, Ergeb. Math. Grenzgeb 3. Folge, Band 2, Springer-Verlag, Berlin and New York, 1984.

[Ga81a] O. Gabber, *Some theorems on Azumaya algebras*, Groupe de Brauer, Séminaire, Les Plans-sur-Bex, Suisse 1980 (M. Kervaire et M. Ojanguren, éd.), Lecture Notes in Math., vol. 844, Springer-Verlag, Berlin and New York, 1981, pp. 129–209.

[Ga81b] _____, *Gersten's conjecture for some complexes of vanishing cycles*, Inst. Hautes Études Sci., 1981, preprint.

[Gi86] H. Gillet, *Gersten's conjecture for the K-theory with torsion coefficients of a discrete valuation ring*, J. Algebra **103** (1986), 377–380.

[Gr/Su88] M. Gros et N. Suwa, *La conjecture de Gersten pour les faisceaux de Hodge-Witt logarithmiques*, Duke Math. J. **57** (1988), 615–628.

[Gr68=GBI,GBII,GBIII] A. Grothendieck, *Le groupe de Brauer* I, II, III, Dix exposés sur la cohomologie des schémas, North-Holland, Amsterdam, 1968.

[Ja/Ro89] B. Jacob and M. Rost, *Degree four cohomological invariants for quadratic forms*, Invent. Math. **96** (1989), 551–570.

[Ja90] U. Jannsen, *Mixed motives and algebraic K-theory*, Lecture Notes in Math., vol. 1400, Springer-Verlag, Berlin and New York, 1990.

[Ka86] K. Kato, *A Hasse principle for two dimensional global fields*, J. Reine Angew. Math. **366** (1986), 142–181.

[Ka73] M. Karoubi, *Périodicité de la K-théorie hermitienne*, Higher Algebraic K-theory III, Hermitian K-Theory and Geometric Applications, Lecture Notes in Math., vol. 343, Springer-Verlag, Berlin and New York, 1973, pp. 301–411.

[Ka75] _____, *Localisation des formes quadratiques*. II, Ann. Sci. École Norm. Sup. 4ème série **8** (1975), 99–155.

[Kn77] M. Knebusch, *Symmetric bilinear forms over algebraic varieties*, Conference on Quadratic Forms, Kingston 1976 (G. Orzech, ed.), Queen's Papers in Pure and Applied Math., vol. 46, Queen's Univ. Press, Kingston, ON, 1977, pp. 103–283.

[Kn91] M. A. Knus, *Quadratic and hermitian forms over rings*, Grundlehren Math. Wiss. **294** (1994).

[Le86] F. Lecomte, *Rigidité des groupes de Chow*, Duke Math. J. **53** (1986), 405–426.

[Me/Su82] A. S. Merkur'ev and A. A. Suslin, *K-cohomology of Severi-Brauer varieties and the norm residue homomorphism*, Izv. Akad. Nauk SSSR **46** (1982), 1011–1046; English transl. in Math. USSR Izv. **21** (1983).

[Mi80] J. S. Milne, *Étale cohomology*, Princeton University Press, Princeton, NJ, 1980.

[Mi80] J. P. Murre, *Un résultat en théorie des cycles de codimension deux*, C. R. Acad. Sci. Paris **296** (1983), 981–984.

[Ni84] Ye. A. Nisnevich, *Espaces homogènes principaux rationnellement triviaux et arithmétique des schémas en groupes réductifs sur les anneaux de Dedekind*, C. R. Acad. Sci. Paris **299** Série I (1984), 5–8.

[Ni89] _____, *Espaces homogènes principaux rationnellement triviaux, pureté et arithmétique des schémas en groupes réductifs sur les extensions d'anneaux locaux réguliers de dimension* 2, C. R. Acad. Sci. Paris **309** Série I (1989), 651–655.

[Oj80] M. Ojanguren, *Quadratic forms over regular rings*, J. Indian Math. Soc. **44** (1980), 109–116.

[Oj82a] _____, *A splitting theorem for quadratic forms*, Comment. Math. Helv. **57** (1982), 145–157.

[Oj82b] _____, *Unités représentées par des formes quadratiques ou par des normes réduites*, Algebraic K-Theory, Oberwolfach, 1980 (K. Dennis, ed.), t. II, Lecture Notes in Math., vol. 967, Springer-Verlag, Berlin and New York, 1982, pp. 290–299.

[Oj84] _____, *On Karoubi's theorem* $W(A) = W(A[t])$, Arch. Math. **43** (1984), 328–331.

[Oj90] _____, *The Witt group and the problem of Lüroth*, Università di Pisa, Dottorato di ricerca in matematica, Ets Editrice Pisa, 1990.

[Pa76] W. Pardon, *The exact sequence of a localization for Witt groups*, Lecture Notes in Math., vol. 551, Springer-Verlag, Berlin and New York, 1976, pp. 336–379.

[Pa82a] _____, *A Gersten conjecture for Witt groups*, Algebraic K-Theory, Part II, Proceedings Oberwolfach 1980 (K. Dennis, ed.), Lecture Notes in Math., vol. 967, Springer-Verlag, Berlin and New York, 1982, pp. 300–315.

[Pa82b] _____, *A relation between Witt groups and zero-cycles in a regular ring*, Algebraic *K*-Theory, Number Theory, Geometry and Analysis, Proceedings, Bielefeld, 1982 (A. Bak, ed.), Lecture Notes in Math., vol. 1046, Springer-Verlag, Boston and New York, 1984, pp. 261–328.

[Pa88] R. Parimala, *Witt groups of affine threefolds*, Duke Math. J. **57** (1988), 947–954.

[Pa89] _____, *Witt groups vis-à-vis Chow groups*, Proceedings of the Indo-French Conference on Geometry (Bombay, 1989) (S. Ramanan and A. Beauville, ed.), National Board for Higher Mathematics, Hindustan Book Agency, Delhi, 1993, pp. 149–154.

[Pe93] E. Peyre, *Unramified cohomology and rationality problems*, Math. Annalen **296** (1993), 247–268.

[Qu73] D. Quillen, *Higher algebraic K-theory: I*, Higher Algebraic *K*-Theory I, Lecture Notes in Math., vol. 341, Springer-Verlag, Berlin and New York, 1973.

[Ra/Ra84] M. S. Ragunathan and A. Ramanathan, *Principal bundles on the affine line*, Proc. Indian Acad. Sci. **93** (1984), 137–145.

[Ra89] M. S. Ragunathan, *Principal bundles on affine space and bundles on the projective line*, Math. Ann. **285** (1989), 309–332.

[Ra93] _____, *Principal bundles admitting a rational section*, Invent. Math. **116** (1994), 409–423.

[Ro90] M. Rost, *Durch Normengruppen definierte birationale Invarianten*, C. R. Acad. Sci. Paris **10**, Série I (1990), 189–192.

[Ro93] _____, *Chow groups with coefficients*, Bonn, 1993, preprint.

[Sa79] H. Saito, *Abelian varieties attached to cycles of intermediate dimension*, Nagoya Math. J. **75** (1979), 95–119.

[Sa91] S. Saito, *On the cycle map for torsion algebraic cycles of codimension two*, Invent. Math. **106** (1991), 443–460.

[Sa92] _____, *Cohomological Hasse principle for a threefold over a finite field*, Algebraic *K*-Theory and Algebraic Topology (Lake Louise, 1991) (P. G. Goerss and J. F. Jardine, ed.), NATO Adv. Sci. Inst. Ser. C, vol. 407 Kluwer, Dordrecht, 1993, pp. 229–241.

[Sa/Su93] S. Saito and R. Sujatha, *A finiteness theorem for cohomology of surfaces over p-adic fields and an application to Witt groups*, these proceedings.

[Sal93] P. Salberger, *Chow groups of codimension two and l-adic realizations of motivic cohomology*, Séminaire de théorie des nombres, Paris (1991/1992) (Sinnou David, éd.), Progr. Math., vol. 116, Birkhäuser, Boston, 1993, pp. 247–277.

[Sa84] D. J. Saltman, *Noether's problem over an algebraically closed field*, Invent. Math. **77** (1984), 71–84.

[Sa85] _____, *The Brauer group and the center of generic matrices*, J. Algebra **97** (1985), 53–67.

[Sa88] _____, *Invariant fields of linear groups and division algebras*, Perspectives in Ring Theory (F. van Oystaeyen and L. Le Bruyn, ed.), Kluwer, Dordrecht, 1988, pp. 279–297.

[Sa90] _____, *Multiplicative field invariants and the Brauer group*, J. Algebra **133** (1990), 533–544.

[Sa93] _____, *Brauer groups of invariant fields, geometrically negligible classes, an equivariant Chow group and unramified H^3*, these proceedings.

[Se59] J.-P. Serre, *On the fundamental group of a unirational variety*, J. London Math. Soc. **34** (1959), 481–484.

[Se65] _____, *Cohomologie galoisienne*, Lecture Notes in Math., vol. 5, Springer-Verlag, Berlin and New York, 1965.

[Se68] _____, *Corps locaux*, Hermann, Paris, 1968.

[SGA4] M. Artin, A. Grothendieck et J.-L. Verdier, *Théorie des topos et cohomologie étale des schémas*, Lecture Notes in Math., vols. 269, 270, 305, Springer-Verlag, Berlin and New York, 1972–73.

[SGA4 1/2] P. Deligne et al., *Cohomologie étale*, Lecture Notes in Math., vol. 569, Springer, Berlin and New York, 1977.

[SGA5] A. Grothendieck et al., *Cohomologie l-adique et fonctions L*, Lecture Notes in Math. vol. 589, Springer, Berlin and New York, 1977.

[Sh89] C. Sherman, *K-theory of discrete valuation rings*, J. Pure Appl. Algebra **61** (1989), 79–98.

[Su83] A. A. Suslin, *On the K-theory of algebraically closed fields*, Invent. Math. **73** (1983), 241–245.

[Su88] _____, *Algebraic K-theory of fields*, Proceedings of the International Congress of Mathematicians, Berkeley 1986, Amer. Math. Soc., Providence, RI, 1988, pp. 222–244.

[Su91] _____, *K-theory and $\mathscr{K}$-cohomology of certain group varieties*, Algebraic K-Theory (A. A. Suslin, ed.), Adv. Soviet Math., vol. 4, Amer. Math. Soc., Providence, RI, 1991, pp. 53–74.

[Th84] R. W. Thomason, *Absolute cohomological purity*, Bull. Soc. Math. France **112** (1984), 397–406.

[vdK76] W. van der Kallen, *The K_2 of a 2-dimensional regular local ring and its quotient field*, Comm. Algebra **4** (1976), 677–679.

[Wi35] E. Witt, *Über ein Gegenbeispiel zum Normensatz*, Math. Zeit. **39** (1935), 462–467.

C. N. R. S., U. R. A. D0752, MATHÉMATIQUES, BÂTIMENT 425, UNIVERSITÉ DE PARIS-SUD, F-91405 ORSAY, FRANCE

Proceedings of Symposia in Pure Mathematics
Volume 58.1 (1995)

K-theory of Simple Algebras

A. S. MERKURJEV

We use the following approach to the study of the K-theory of a simple algebra D of finite dimension over its center F. We define the *reduced norm homomorphism*

$$\mathrm{Nrd}_n : K_n D \to K_n F$$

and study separately the kernel $SK_n D$ and the image $\mathrm{Nrd}_n D$ of this map.

The reduced norm homomorphism is compatible with the usual norm map for a field extension in the following sense. If L is a subfield in D containing F such that $[L : F] = \deg D$, then the following diagram commutes

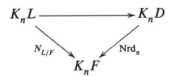

The group $K_0 D$ is infinite cyclic. Some interesting results can be obtained from the study of the behaviour of this group under certain field extensions (function fields of algebraic varieties).

The group $SK_1 D$ is not trivial in general and appears to be the obstruction to the rationality property of the algebraic group $\mathrm{SL}_1(D)$. We prove Rost's theorem giving a cohomological description of $SK_1 D$ for a biquaternion algebra D. In certain cases we give the description of the group of reduced norms $\mathrm{Nrd}_1 D$.

Much less is known about $K_2 D$. Even the definition of Nrd_2 needs certain efforts. The only case where we have a satisfactory description of this group is the case of a quaternion algebra.

Finally, we show that for $n = 3$ (and therefore, for $n \geq 3$) the reduced norm homomorphism does not exist.

1991 *Mathematics Subject Classification*. Primary 11R34, 11R52, 16A54; Secondary 19B20, 19D99.
This paper is in final form and no version of it will be submitted for publication elsewhere.

I thank J.-P. Tignol for his notes which were very helpful for the preparation of this paper.

1. K_0 of simple algebras

Let D be a central simple algebra over a field F. The function $P \mapsto \dim_F P$ on the set of finitely generated D-modules gives rise to the homomorphism $d\colon K_0 D \to \mathbb{Z}$. By Wedderburn's theorem, $D \simeq M_k(T)$ for some division algebra T over F. Since $M = T^k$ is the unique (up to an isomorphism) irreducible D-module, we have $K_0 D = \mathbb{Z} \cdot [M]$. Hence, d is an injection and $\operatorname{im}(d) = \frac{\dim D}{k} \cdot \mathbb{Z}$. Actually, the reduced norm map

$$\operatorname{Nrd}_0\colon K_0 D \to K_0 F = \mathbb{Z}$$

is equal to d divided by $\deg D$ and $\operatorname{im}(\operatorname{Nrd}_0) = \operatorname{ind} D \cdot \mathbb{Z}$.

Now let L/F be any field extension, $D_L = D \otimes_F L$. How can we determine $\operatorname{ind} D_L$? The following idea is due to Schofield and Van den Bergh [24]. Assume L is a function field of some irreducible variety X defined over F. Consider the category $M(X, D)$ of coherent sheaves over X which are left D-modules and the full subcategory $P(X, D)$ of locally free sheaves. It is clear that the composition

$$K_0(P(X, D)) \xrightarrow{i} K_0(M(X, D)) \xrightarrow{j} K_0(D_L) \xrightarrow{d} K_0 L = \mathbb{Z},$$

where j is induced by the functor $\mathscr{F} \mapsto \mathscr{F}_\xi$ ($=$ stalk at generic point) coincides with the map $[\mathscr{F}] \mapsto \operatorname{rank} \mathscr{F}$.

Since any finitely generated D_L-module can be extended to a coherent sheaf of D-modules over X, j is a surjective map. If the variety X is nonsingular, then any coherent sheaf of D-modules has a finite resolution in the category $P(X, D)$, and hence, i is an isomorphism. Comparing the images of d and the composition $d \circ j \circ i$ we get the following

THEOREM 1. *Let D be a central division algebra of finite dimension over a field F, and let X be a nonsingular irreducible variety over F. Then*

$$\operatorname{ind} D_{F(X)} = \tfrac{1}{\deg D} \cdot \gcd\{\operatorname{rank} M, \quad M \in P(X, D)\}.$$

The group $K_0(P(X, D))$ was computed by I. Panin for all homogeneous varieties X, i.e., projective varieties which carry a transitive action of some connected linear algebraic group. This class of varieties contains, for example, grassmannians, quadrics, and also their twisted forms, products and transfers. Therefore, for any homogeneous variety X one can get a formula for $\operatorname{ind} D_{F(x)}$ which is known as the index reduction formula.

We consider in detail the case of quadric hypersurface. Let $\operatorname{char} F \neq 2$, let q be a nondegenerate quadratic form over F, and let X_q be a projective quadric hypersurface defined by the equation $q = 0$. Denote by $F(q) = F(X_q)$ the function field of X_q. It turns out that for any central simple algebra D over F the index of $D_{F(q)}$ is equal either to $\operatorname{ind} D$ or $\tfrac{1}{2} \operatorname{ind} D$.

Using the computation of $K_0(P(X, D))$ [27] one can get the following

THEOREM 2. *Let D be a division algebra over F, let q be a quadratic form over F, and let $C_0(q)$ be the even Clifford algebra of q. Then the following conditions are equivalent*:

(1) $D_{F(q)}$ *is not a division algebra, i.e., $\operatorname{ind} D_{F(q)} = \frac{1}{2} \operatorname{ind} D$.*

(2) *There exists an F-algebra homomorphism $C_0(q) \to D$.*

This theorem stated in a more complicated way was proved in [9]. The present formulation and the "elementary" proof below are due to J.-P. Tignol [29].

Assume $q = \langle 1, -a_1, -a_2, \ldots, -a_n \rangle$ then $C_0(q) = C(\langle a_1, a_2, \ldots, a_n \rangle)$ is generated by $u_1, u_2, \ldots, u_n$ such that $u_i^2 = a_i$ for $i = 1, 2, \ldots, n$ and $u_i u_j = -u_j u_i$ for $i \neq j$ [7].

One can modify the condition (2) as follows:

(2′) There exist $d_1, d_2, \ldots, d_n \in D$ such that $d_i^2 = a_i$ for $i = 1, 2, \ldots, n$ and $d_i d_j = -d_j d_i$ for $i \neq j$.

We have to to show (1) $\Leftrightarrow$ (2′).

(2′) $\Rightarrow$ (1) is easy. We write $F(q) = F(x_1, x_2, \ldots, x_n)$ with $1 - a_1 x_1^2 - a_2 x_2^2 - \cdots - a_n x_n^2 = 0$ and consider $u = \sum x_i d_i \in D_{F(q)}$. Then $u^2 = \sum x_i^2 d_i^2 = \sum x_i^2 a_i = 1$, hence $(u-1)(u+1) = 0$, and therefore, $D_{F(q)}$ is not a division algebra.

(1) $\Rightarrow$ (2′): induction on n. Let $L = F(q)$. If $n = 1$, then $q = \langle 1, -a \rangle$, $L = F(\sqrt{a})$, and since D_L is not a division algebra the quadratic extension $C_0(q) = L$ can be imbedded to D [3].

$n - 1 \Rightarrow n$. Consider the field of rational functions $F' = F(t)$. It is clear that $D' = D(t) = D \otimes_F F'$ is a division algebra over F'. The function field $L' = F'(q')$ of a quadratic form $q' = \langle 1, -a_1, \ldots, -a_{n-2}, -a_{n-1} - t^2 a_n \rangle$ over F' given by the equation $1 - a_1 y_1^2 - \cdots - a_{n-2} y_{n-2}^2 - (a_{n-1} + t^2 a_n) y_{n-1}^2 = 0$ is isomorphic to $F(q)$ (we can set $x_i = y_i$ for $i = 1, 2, \ldots, n-1$ and $x_n = t y_{n-1}$). Since $D' \otimes_{F'} L' = D \otimes_F L$ is not a division algebra, we can find by induction $d_1, d_2, \ldots, d_{n-1} \in D(t)$ such that $d_1^2 = a_1, \ldots, d_{n-2}^2 = a_{n-2}$, $d_{n-1}^2 = a_{n-1} + t^2 a_n$ and $d_i d_j = -d_j d_i$.

For the last (crucial) step, modifying $d_1, \ldots, d_{n-1}$ so that they are in $D[t]$, we need the following

LEMMA 1. *Let Λ be a subring in $D(t)$ containing $F[t]$. Then the following conditions are equivalent*: (1) Λ *is a finitely generated $F[t]$-module*, (2) *There exists $d \in D(t)^*$ such that $d \Lambda d^{-1} \subset D[t]$.*

PROOF. (2) $\Rightarrow$ (1) is easy, since $D[t]$ is finitely generated $F[t]$-module.

(1) $\Rightarrow$ (2). By assumption, there exists a nonzero polynomial $f \in F[t]$ such that $f \Lambda \subset D[t]$. Since $D[t]$ is a P.I.D. and $Df\Lambda$ is a left ideal in $D[t]$, $Df\Lambda = D[t] \cdot d$ for some $d \in D(t)^*$. Finally, we have $d\Lambda \subset D[t] d\Lambda = Df\Lambda\Lambda = Df\Lambda = D[t]d$; hence, $d\Lambda d^{-1} \subset D[t]$. $\square$

Coming back to the proof of the theorem we consider the ring $\Lambda = F[t, d_1, d_2, \ldots, d_{n-1}] \subset D(t)$. The relations show that Λ is generated as $F[t]$-module by $d_1^{\varepsilon_1} d_2^{\varepsilon_2} \cdots d_{n-1}^{\varepsilon_{n-1}}$, where $\varepsilon_i = 0$ or 1. By Lemma 1, one can find $d \in D(t)^*$ such that $d \Lambda d^{-1} \subset D[t]$. Replacing the d_i by $dd_i d^{-1}$ we may assume $d_i \in D[t]$. Therefore, $d_1, \ldots, d_{n-2} \in D$ and $d_{n-1} = d' + d''t$ for $d', d'' \in D$. Then the relations $d_i d_{n-1} = -d_{n-1} d_i$ imply $d_i d' = -d' d_i$ and $d_i d'' = -d'' d_i$. Finally, the equality $d_{n-1}^2 = a_{n-1} + t^2 a_n$ implies $d'^2 = a_{n-1}$, $d''^2 = a_n$, and $d'd'' = -d''d'$. Therefore, $d_1, \ldots, d_{n-2}, d', d''$ yield the required elements.

If $q \notin I^2 F$, then $C_0(q)$ is a simple algebra; therefore, the condition (2) in the theorem is equivalent to $C_0(q) \hookrightarrow D$.

If $q \in I^2 F$, then $C_0(q) \simeq C'(q) \times C'(q)$ for some central simple algebra $C'(q)$. Hence, (2) is equivalent to $C'(q) \hookrightarrow D$.

COROLLARY. *D remains a division algebra over $F(q)$ at least in the following cases*:
(1) *If $q \notin I^2 F$ and $\dim D < 2^{\dim q - 1}$ or $C_0(q)$ is not a division algebra.*
(2) *If $q \in I^2 F$ and $\dim D < 2^{\dim q - 2}$ or $C'(q)$ is not a division algebra.*

EXAMPLE. $q = \langle 1, -a, -b, ab, -c \rangle$, then $C_0(q) = M_2\left(\left(\frac{a,b}{F}\right)\right)$ is not a division algebra, so any division algebra over F remains a division algebra over $F(q)$.

The Corollary can be applied to the study of the values of the u-invariant of a field:

$$u(F) = \sup\{\dim p : p \text{ is an anisotropic quadratic form over } F\}.$$

THEOREM 3 [9]. *For any $n \geq 1$ there exists a field F with $u(F) = 2n$.*

The proof is based on the following observations:
(1) If $p \in I^2 F$ and $C'(p)$ is a division algebra, then p is anisotropic. (For if $p \simeq r \perp \mathbb{H}$, then $C'(p) \simeq M_2(C'(r))$.)
(2) If $\dim p = 2n$ and q is a quadratic form of dimension $2n + 1$, then $C'(p)$ remains a division algebra over $F(q)$. (We may apply the corollary since $\dim C'(p) = 2^{2n-2} < 2^{\dim q - 2}$.)

We start with some field F_0 and $p \in I^2 F_0$, $\dim p = 2n$ such that $C'(p)$ is a division algebra. Let F_1 be a free composite of all function fields $F_0(q)$ where q ranges over all quadratic forms over F_0 of dimension $2n + 1$. The field F_1 satisfies the following properties:
(a) All the forms q over F_0 of dimension $\geq 2n + 1$ are isotropic over F_1.
(2) $C'(p)$ remains a division algebra over F_1.

Iterating this process one gets a sequence of fields $F_0 \subset F_1 \subset F_2 \subset \cdots$. Property (a) implies that the u-invariant of the field $F = \varinjlim F_i$ is at most

$2n$. But it follows from (b) that $C'(p)$ remains a division algebra over F, so p is anisotropic over F and therefore, $u(F) = 2n$.

REMARKS. Slightly modifying the construction (splitting all the forms $\langle 1, -a, -b, ab, -c\rangle$ and $2n$-dimensional forms q such that $q \in I^2$ and $C'(q) \not\simeq C'(p)$)) one can get additional properties for the field F in the theorem:

(1) F has no odd-degree extensions.

(2) The cohomological dimension cd F is at most 2 (cd $F = 2$ if $u(F) \geq 4$).

(3) p is the unique anisotropic quadratic form of dimension $2n$.

2. K_1 of simple algebra

Let D be a central simple algebra over a field F. The reduced norm map Nrd: $D \to F$ can be defined by the commutative diagram

$$\begin{array}{ccc} D & \xrightarrow{\mathrm{Nrd}} & F \\ \uparrow & & \uparrow \\ D_L & \xrightarrow{\det} & L \end{array}$$

where L is a splitting field [3].

The group $K_1 D$ is known to be $D^*/[D^*, D^*]$ [3]. One can define the reduced norm homomorphism

$$\mathrm{Nrd}_1 : K_1 D = D^*/[D^*, D^*] \xrightarrow{\mathrm{Nrd}_1} F^* = K_1 F.$$

The group $SK_1 D = \ker(\mathrm{Nrd}_1) = \{d \in D^* : \mathrm{Nrd}(d) = 1\}/[D^*, D^*]$ has the following properties [3]:

(1) If D is split, then $SK_1 D = 0$.

(2) There is a canonical isomorphism $SK_1(M_n(D)) \simeq SK_1 D$.

(3) Functorial property: for any field extension L/F one has a homomorphism $SK_1 D \to SK_1(D_L)$.

(4) If L/F is a finite extension, then there exists a norm homomorphism $SK_1(D_L) \to SK_1 D$.

(5) ind $D \cdot SK_1 D = 0$ (follows from (3) and (4)).

(6) If the degree $[L : F]$ is prime to ind D, then $SK_1 D \hookrightarrow SK_1(D_L)$.

(7) If $D = D_1 \otimes D_2 \otimes \cdots \otimes D_K$ and the ind D_i are coprime, then $SK_1 D \simeq \coprod SK_1(D_i)$.

(8) Homotopy invariance : $SK_1(D(t)) \simeq SK_1 D$ [20].

(9) If ind D is square free, then $SK_1 D = 0$ [31].

(10) If cd $F \leq 2$, then $SK_1 D = 0$ [25], [32].

In general the group $SK_1 D$ is not trivial [19]. V. Platonov has constructed the following example: let a field k contain a primitive nth root of unity ξ, $a, b \in k^*$, $F = k((X))((Y))$ be the iterated formal Laurent series field, and let $D = A_\xi(a, X) \otimes_F A_\xi(b, Y)$ be a tensor product of two cyclic algebras of degree n. Platonov has shown that

$$SK_1 D \simeq Br(E/k)/(Br(L/k) + Br(M/k)),$$

where $L = F(\sqrt[n]{a})$, $M = F(\sqrt[n]{b})$, $E = L \cdot M$, and for some choice of k, a, b this group is nontrivial.

A. Suslin has conjectured [26] that for any D of degree m there exists a homomorphism

$$\lambda: SK_1 D \to H^4(F, \mu_m^{\otimes 3})/[D] \cup H^2(F, \mu_m^{\otimes 2})$$

such that for Platonov's example we have a commutative diagram

$$
\begin{array}{ccc}
SK_1 D & \xrightarrow{\;\;\sim\;\;} & Br(E/k)/(Br(L/k) + Br(M/k)) \\
\lambda \downarrow & & \big\uparrow \\
H^4(F, \mu_{n^2}^{\otimes 3})/[D] \cup H^2(F, \mu_{n^2}^{\otimes 2}) & \xrightarrow{\partial\partial'} & H^2(k, \mu_{n^2})/(a) \cup H^1(k, \mu_{n^2}) + (b) \cup H^1(k, \mu_{n^2})
\end{array}
$$

where $\partial\partial'$ is a double residue.

Suslin has constructed a homomorphism to a slightly different target group, which is 2 times the desired map [26].

In the case when $D = \left(\frac{a,b}{F}\right) \otimes_F \left(\frac{c,d}{F}\right)$ is a tensor product of two quaternion algebras, M. Rost has proved the following

THEOREM 4. *There exists a natural exact sequence*

$$0 \to SK_1 D \to H^4 F \to H^4 F(q),$$

where $H^4 F = H^4(F, \mathbb{Z}/2)$ and $q = \langle a, b, -ab, -c, -d, cd \rangle$ is an Albert form of D.

REMARK. If D is not a division algebra (or, equivalently, q is isotropic), we have $SK_1 D = 0$ and $H^4 F \hookrightarrow H^4 F(q)$. So we may assume D is a division algebra.

EXAMPLE. Let $\sqrt{-1} \in F^*$, so $\sqrt{-1} \in SK_1 D$. It turns out that the image of $\sqrt{-1}$ in $H^4 F$ equals $u = (a, b, c, d) \in H^4 F$. This immediately gives examples with nontrivial $SK_1 D$. The element u can be derived directly from D or q:

(1) $u = [D]^{(2)}$ is a divided square of D (this observation is due to B. Kahn);

(2) $u = Sw_4(q)$ is the 4th Stiefel-Whitney class of q [16].

PROOF OF THEOREM 4. The proof of the theorem is divided into 3 steps.

Step 1. Let X be any algebraic variety over a field F, by X^i we denote the subset of X consisting of all points of codimension i. The homology of the complex

$$\coprod_{x \in X^{i-1}} K_{n+1} F(x) \to \coprod_{x \in X^i} K_n F(x) \to \coprod_{x \in X^{i+1}} K_{n-1} F(X)$$

where the K_n's are the Milnor K-groups, we denote by $H^i(X, K_{n+i})$ [6] (we assume $K_i = 0$ if $i < 0$). Let $L = F(\sqrt{\alpha})/F$ be a quadratic extension such that X has a rational point over L. Consider the following commutative diagram:

$$
\begin{array}{ccccccc}
H^4L & \longrightarrow & H^4L(X) & & & & \\
\uparrow & & \uparrow & & & & \\
H^4F & \longrightarrow & H^4F(X) & & & & \\
\uparrow{\scriptstyle h} & & \uparrow{\scriptstyle h} & & & & \\
K_3F & \longrightarrow & K_3F(x) & \longrightarrow & \coprod_1 K_2F(x) & & \\
\uparrow{\scriptstyle N} & & \uparrow{\scriptstyle N} & & \uparrow{\scriptstyle N} & & \\
K_3L & \longrightarrow & K_3L(X) & \longrightarrow & \coprod_1 K_2L(x) & \longrightarrow & \coprod_2 K_1L(x) \\
& & \uparrow{\scriptstyle 1-\sigma} & & \uparrow{\scriptstyle 1-\sigma} & & \uparrow{\scriptstyle 1-\sigma} \\
& & K_3L(X) & \longrightarrow & \coprod_1 K_2L(x) & \longrightarrow & \coprod_2 K_1L(x) & \longrightarrow & \coprod_3 K_0L(x) \\
& & & & \uparrow & & \uparrow & & \uparrow \\
& & & & \coprod_1 K_2F(x) & \longrightarrow & \coprod_2 K_1F(x) & \longrightarrow & \coprod_3 K_0F(x)
\end{array}
$$

where h is the composition of the norm residue map $K_3F \to H^3F$ and the cup product by $(d) \in H^1F$.

By a diagram chase one gets a map

$$\theta: \ker(H^4F \to H^4F(X)) \to \ker(H^2(X, K_3) \to H^2(X_L, K_3))$$

provided the following sequences are exact:

(1) $0 \to H^4L \to H^4L(X)$,

(2) $K_3F \overset{h}{\to} H^4F \to H^4L$,

(3) $K_3L(X) \overset{N}{\to} K_3F(X) \overset{h}{\to} H^4F(X)$,

(4) $K_2L(x) \overset{1-\sigma}{\to} K_2L(x) \overset{N}{\to} K_2F(x)$, $x \in X^1$.

Since X has a rational point over L, (1) is exact. Then both (2) and (3) follow from the commutative diagram

$$
\begin{array}{ccc}
K_3L/2 & \overset{N}{\longrightarrow} & K_3F/2 \\
{\scriptstyle \cong}\downarrow & & {\scriptstyle \cong}\downarrow \quad{}^{h}\searrow \\
H^3L & \underset{N}{\longrightarrow} H^3F & \overset{\cup(d)}{\longrightarrow} H^4F \longrightarrow H^4L
\end{array}
$$

the exactness of the bottom row [2] and the bijectivity of the norm residue symbol of degree 3 [14], [21] for (3) one should replace F by $F(X)$, L by $L(X)$ and use $\operatorname{im} N \supset 2K_3F(X))$.

Exactness of (4) is the K_2-analogue of Hilbert's theorem 90 [13].

Assume now that $X = X(\varphi)$ is a projective quadric, $\dim X \geq 3$, and φ is not similar to a subform of an anisotropic 3-fold Pfister form. The diagram chase shows that the injectivity of θ follows from:

(5) There is a natural isomorphism $K_2L \xrightarrow{\sim} H^1(X_L, K_3)$,
and the exactness of the following sequences:

(6) $K_3L \to K_3L(X) \to \coprod_1 K_2L(x)$,

(7) $0 \to K_3F/2 \to K_3F(X)/2$.

The statement (5) follows from

LEMMA 2. *The homomorphism $K_2F \to H^1(X, K_3)$ given by multiplication by the class of hyperplane section in $H^1(X, K_1)$ is injective and if X has a rational point then it is an isomorphism.*

PROOF. Suppose first that X has a rational point. Choose an open subset $U \subset X$ isomorphic to an affine space, a rational point $p \in Y = X - U$, and a fibration $\pi: Y - p \to Z$ with affine line fibers and a quadric Z [5]. Then clearly $\dim Z \geq 1$ and $H^1(X, K_3) \simeq H^0(Y, K_2) = H^0(Y - p, K_2) = H^0(Z, K_2) = K_2F$ [25] and it is easy to see that the composition is the desired isomorphism.

In the general case the assertion follows from the commutative diagram

$$
\begin{array}{ccc}
K_2F & \longrightarrow & H^2(X, K_3) \\
\downarrow{\scriptstyle i} & & \downarrow \\
K_2F(X) & \xrightarrow{\ \sim\ } & H^1(X_{F(X)}, K_3)
\end{array}
$$

since i is injective [25].

The sequence (6) is simply exact since X_L has a rational point. For (7), according to the bijectivity of the norm residue homomorphism of degree 3, it suffices to show that $H^3F \hookrightarrow H^3F(X)$. But this is true if φ is not a subform of a 3-fold Pfister form [2].

It is clear from the diagram chase that the surjectivity of θ follows from the exactness of the sequences

(8) $H^1(X_L, K_3) \xrightarrow{1-\sigma} H^1(X_L, K_3) \xrightarrow{N} H^1(X, K_3)$,

(9) $K_3F \to K_3F(X) \to \coprod_1 K_2F(x)$.

The exactness of (8) follows from Lemma 2 and the commutative diagram

$$
\begin{array}{ccccc}
K_2L & \xrightarrow{\ 1-\sigma\ } & K_2L & \xrightarrow{\ N\ } & K_2F \\
\downarrow{\scriptstyle \cong} & & \downarrow{\scriptstyle \cong} & & \uparrow \\
H^1(X_L, K_3) & \xrightarrow{\ 1-\sigma\ } & H^1(X_L, K_3) & \xrightarrow{\ N\ } & H^1(X, K_3),
\end{array}
$$

and the K_2-analogue of Hilbert's theorem 90.

For (9) let $u \in \ker(K_3F(X) \xrightarrow{(\partial_x)} \coprod_1 K_2F(x))$. Since (9) is exact over L we have: $2u \in \operatorname{im}(K_3F \to K_3F(X))$. Using (7) one gets $v \in K_3F$ such that $2u = 2v_{F(x)}$. Modifying u by $v_{F(x)}$ we may assume $2u = 0$, i.e., $u = \{-1\} \cdot w$ for some $w \in K_2F(X)$ [14], [21]. Since $0 = \partial_x(u) = \{-1\} \cdot \partial_x(w)$ for all $x \in X^1$ there are $b_x \in F(x)^*$ and $a_x \in F(x)_0^*$ ($F(x)_0$ is the subfield in $F(x)$ of all algebraic over prime field elements) such that $\partial_x(u) = a_x \cdot b_x^2$ and $\{-1, a_x\} = 0 \in K_2 F(x)_0$ [25].

Let $F_x = F \cdot F(x)_0 \subset F(x)$ for any $x \in X^1$. Since $F(x)$ is a finitely generated extension of F and F_x is algebraic over F, then F_x is a finite extension of F. Let y be the generic point of some hyperplane section of X. Since $\operatorname{Pic} X$ is generated by y, we can find $f_x \in F_x(X)^*$ such that $\operatorname{div}(f_x) = x' - m_x \cdot y$, where x' is a point of X_{F_x} lying over x and $m_x \in \mathbb{Z}$ for all $x \neq y$. Since the elements $t_x = N_{F_x/F}(\{f_x, a_x\})$ in $K_2 F(X)$ satisfy $\partial_x(t_x) = a_x$ and $\{-1\} \cdot t_x = N_{F_x/F}(\{f_x, -1, a_x\}) = 0$, modifying w by the sum of t_x for all x such that $a_x \neq 1$, $x \neq y$, we may assume that $a_x = 1$ for all $x \neq y$.

Since a_y is algebraic element, $\nu(a_y) = 0$ for any discrete valuation ν of the field $F(y)$ over F. Hence,

$$a = (a_x) \in \ker \left(\coprod_1 K_1 F(x) \xrightarrow{\delta} \coprod_2 K_0 F(x) \right)$$

and, therefore, a defines an element $\overline{a} \in H^1(F, K_2)$. The image $\delta(b^2)$ for $b = (b_x)$ equals $\delta(\partial(w) - a) = -\delta(a) = 0$, hence $\delta(b) = 0$ and b also defines an element $\overline{b} \in H^1(F, K_2)$.

The homomorphism $K_1 F \to H^1(X, K_2)$ given by multiplication by the class of a hyperplane section is an isomorphism [5]; hence, modifying a_y, b_y by an element in F^* we may assume that $\overline{b} = 0 \in H^1(X, K_2)$, i.e., $b = \partial_x(r)$ for some $r \in K_2 F(X)$. Since $\overline{a} + 2\overline{b} = 0 \in H^1(X, K_2)$, $\overline{a} = 0$. But $a_y \in F(y)_0^* \subset F^*$; therefore, $a_y = 1$, so, $a_x = 1$ for all $x \in X^1$. Finally, $\partial_x(w) = b_x^2 = \partial_x(2r)$; hence, $w - 2r \in H^0(X, K_2) = K_2 F$ [25] and $u = \{-1\} \cdot w = \{-1\} \cdot (w - 2r) \in \{-1\} \cdot K_2 F \subset \operatorname{im}(K_3 F \to K_3 F(X))$.

This proves the bijectivity of θ. Since X_L has a rational point, clearly $H^2(X_L, K_3) \hookrightarrow H^2(X_{\text{sep}}, K_3)$, where $X_{\text{sep}} = X_{F_{\text{sep}}}$ and F_{sep} is the separable closure of F. So, we have proved the following

PROPOSITION 1. *Let $X = X(\varphi)$ be a projective quadric; assume $\dim X \geq 3$ and φ is not similar to a subform of the anisotropic 3-fold Pfister form. Then there exists a natural isomorphism*

$$\theta: \ker(H^4 F \to H^4 F(X)) \xrightarrow{\sim} \ker(H^2(X, K_3) \to H^2(X_{\text{sep}}, K_3)).$$

Step 2. Let X be a quadric of dimension at most 4. We consider the Brown-Gersten-Quillen spectral sequence [18]:

$$H^p(X, K_{-q}^Q) \Rightarrow K_{-p-q}(X)$$

converging to the Quillen K-theory of X with the topological filtration. Since $K_n^Q F = K_n F$ for a field F and $n = 0, 1, 2$, we have $H^p(X, K_{n+p}^Q) = H^p(X, K_{n+p})$ for $n = 0, 1$ and $H^0(X, K_n^Q) = H^0(X, K_n)$ for $n = 0, 1, 2$.

The restriction on the dimension of X implies that the term $E_\infty^{2,-3} = K_1 X^{(2/3)}$ equals to the homology group of the complex:

$$H^0(X, K_2) \xrightarrow{d} H^2(X, K_3) \xrightarrow{d'} H^4(X, K_4)$$

where d and d' are the differentials in the spectral sequence. But the composition $K_2 F \to K_2 X \xrightarrow{\alpha} H^0(X, K_2)$ is an isomorphism [25]; hence, α is surjective and, therefore, $d = 0$. On the other hand, $H^4(X, K_4) = CH^4 X = K_0 X^{(4)} \simeq \mathbb{Z}$ if $\dim X = 4$ and $= 0$ if $\dim X < 4$ [5], [28], so $d' = 0$. We have proved

PROPOSITION 2. *If X is a quadric of dimension at most 4, then $H^2(X, K_3) \simeq K_1 X^{(2/3)}$.*

Step 3. We compute the topological filtration on $K_1 X$ for a quadric $X = X(q)$, where q is the Albert form, $\dim X = 4$. If q is hyperbolic, then [5]

$$K_1 X = F^* \cdot 1 \oplus F^* \cdot h \oplus F^* \cdot P_1 \oplus F^* \cdot P_2 \oplus F^* \cdot l \oplus F^* \cdot p,$$

where h, P_1, P_2, l, $p \in K_0 X$ are respectively the classes of a hyperplane section, two nonequivalent planes in X, line and a rational point. Using the equalities $h^2 = P_1 + P_2 - l$ and $h^3 = 2l - p$ in $K_0 X$, one can rewrite the formula

$$K_1 X = F^* \cdot 1 \oplus F^* \cdot h \oplus F^* \cdot h^2 \oplus F^* \cdot h^3 \oplus F^* \cdot P_1 \oplus F^* \cdot P_2.$$

The powers of h (but not P_1 and P_2) are defined in the nonsplit case. Now let q be an anisotropic form or, equivalently, let D be a division algebra. For any $u \in D^*$ choose a maximal subfield E containing u. Since the algebra D_E is split, q_E is hyperbolic and one can define classes $P_{1,2} \in K_0 X_E$ and $u \cdot P_{1,2} \in K_1 X_E$. Consider two homomorphisms $K_1 D \to K_1 X$ given by $u \mapsto N_{E/F}(u \cdot P_i)$. The images of u we will simply denote by $u P_1 \in K_1 X$ (they are independent of the choice of a field E). Hence, $K_1 X$ contains the sum

$$F^* \cdot 1 + F^* h + F^* h^2 + F^* \cdot h^3 + K_1 D \cdot P_1 + K_1 D \cdot P_2.$$

On the other hand, the group $K_1 X$ was computed in [27]: $K_1 X \simeq (F^*)^4 \oplus K_1 C_0(q)$. Comparing the computations and using the isomorphism $C_0(q) \simeq D \times D$, one gets the following formula for $K_1 X$:

$$K_1 X = F^* 1 \oplus F^* \cdot h \oplus F^* \cdot h^2 \oplus F^* \cdot h^3 \oplus K_1 D \cdot P_1 \oplus K_1 D \cdot P_2.$$

Let us compute the topological filtration. Since

$$K_1 X^{(1)} = \ker(K_1 X \to K_1 F(X))$$

one easily sees that

$$K_1 X^{(1)} = F^* \cdot h \oplus F^* \cdot h^2 \oplus F^* \cdot h^3 \oplus K_1 D \cdot P_1 \oplus K_1 D \cdot P_2.$$

Since all the terms except $F^* \cdot h$ belong to $K_1 X^{(2)}$ and the composition

$$F^* \xrightarrow{\cdot h} K_1 X^{(1/2)} = H^1(X, K_2) \simeq F^*$$

is the identity [5],

$$K_1 X^{(2)} = F^* \cdot h^2 \oplus F^* \cdot h^3 \oplus K_1 D \cdot P_1 \oplus K_1 D \cdot P_2.$$

The determination of $K_1 X^{(3)}$ is more delicate. In the splitting case $P_1 + P_2 - h^2 = l \in K_0 X^{(3)}$; therefore, $u \cdot P_1 + u \cdot P_2 - u \cdot h^2 \in K_1 X^{(3)}$ for any $u \in D^*$. Hence, in the non-split case the element $\alpha(u) = u \cdot P_1 + u \cdot P_2 - \mathrm{Nrd}(u) \cdot h^2$ belongs to $K_1 X^{(3)}$ for any $u \in D^*$.

Let A be a subgroup in $K_1 X^{(3)}$ generated by $F^* \cdot h^3$ and $\alpha(u)$ for all $u \in D^*$. Consider an isomorphism

$$K_1 X^{(2)}/A \xrightarrow{\sim} F^* \oplus K_1 D$$

defined by $ah^2 + bh^3 + u_1 P_1 + u_2 P_2 \bmod A \mapsto (\mathrm{Nrd}\, u_1 \cdot a^{-1}) \oplus (u_1 \cdot u_2^{-1})$. Since in the split case A clearly coincides with $K_1 X^{(3)}$, we have the following diagram

$$
\begin{array}{ccccc}
K_1 X^{(2/3)} & \twoheadleftarrow & K_1 X^{(2)}/A & \xrightarrow{\;\sim\;} & F^* \oplus K_1 D \\
\downarrow & & \downarrow & & \downarrow{\scriptstyle\beta} \\
K_1 X_{\mathrm{sep}}^{(2/3)} & \xleftarrow{\;\sim\;} & K_1 X_{\mathrm{sep}}^{(2)}/A_{\mathrm{sep}} & \xrightarrow{\;\sim\;} & F_{\mathrm{sep}}^* \oplus K_1 D_{\mathrm{sep}} = F_{\mathrm{sep}}^* \oplus F_{\mathrm{sep}}^*.
\end{array}
$$

The kernel of β equals $SK_1 D$; therefore, there is a surjective

$$SK_1 D \twoheadrightarrow \ker(K_1 X^{(2/3)} \longrightarrow K_1 X_{\mathrm{sep}}^{(2/3)}).$$

Combining all the steps (q is anisotropic and, therefore, is not similar to a subform of 3-fold Pfister form) one obtains an exact sequence

$$SK_1 D \longrightarrow H^4 F \longrightarrow H^4 F(q).$$

It remains to show that the first map is injective. In fact, injectivity is equivalent to the equality $K_1 X^{(3)} = A$.

Let $\varphi \subset q$ be a subform of dimension 5, $X = X(\varphi)$, $\dim X = 3$. In the splitting case,

$$K_1 X = F^* \cdot 1 \oplus F^* \cdot h \oplus F^* \cdot l \oplus F^* \cdot p = F^* \cdot 1 \oplus F^* \cdot h \oplus F^* \cdot h^2 \oplus F^* \cdot l,$$

since $h^2 = 2l - p \in K_0 X$.

Assume q is anisotropic and $u \in E^* \subset D^*$, where E is a maximal subfield in D. The field E splits D and therefore X completely. As above we consider the homomorphism $K_1 D \to K_1 X$, $u \mapsto N_{E/F}(u \cdot l) \stackrel{\mathrm{def}}{=} u \cdot l$. The group $K_1 X$ contains $F^* \cdot 1 + F^* \cdot h + F^* \cdot h^2 + K_1 D \cdot l$. Using the Swan's computation of $K_1 X$ and the isomorphism $C_0(\varphi) \simeq D$, one gets the equality

$$K_1 X = F^* \cdot 1 \oplus F^* \cdot h \oplus F^* \cdot h^2 \oplus K_1 D \cdot l.$$

As above, the first two terms of the topological filtration are easy to compute:

$$K_1 X^{(1)} = F^* \cdot h \oplus F^* \cdot h^2 \oplus K_1 D \cdot l,$$
$$K_1 X^{(2)} = F^* \cdot h^2 \oplus K_1 D \cdot l.$$

The commutative diagram

$$
\begin{array}{ccccc}
K_1 X^{(2)} & = & F^* \cdot h^2 \oplus K_1 D \cdot l & \longrightarrow & K_1 1 X^{(2/3)} \\
\downarrow & & \downarrow & & \downarrow \\
K_1 X^{(2)}_{\text{sep}} & = & F^*_{\text{sep}} \cdot h^2 \oplus K_1 D_{\text{sep}} \cdot l & \longrightarrow & K_1 X^{(2/3)}_{\text{sep}}
\end{array}
$$

gives a map

$$SK_1 D \to \ker(K_1 X^{(2/3)} \to K_1 X^{(2/3)}_{\text{sep}}) \hookrightarrow H^4 F.$$

Actually, this is the same map as before. To prove the injectivity we need to show that

$(*)$ $\qquad\qquad\qquad SK_1 D \cap K_1 X^{(3)} = 0 \in K_1 X.$

But since $\dim X = 3$, there is a surjection $H^3(X, K_4) \twoheadrightarrow K_1 X^{(3)}$ and one easily checks that the composition

$$H^3(X, K_4) \twoheadrightarrow K_1 X^{(3)} \hookrightarrow K_1 X^{(2)} = F^* \cdot h^2 \oplus K_1 D \cdot l \xrightarrow{\text{pr}} F^* \cdot h^2$$

coincides up to sign with the norm map $N: H^3(X, K_4) \to F^*$ given by the usual norms $F(x)^* \to F^*$ for any $x \in X^3$,

Therefore, the equality $(*)$ follows from another theorem of Rost [23]: for any quadric X, $d = \dim X \le 3$, the norm map $N: H^d(X, K_{d+1}) \to F^*$ is injective. Thus, the proof of the theorem is complete. $\square$

The immediate consequence is the following

COROLLARY. *If* $\mathrm{cd} F \le 3$, *then* $SK_1 D = 0$.

Suslin has conjectured that this statement holds for any simple algebra instead of biquaternion algebras.

We consider another application of the Rost's theorem.

Let D be a central simple algebra over a field F. We consider a linear group $G = \mathrm{SL}_1(D)$. The group of L-points for any field extension L/F equals

$$G(L) = \{x \in D_L^* : \mathrm{Nrd}_L(d) = 1\}.$$

Therefore, there is a canonical subjection $G(L) \twoheadrightarrow SK_1 D_L$. The image of the generic point $\xi \in G(F(G))$ in $SK_1(D_{F(G)})$ we will denote by $x_D \in SK_1(D_{F(G)})$ and call it the *generic element in* SK_1.

PROPOSITION 3. (1) *The generic element* x_D *is trivial iff* $SK_1(D_L) = 0$ *for any field extension* L/F.

(2) *If the variety* G *is rational, then the generic element* x_D *is trivial.*

PROOF. (1) The implication " $\Leftarrow$ " is trivial. For the proof " $\Rightarrow$ " we choose a basis $d_1, d_2, \ldots, d_n$ of D over F. Then $\xi = \sum x_i d_i \in D^*_{F(G)}$ is the generic element in $D_{F(G)}$, where the x_i are coordinate functions of G. Suppose $\xi = [f_1, g_1] \cdots [f_n, g_n]$ for some $f_i, g_i \in D^*_{F(G)}$. There is an open subset $U \subset G$ such that we can specialize the f_i and g_i. This implies that for any field extension L/F we have $U(L) \subset [D^*_L, D^*_L]$, and $G(L) = U(L) \cdot U(L) = [D^*_L, D^*_L]$; therefore, $SK_1(D_L) = 0$.

(2) One can apply the theorem of Voskresenskii [30]; G_L/R-equivalence $\simeq SK_1 D_L$. $\square$

COROLLARY. *If there is a field extension* L/F *such that* $SK_1(D_L) \neq 0$, *then* $x_D \neq 0$ *and the variety* G *is not rational.*

EXAMPLES. (1) $D = M_n(F)$; $SL_1(D) = SL_n(F)$ is a rational variety

(2) $D = \left(\frac{a,b}{F}\right)$ is a quaternion algebra. The variety $SL_1(D)$ is given by the quadratic equation $x^2 - ay^2 - bz^2 + abt^2 = 1$; it is rational since it has a rational point.

(3) If $\operatorname{ind} D$ is square free, then $SK_1(D_L) = 0$ for any field extension L/F and $x_D = 0$.

CONJECTURE (Suslin [26]). *If* $\operatorname{ind} D$ *is not square free, then* $x_D \neq 0$. *In particular,* $G = SL_1(D)$ *is not rational.*

The following theorem proves the conjecture in a particular case.

THEOREM [10]. *Suppose* $\operatorname{char} F \neq 2$ *and* $\operatorname{ind} D$ *is divisible by* 4. *Then* $x_D \neq 0$, *and therefore, the variety* $G = SL_1(D)$ *is not rational.*

PROOF. The idea of the proof is to produce a field extension L/F such that $SK_1(D_L) \neq 0$. In the (crucial) case: $\operatorname{ind} D = 4$, $\exp D = 2$, we apply Rost's theorem.

LEMMA 3. *Let* $a, b, c \ d \in F^*$. *The following conditions are equivalent:*
(1) $\{a, b, c, d\} \in 2K_4 F$;
(2) *The 4-fold Pfister form* $\langle\langle a, b, c, d, \rangle\rangle$ *is hyperbolic;*
(3) $(a, b, c, d) = 0 \in H^4 F$.

PROOF. $(1) \Rightarrow (2)$ is known [4]. The norm residue homomorphism of degree 4 shows that $(1) \Rightarrow (3)$.

To prove $(3) \Rightarrow (1)$ we consider a commutative diagram

$$
\begin{array}{ccccc}
K_3 E/2 & \xrightarrow{N} & K_3 F/2 & \xrightarrow{\cdot \{d\}} & K_4 F/2 \\
\Vert \downarrow & & \Vert \downarrow & & \downarrow \\
H^3 E & \xrightarrow{N} & H^3 F & \xrightarrow{(d)} & H^4 F
\end{array}
$$

where $E = F(\sqrt{d})$ and one should argue as in the proof of Rost's theorem. $\square$

Let $q \in I^2 F$, $C(q) = M_2(C'(q))$.

LEMMA 4. *Let* $n \in \mathbb{N}$ *such that* $2^{n-1} \geq \dim q$. *If* $C'(q)$ *is a division algebra, then there exist a field extension* L/F *and an anisotropic* n-fold *Pfister form* f *over* L *such that* q_L *is isomorphic to a subform of* f.

PROOF. Adjoin indeterminates (if necessary) in order to get an anisotropic n-fold Pfister form f. Choose an extension L/F such that:

(1) f_L is anistropic

(2) $C'(q_L) = C'(q)_L$ is a division algebra,

(3) the Witt index of $f_L \perp (-q_L)$ is maximal.

Let $f_L \perp (-q_L) = h \perp k \cdot \mathbb{H}$ where h is anisotropic. We claim that $k \geq \dim q$.

Suppose on the contrary that $k < \dim q$, then $\dim h = \dim f + \dim q - 2k = 2^n + \dim q - 2k \geq 3 \dim q - 2k > \dim q$. We replace a field L by $L(h)$. Since $\dim C'(q_L) = 2^{\dim q - 2} < 2^{\dim h - 2}$, then by Theorem 2 $C'(q_L)$ remains a division algebra over $L(h)$. The Witt index of $f_L \perp (-q_L)$ over $L(h)$ is bigger than over L; therefore, by the choice of L we must have $f_{L(h)}$ isotropic and hence hyperbolic. This implies that $f_L \simeq \langle x \rangle \cdot h \perp p$ for some $x \in L^*$ and a form p over L [33]. Note that f_L, q_l, h, $p \in I^2 L$. The equality $x \cdot h + p - q_L = h$ in the Witt ring $W(L)$ implies that $p - q_L = \langle 1, -x \rangle h \in I^3 L$; therefore, $C(p) \sim C(q_L)$ and $C'(p) \sim C'(q_L)$. But $C'(q_L)$ is a division algebra, so $\dim C'(p) \geq \dim C'(q_L)$ and $\dim p \geq \dim q$. Finally, $\dim p = \dim f - \dim h = 2k - \dim q < \dim q$: contradiction. Thus, $k \geq \dim q$. Then $f_L \perp (\dim q) \cdot \mathbb{H} = f_L \perp (-q_L \perp q_L) = h \perp k\mathbb{H} \perp q_L$ and cancelling hyperbolic planes we get $f_L \simeq h \perp q_L \perp (k - \dim q) \cdot \mathbb{H}$, therefore, $q_L \subset f_L$. $\square$

We return to the proof of the theorem. We wish to find a field extension L/F such that $SK_1(D_L) \neq 0$.

Consider the first case: $\operatorname{ind} D = 4$, $\exp D = 2$, i.e., D is a biquaternion algebra. Applying Lemma 4 to $n = 4$, q being an Albert form of D, we can obtain a field extension L/F and an anisotropic 4-fold Pfister form $f = \langle\langle x, y, z, t \rangle\rangle$ over L such that $q_L \subset f$. Since f is clearly hyperbolic over $L(q)$, Lemma 3 implies that (x, y, z, t) is a nontrivial element in $\ker(H^4 L \to H^4 L(q))$. By the theorem of Rost $SK_1(D_L) \neq 0$.

Now consider the case where $\operatorname{ind} D = 4 = \exp D$. The tensor square $D^{\otimes 2}$ is similar to some quaternion algebra Q; let C be the corresponding conic curve. We claim that D remains a division algebra over a function field $F(C)$. For if $D_{F(C)}$ is not a division algebra, then Q can be imbedded to D (Theorem 2); hence, $D \simeq Q \otimes_F Q'$ for some quaternion algebra Q' and $\exp D \leq 2$: contradiction.

Since $\operatorname{ind} D_{F(C)} = 4$ and $\exp D_{F(C)} = 2$, we are reduced to the preceding case.

Assume now that F has no odd-degree extensions and D is a division algebra. Let E be a maximal subfield in D and let L be a field such that $F \subset L \subset E$ and $[E : L] = 4$, then D_L has index 4; therefore, $\exp D_L = 2$ or 4 and we are back to previous cases.

For the general case, one should take a maximal extension of odd degree and return to the previous case. $\square$

Let X be an algebraic variety of dimension d. The group

$$H^d(X, K_{d+1}) = \operatorname{coker}\left(\coprod_{\dim x=1} K_2 F(x) \to \coprod_{\dim x=0} K_1 F(x) \right)$$

we denote by $H_0(X, K_1)$. For a projective X the norm maps $K_1 F(x) \to K_1 F$ induce the norm map $N: H_0(X, K_1) \to K_1 F$.

THEOREM [15]. *Let D be a central simple algebra over a field F, let $X = X(D)$ be a corresponding Severi-Brauer variety [1]. Then there exists a commutative diagram*

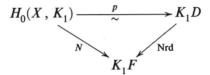

with a natural isomorphism p. In particular, $\ker(H_0(X, K_1) \xrightarrow{N} K_1 F) \simeq SK_1 D$.

PROOF. Assume D is a division algebra of dimension n^2. The residue field $F(x)$ for any closed point $x \in X$ splits D; hence, $[F(x) : F] = k \cdot n$ for some k. Therefore, $F(x)$ can be imbedded to $M_k(D)$ and we get a map

$$K_1 F(x) \to K_1(M_k(D)) \simeq K_1 D,$$

and it is easy to check that we have a commutative diagram

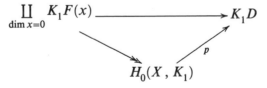

with a surjective homomorphism p.

For the proof of the injectivity of p, one constructs a map in the opposite direction. But first we wish to understand the structure of the set of closed points of X.

For any field extension L/F the set $X(L) = \operatorname{Mor}(\operatorname{Spec} L, X)$ of L-points of X is naturally bijective to the set of right ideals in D_L of dimension n.

Let $L \subset D$ be a maximal subfield. The L-space $\operatorname{Hom}_L(D, L)$ is a right ideal in $\operatorname{End}_L D \simeq D_L$ of dimension n and hence defines a morphism $\operatorname{Spec} L \to X$. Let x_L be a point of its image. Clearly, $F(x_L) \hookrightarrow L$ and

$[F(x) : F] \geq n$ since $F(x)$ splits D. Hence, $F(x_L) = L$ and $\deg x = n$. Therefore, we have a map $L \mapsto x_L$ from the set of maximal subfields in D to the set of closed points in X of the smallest degree n. In fact, this is 1-1 correspondence [11].

Consider a Zariski-dense subset $S = \{u \in D : F[u]$ is a maximal subfield in $D\}$ in D^*. For $u \in S$, $F[u]$ is a maximal subfield in D; hence, it defines a closed point $x_{F[u]} \in X$ with $F(x_{F[u]}) = F[u]$. The map opposite to p we define first on the set S:

$$q: S \to H_0(X, K_1), \qquad u \mapsto u \cdot x_{F[u]},$$

and then extend it to D^*. The difficult part of the proof is to show that q is a homomorphism. If u, $v \in S$ such that $uv \in S$, then

$$(u \cdot x_{F[u]}) + (v \cdot x_{F[v]}) = (uv, x_{F(uv)}) \in H_0(X, K_1).$$

This can be shown by the specialization argument. Consider the field $F' = F(t)$ of rational functions, an algebra $D' = D(t)$, and the elements $u \in S$, $v(t) = vt + 1 - t \in D(t)^*$, where $v \in S$; $v(0) = 1$, $v(1) = v$. Let

$$w = (uv(t) \cdot x_{F'[uv(t)]}) - (u \cdot x_{F'[u]}) - (v(t) \cdot x_{F'[u(t)]}) \in H_0(X_{F(t)}, K_1).$$

We have $w(0) = 0$ and $w(1) = (uv \cdot x_{F[uv]}) - (u \cdot x_{F[u]}) - (v \cdot x_{F[v]})$ and we wish to show that $w(1) = 0$.

Consider a commutative diagram

$$
\begin{array}{ccccc}
H_0(X, K_1) & \xrightarrow{\;i\;} & H_0(X_{F(t)}, K_1) & \longrightarrow & \displaystyle\coprod_{y \in \mathbb{A}^1} H_0(X_{F(y)}, K_0) \\
& & \downarrow{\scriptstyle p} & & \downarrow{\scriptstyle p_0} \\
& & K_1 D' & \longrightarrow & \displaystyle\coprod_{y \in \mathbb{A}^1} K_0 D_{F(y)} \coprod_{y \in \mathbb{A}^1} K_0 D_{F(y)}
\end{array}
$$

where the upper row is a part of the exact localization sequence. The Chow group $CH_0 X = H_0(X, K_0)$ is known to have no torsion [17]; hence, p_0 is injective. The diagram chase shows that $w \in \operatorname{im}(i)$, and hence, the specialization of w at all the points coincide; therefore, $w(1) = 0$. $\square$

Now we are going to study the image $\operatorname{Nrd} D$ of the reduced norm map $\operatorname{Nrd}_1 : K_1 D \to K_1 F = F^*$.

Let n be divisible by $\exp D$; any $a \in F^*$ defines an element $(a) \in H^1(F, \mu_n)$. Taking the cupproduct with $[D] \in H^2(F, \mu_n)$ we get an element $(a) \cup [D]$ in $H^3(F, \mu_n^{\otimes 2})$. If $a \in \operatorname{Nrd} D$, then $a = N_{E/F}(u)$ for some splitting field E and $u \in E$. Therefore, $(a) \cup [D] = (N_{E/F}(u)) \cup [D] = N_{E/F}((u) \cup [D_E]) = 0$ since D_E is split.

Let $A(D) = \{a \in F^* : (a) \cup [D] = 0 \in H^3(F, \mu_n^{\otimes 2})\}$ be a subgroup in F^*. We have shown that $\operatorname{Nrd}(D) \subset A(D)$.

For any $i > 0$ and $b \in \operatorname{Nrd}(D^{\otimes i})$ we have $(b^i) \cup [D] = (b) \cup [D^{\otimes i}] = 0$; therefore, $\operatorname{Nrd}(D^{\otimes i})^i \subset A(D)$.

THEOREM 7 [12], [25]. *If* $n = \exp D$ *and* $\operatorname{ind} D$ *is almost square free; i.e., is not divisible by* 8 *and square of odd prime number, then* $A(D)$ *is equal to the product of* $\operatorname{Nrd}(D^{\otimes i})^i$ *for* $i > 0$.

COROLLARY. (1) *If* $\operatorname{ind} D$ *is a prime, then* $A(D) = \operatorname{Nrd} D$.

(2) *If* $\operatorname{ind} D = 4$, *then* $A(D) = \operatorname{Nrd}(D^{\otimes 2})^2 \cdot \operatorname{Nrd} D$. *In particular, if* D *is a biquaternion algebra, then* $A(D) = F^{*2} \cdot \operatorname{Nrd} D$.

REMARK. If $\operatorname{ind} D$ is not almost square free, then the statement of the Theorem 7 is not true in general [12]. Similar examples were found by Tignol.

3. K_2 and K_3 of simple algebra

To define a reduced norm map $\operatorname{Nrd}_2 : K_2 D \to K_2 F$ for a central simple algebra D over a field F we consider the following commutative diagram

$$
\begin{array}{ccccc}
K_2 D & \longrightarrow & K_2 D(X) & \longrightarrow & \coprod_{x \in X^1} K_1 D(x) \\
& & \downarrow \wr & & \downarrow \wr \\
0 \longrightarrow K_2 F & \longrightarrow & K_2 F(X) & \longrightarrow & \coprod_{x \in X^1} K_1 F(x)
\end{array}
$$

where X is a Severi-Brauer variety corresponding to D, $D(X) = D \otimes_F F(X)$. The vertical maps are isomorphisms since $F(x)$ splits D for any point $x \in X$. Since the bottom row is exact [25], the diagram defines a reduced norm map $\operatorname{Nrd}_2 : K_2 D \to K_2 F$. As before $SK_2 D = \ker(\operatorname{Nrd}_2)$.

CONJECTURE. *If* $\operatorname{ind} D$ *is square free, then* $SK_2 D = 0$.

THEOREM 8 [8], [22]. *If* D *is a quaternion algebra, then* $SK_2 D = 0$.

Denote by h the composition of the norm residue map $K_2 F \to H^2 F$ and the homomorphism $H^2 F \to H^4 F$ given by multiplication by $[D] \in H^2 F$.

THEOREM 9 [8]. *If* D *is a quaternion algebra, then* $\operatorname{Nrd}_2 D = \ker(K_2 F \xrightarrow{h} H^4 F)$.

In the last part of the paper, we prove

PROPOSITION 4. *The reduced norm sup* $\operatorname{Nrd}_3 : K_3 D \to K_3 F$ *does not exist.*

PROOF. We work with Quillen K-groups now. Let t be an indeterminate, $a \in F^*$, $L = F(\sqrt{a})$. Consider a skew-polynomial ring $T = L[t, j]$, $j^2 = t$, $j \cdot \sqrt{a} = -\sqrt{a} \cdot j$. The imbedding $L \hookrightarrow T$ induces an isomorphism $K_* L \xrightarrow{\sim} K_* T$ [18].

Assume that the reduced norm for K_3 exists. Since the quaternion algebra $D = \left(\frac{a, t}{F(t)}\right)$ is a localization of T, then an exact localization sequence can be included in the commutative diagram:

$$
\begin{array}{ccccccc}
K_3T & = & K_3L & \xrightarrow{\ i\ } & K_3D & \xrightarrow{\ \partial\ } & \left(\displaystyle\coprod_{\substack{x\in \mathbf{A}^1;\\ x\neq 0}} K_2D(x) \right) \oplus K_2L \\
 & & \Big\downarrow{\scriptstyle N_{L/F}} & & \Big\downarrow{\scriptstyle \mathrm{Nrd}_3} & & \\
 & & K_3F & \xrightarrow{\ j\ } & K_3F(t). & &
\end{array}
$$

Consider $u = \{-1, a, j\} \in K_3D$. It is clear that $\partial(u) = 0 \oplus \{-1, a\}$. But $\{-1, a\} = 2\{-1, \sqrt{a}\} = 0 \in K_2L$; hence, $u \in \mathrm{im}(i)$. Therefore, $\mathrm{Nrd}(u) = \{-1, a, -t\} \in \mathrm{im}(j)$. On the other hand, the residue of $\{-1, a, -t\}$ at t is $\{-1, a\} \in K_2F$ which is not necessarily zero. □

References

1. M. Artin, *Brauer-Sevri varieties*, Lecture Notes in Math., vol. 917, Springer-Verlag, Berlin and New York, 1982, pp. 194–210.

2. J. K. Arason, *Cohomologische Invarianten quadratischer Fromen*, J. Algebra **36** (1975), 448–491.

3. P. K. Draxl, *Skew fields*, Read. Math., Univ. of Bielefeld, Cambridge Univ. Press, 1983.

4. R. Elman and T.-Y. Lam, *Pfister forms and K-theory of fields*, J. Algebra **23** (1972), 181–213.

5. N. A. Karpenko, *Algebro-geometric invariants of quadratic forms*, Leningrad Math. J. **2** (1991), 141–162.

6. K. Kato, *Milnor K-theory and the Chow group of zero cycles*, Contemp. Math., vol. 55, Amer. Math. Soc., Providence, RI, 1986, pp. 241–254.

7. T.-Y. Lam, *The algebraic theory of quadratic forms*, Benjamin, Reading, Mass, 1973.

8. A. S. Merkurjev, *Group SK_2 for quaternion algebras*, Izv. Akad. Nauk SSSR **52** (1988), 310–335.

9. _____, *Simple algebras and quadratic forms*, Izv. Akad. Nauk SSSR **55** (1991), 218–224.

10. _____, *Generic element in SK_1 for simple algebras*, K-theory, vol. 7, 1993, pp. 1–3.

11. _____, *Closed points of Severi-Brauer varieties*, Vestnik SPGU, Mathematics, Sankt Petersburg, vol. 2, no. 8, 1993, pp. 51–53 (Russian).

12. _____, *Certain K-cohomology groups of Severi-Brauer varieties*, these Proceedings.

13. A. S. Merkurjev and A. A. Suslin, *K-cohomology of Severi-Brauer varieties and norm residue homomorphism*, Izv. Akad. Nauk SSSR **46** (1982), 1011–1046.

14. _____, *The norm residue homomorphism of degree* 3, Izv. Akad. Nauk SSSR **54** (1990), 339–356.

15. _____, *The group of K_1-zero-cycles on Severi-Brauer varieties*, Nova Journal of Algebra and Geometry, vol. 1, no. 3, 1992, pp. 297–315.

16. J. Milnor, *Algebraic K-theory and quadratic forms*, Invent. Math **3** (1970), 318–344.

17. I. A. Panin, *Application of K-theory in algebraic geometry*, Ph.D. Thesis, LOMI, Leningrad, 1984.

18. D. Quillen, *Higher K-theory*. I, Lecture Notes in Math., vol. 341, Springer, Berlin, 1973, pp. 83–147.

19. V. P. Platonov, *The Tannaka-Artin problem and reduced K-theory*, Izv. Akad. Nauk USSR **40** (1976), 227–261.

20. _____, *Birational properties of the reduce Whitehead group*, Doklady BSSR **21** (1977), 197–198.

21. M. Rost, *Hilbert 90 for K_3 for degree-two extensions*, Regensburg, Germany, 1986, preprint.

22. _____, *Injectivity of $K_2D \to K_2F$ for quaternion algebra*, Regensburg, Germany, 1986, preprint.

23. ____, *The group* $H_0(X, K_1)$ *for quadrics*, Regensburg, Germany, 1989, preprint.

24. A. H. Schofield and M. Van den Bergh, *The index of a Brauer class on a Severi-Brauer variety*, Trans. Amer. Math. Soc. **333** (1992), 729–739.

25. A. A. Suslin, *Algebraic K-theory and the norm-residue homomorphism*, J. Soviet Mat. **30** (1985), 2556–2611.

26. ____, SK_1 *of division algebras and Galois cohomology*, Adv. in Soviet Math. **4** (1991), 75–99.

27. R. G. Swan, *K-theory of quadric hypersurfaces*, Ann. of Math. **121** (1985), 113–153.

28. ____, *Zero cycles on quadric hypersurfaces*, Proc. Amer. Math. Soc. **107** (1989), 43–46.

29. J.-P. Tignol, *Reduction de l'indice d'une algebre simple centrale sur le corps des fonctions d'une quadrique*, Bull. Soc. Math. Belgique **42** (1990), 735–745.

30. V. E. Voskresenskii, *Algebraic tori*, Nauka, Moscow, 1977.

31. S. Wang, *On the commutator group of a simple algebra*, Amer. J. Math. **72** (1950), 323–334.

32. V. I. Yanchevskii, *The commutator subgroups of simple algebras with surjective reduced norms*, Soviet Math. Dokl. **16** (1975), 492–495.

33. W. Scharlan, *Quadratic and hermitian forms*, Grundelhren Math. Wiss., vol. 270, Springer-Verlag, Berlin, 1985.

DEPARTMENT OF MATHEMATICS AND MECHANICS, ST. PETERSBURG STATE UNIVERSITY, 198904 ST. PETERSBURG, RUSSIA

E-mail address: merkurev@math.lgu.spb.su

Proceedings of Symposia in Pure Mathematics
Volume **58.1** (1995)

Abelian Class Field Theory of Arithmetic Schemes

WAYNE RASKIND

Dedicated to the memory of my mother, Gertrude Raskind (1916–1993)

Contents

1991 *Mathematics Subject Classification.* Primary 19F05; Secondary 11-02, 11G99, 14G15, 14C35, 14G20, 14G25, 19E15.

Research supported by grants from the National Science Foundation, the National Security Agency and the Alexander von Humboldt-Stiftung.

This paper is in final form and no version of it will be submitted for publication elsewhere.

Introduction

In this article I shall survey the abelian class field theory of absolutely finitely generated fields and their various localizations. This was the subject of my five lectures at the Institute, and the presentation here is a greatly expanded version of the lectures.

An abelian extension of a field k is a Galois field extension with abelian Galois group. Abelian class field theory was originally concerned with abelian extensions of number fields, function fields in one variable over finite fields and local fields. The beginnings of the subject were in the theory of quadratic forms over numbers fields and quadratic reciprocity. Kronecker asserted and Weber gave the first complete proof that every abelian extension of **Q** is contained in a cyclotomic field [**Kr**], [**We1**]. The word "class field" seems to appear for the first time in the work of Weber [**We2**], although the notion was certainly in the mind of Kronecker. Given any number field k, Hilbert conjectured [**Hi,** §16] the existence of what is now called the Hilbert class field; that is, a normal field extension of k which is everywhere unramified and whose Galois group is canonically isomorphic to the ideal class group of k. This "reciprocity law for unramified class field theory" was proved by Furtwängler [**Furt**]. Takagi [**Ta1**], [**Ta2**] proved the general reciprocity law for abelian extensions of number fields by an indirect argument which did not immediately apply to L-series. Artin [**A1**], [**A2**] reformulated the reciprocity law to relate his nonabelian L-series with the abelian L-series associated to Grössencharaktere as defined by Hecke [**He**]. After class field theory for number fields had been established, the theory in local fields was developed by Hasse [**Ha1**], F. K. Schmidt [**Schmidt**] and Chevalley [**C1**]. With the introduction of idèles by Chevalley [**C2**], the global theory took on an analytic meaning which was developed further by Tate in his thesis [**T1**] to give a more conceptual proof of the analytic continuation of the zeta function of any number field and Hecke L-series. This approach made possible the philosophy of Langlands [LL] relating nonabelian representations of the absolute Galois group of a global field with automorphic representations of GL_n.

A function field in one variable over a finite field has many of the same local and global properties as a number field, and this suggested that there should be a class field theory for such fields. This was developed in the 1930's by Hasse, Schmid, Schmidt, and Witt [Ha2], [Schmid], [Schmidt2], [Wi]. Viewing the field as the function field of a smooth projective curve over a finite field, Weil used his theory of the Jacobian to express the reciprocity law for unramified class field theory in the form of an isomorphism between the Galois group of the maximal abelian geometric unramified covering of X and the group of rational points of the Jacobian of X [W1]. Lang [L2] then showed how to use Rosenlicht's theory of generalized Jacobians [Ro] to prove an analogous statement for all abelian extensions (see also [S1]).

In the 1940's and 50's, the proofs of the basic results in class field theory were further simplified by cohomological methods [HN], [T2], [AT], [W2], and in the 1980's Neukirch found a very elegant group-theoretic approach [N1], [N3]. Although many fascinating and difficult unsolved problems remain (e.g., construction of explicit class fields and reciprocity laws, Leopoldt's conjecture), it might be fair to say that the general theory of abelian extensions of local and global fields has reached a certain completeness.

Now let K be a field which is finitely generated over the prime subfield and denote by K^{ab} the maximal abelian extension of K. We regard K as the function field of a proper scheme X of finite type over $\mathbf{Z}$. By the dimension of K, we mean the dimension of X as a scheme (often called the Kronecker dimension). The abelian class field theory of K is concerned with the description of how closed points of X decompose in abelian extensions of K and a description of $\operatorname{Gal}(K^{ab}/K)$ in terms of data about X alone. As for nonabelian class field theory, there is little known at present in dimension greater than one, but Kapranov has recently made some striking analogies between the Langlands correspondence and topological quantum field theory [Ka].

The beginning of higher-dimensional class field theory was in the 1950's with the work of Lang [L1], [L2], who treated the case of unramified extensions of Albanese type of a normal projective variety X over a finite field. If Y is an abelian unramified cover of X, he defined the reciprocity map from the group of zero-cycles $Z_0(X)$ to the Galois group $\operatorname{Gal}(Y/X)$ (the Artin map given by Frobenius substitutions). The reciprocity law may be expressed by saying that if the covering arises by pullback from the Albanese variety of X, the kernel of this map consists of norms down to X of zero-cycles on Y and cycles on X which are Albanese equivalent to zero. Lang also outlined how to generalize the theory to ramified extensions [L2], and this is explained in Serre's book [S1]. The Galois group of the maximal abelian extension of K is described by means of an inverse limit of rational maps of X into principal homogeneous spaces under algebraic groups (generalized Albanese varieties).

The situation pretty much remained this way until Parshin made a great advance in the 1970's by seeing the connection between algebraic K-theory and class field theory [P1]. First, he saw the generalization of local class field theory to higher-dimensional local fields ([P1], [P4]), and this was done independently by Kato ([K1]). Then he worked toward the reciprocity law for unramified class field theory, and this was accomplished by Bloch, Kato and Saito ([B2], [KS2], [Sa3]). During this time, Bloch, Coombes and Saito ([B2], [Co], [Sa2]) were developing the class field theory of curves over local fields and Saito [Sa7] was studying the class field theory of two-dimensional local rings. Kato and Saito [KS1] proved the general reciprocity law for the abelian class field theory of absolutely finitely generated fields of dimension 2 and then for fields of any dimension [KS3]. We will give a more streamlined statement and proof of their beautiful result which is based on their unpublished notes. The reader should note that many of the important techniques in the subject were introduced by Bloch in his paper mentioned above.

The main theorem says roughly that there is an isomorphism between the inverse limit over moduli of relative Chow groups of zero cycles modulo rational equivalence and the Galois group of the maximal abelian extension of K. More precisely, let X be a proper integral scheme of finite type over $\mathbf{Z}$ with structure sheaf $\mathscr{O}_X$ and function field K. Denote by $\mathscr{K}_d^M(\mathscr{O}_X)$ the sheaf associated to the Zariski presheaf of Milnor K-groups (see below). If I is a nonzero coherent ideal of $\mathscr{O}_X$, we put (see Chapter 1)

$$\mathscr{K}_d^M(\mathscr{O}_X, I) := \mathrm{Ker}[\mathscr{K}_d^M(\mathscr{O}_X) \to \mathscr{K}_d^M(\mathscr{O}_X/I)],$$

where $d = \dim X$. Let Σ denote the set of archimedean places of the algebraic closure of the prime field in K. We consider the Zariski cohomology group with compact support $H_\Sigma^d(X, \mathscr{K}_d^M(\mathscr{O}_X, I))$, which should be viewed as a relative Chow group or a class group with modulus (if char. $K \neq 0$, the Σ plays no role). Indeed, in the case where $d = 1$, this cohomology group is a ray class group with modulus I. In Chapter 6 we shall define a reciprocity map

$$\varprojlim_{I \neq 0} H_\Sigma^d(X, \mathscr{K}_d^M(\mathscr{O}_X, I)) \to \mathrm{Gal}(K^{\mathrm{ab}}/K).$$

The reciprocity law may be stated as follows (see Theorems 6.1, 6.2):

THEOREM. (1) *If* char. $K = 0$, *the reciprocity map is an isomorphism.*

(2) *If* K *is of characteristic* $p > 0$, *let* $\mathbf{F}$ *be the algebraic closure of* $\mathbf{F}_p$ *in* K. *Then the reciprocity map is injective with image equal to the subgroup of* $\mathrm{Gal}(K^{\mathrm{ab}}/K)$ *consisting of all elements whose image in* $\mathrm{Gal}(\overline{\mathbf{F}}/\mathbf{F}) = \widehat{\mathbf{Z}}$ *is an integral power of the Frobenius.*

It is possible (at least in dimension two) to state the reciprocity law in terms of idèles, but this formulation is much more complicated (see [KS1]).

Also, there are good reasons to believe that there will be no corresponding analytic theory as in the 1-dimensional case (see Remark 6.3).

These notes are organized as follows: in Chapter 1, we review much of the formalism from algebraic K-theory and étale cohomology that we require. In particular, we review the duality theorems which play an important part in the definition of the reciprocity maps. All of this material may be found elsewhere, but we thought it might be helpful to at least have the statements of some of the results needed. In Chapter 2, we discuss the class field theory of n-dimensional local fields. We give the basic outline of the theory of Parshin and Kato, then give some recent work of Fesenko. In Chapter 3, we discuss the class field theory of two-dimensional local rings following work of Saito and Koya. In Chapter 4, we do the class field theory of varieties over local fields following work of Bloch, Coombes and Saito. In Chapter 5, we do unramified class field theory of arithmetic schemes following the work of Bloch, Kato, Saito, Lang, Milne, Colliot-Thélène/Sansuc/Soulé and Colliot-Thélène/Raskind. In Chapter 6, we prove the general reciprocity law for abelian extensions of an absolutely finitely generated field. The proof given is taken from unpublished notes of Kato-Saito. It actually requires very little in Chapters 2–5. Finally, in Chapter 7 we give some applications of the theory to "Hasse principles" for higher-dimensional global fields following work of Kato, Jannsen, Colliot-Thélène and Saito.

A word about the style of these notes. I have tried to write them in such a way that it is possible for someone to either browse through them to see the main results of the theory or really learn some of the details of the proofs. I also hope that they might be accessible to a wide range of mathematicians. The main results are stated near the beginning of each chapter before we get into the details. My aim is to really give the reader a feel for the subject. There are many techniques used, but there are a few very important themes which I have tried to stress. In particular, I have attempted to give proofs which are as geometric as possible. Some of the proofs are new or have been simplified.

Limitations of time and space required me to leave out several important topics, perhaps the most important of which is higher-dimensional ramification theory (for a short survey, see [K9]). I also regret that I did not have the time to work out more examples.

The list of references is meant to be an extensive but not exhaustive list of papers and books relevant to the subject. For one-dimensional class field theory, I have taken much of the history and many of the references from a survey by Hasse [Ha4] and a commentary by Iwasawa on the work of Takagi [Iw2].

Acknowledgments

I have had the good fortune of being able to discuss higher-dimensional class field theory extensively with K. Kato and S. Saito, who have also allowed me to use some of their unpublished work. In particular, Sections 4–7 of Chapter 6 rely heavily on their unpublished notes. Much of this paper is just a reorganization and consolidation of their work. J.-L. Colliot-Thélène introduced me to the subject several years ago, and much of what I know I learned from him. I would like to thank I. Fesenko for his help with the bibliography and for informing me of his recent work.

Parts of this paper were written while I enjoyed the hospitality of Universitat Autònoma de Barcelona and Universität zu Köln.

B. Jacob and A. Rosenberg did an excellent job of organizing the Institute. I would like to thank them for inviting me to give these lectures and for waiting very patiently for me to write them up. The audience was very attentive and laughed politely at my jokes. Finally, I would like to thank my wife for agreeing to delay our honeymoon so that I could give these lectures and for putting up with me during the final stages of preparation of this paper.

Notation

In Chapter 1 below, we give some preliminaries on algebraic K-theory and étale cohomology. Here we simply list some of the notation we shall use for easy access.

For X a noetherian scheme, we denote by X^i the set of points of X such that the local ring $\mathcal{O}_{X,x}$ has dimension i. The residue field of a point x of X will be denoted by $\kappa(x)$. We denote by $CH^i(X)$ the group of codimension i-cycles modulo rational equivalence.

If k is a field, we denote by k^{sep} a separable closure of k, by $\overline{k}$ an algebraic closure and $G = \mathrm{Gal}(k^{\mathrm{sep}}/k)$. If M is discrete G-module, we denote by $H^i(k, M)$ the Galois cohomology group of G with values in M.

If X is defined over k and L is any field extension of k, we write $X_L := X \times_k L$. In particular, $\overline{X} := X \times_k \overline{k}$. If k is a global field and v is a place of k, k_v will denote the completion of k with respect to v. If A is a local ring with maximal ideal m, we denote by $\widehat{A}$ the completion of A in the m-adic topology and A^h the henselization of A with respect to m.

The Zariski sheaf of Quillen K_i-groups on X will be denoted by $\mathcal{K}_i$ and the sheaf of Milnor K groups will be denoted by $\mathcal{K}_i^M(\mathcal{O}_X)$. If I is an ideal of $\mathcal{O}_X$, let $Y = \mathrm{Spec}(\mathcal{O}_X/I)$ with the reduced scheme structure and denote by $i : Y \to X$ the inclusion. By abuse of notation, we write

$$\mathcal{K}_i^M(\mathcal{O}_X/I)$$

instead of

$$i_*\mathcal{K}_i^M(\mathcal{O}_X/I).$$

Set

$$\mathcal{K}_i^M(\mathcal{O}_X, I) = \mathrm{Ker}[\mathcal{K}_i^M(\mathcal{O}_X) \to \mathcal{K}_i^M(\mathcal{O}_X/I)].$$

These sheaves will play a crucial role in Chapter 6.

If n is a positive integer prime to any residue characteristic of X, we let $\mathbf{Z}/n(1) = \mu_n$, the sheaf of nth roots of unity. Then for $j > 0$, $\mathbf{Z}/n(j) = \mathbf{Z}/n(1)^{\otimes j}$ and for $j < 0$, $\mathbf{Z}/n(j) = \mathrm{Hom}(\mathbf{Z}/n(-j), \mathbf{Z}/n)$. Cohomology with values in such sheaves will always be étale cohomology. If M is a Galois module which is killed by n, we denote by $M(i) = M \otimes \mathbf{Z}/n(i)$ the twist by i of M. When M is a $\mathbf{Z}_l$-module, we put $M(i) = M \otimes \mathbf{Z}_l(i)$. If $\mathscr{F}$ is an étale sheaf on X, we denote by $\mathscr{H}^i(\mathscr{F})$ the Zariski sheaf associated to the presheaf:

$$U \mapsto H^i(U_{\text{ét}}, \mathscr{F}_U).$$

In practice $\mathscr{F}$ will be something like $\mathbf{Z}/n(j)$ or the extension by zero of $\mathbf{Z}/n(j)$ from an open subscheme of X.

If some residue characteristic divides n, we will consider μ_n as a sheaf in the flat topology. If X is defined over a field of characteristic $p > 0$, we denote by $H^i(X, \mathbf{Z}/p^n(j))$ the group $H^{i-j}(X, W_n\Omega^j_{X,\log})$.

For an abelian group A, we denote by A_n the subgroup of A consisting of elements killed by n and by A/n the quotient A/nA. If l is a prime number, we denote by $T_l(A)$ the group

$$\varprojlim_m A_{l^m}$$

and call it the l-Tate module of A. We set $V_l(A) = T_l(A) \otimes \mathbf{Q}_l$ and $A^* = \mathrm{Hom}(M, \mathbf{Q}/\mathbf{Z})$.

If $X = \mathrm{Spec}\, A$ is an affine scheme, we denote by μ_n the group-scheme of nth roots of unit $\mathrm{Spec}\, A[x]/(x^n - 1)$. If p is a prime number, then α_p denotes the group-scheme $\mathrm{Spec}\, A[x]/x^p$.

Chapter 1. Preliminaries on algebraic K-theory and étale cohomology

1.1. Algebraic K-theory. Let X denote a scheme with structure sheaf $\mathscr{O}_X$. By a *vector bundle* on X, we mean a locally free $\mathscr{O}_X$-module of finite rank. We denote the category of vector bundles on X by $\mathscr{P}(X)$; if X is noetherian, we denote the category of coherent $\mathscr{O}_X$-modules by $\mathscr{M}(X)$. These are exact categories in the sense of Quillen [Q], and his Q-construction may be applied to them to yield categories which will be denoted by $Q\mathscr{P}(X)$ and $Q\mathscr{M}(X)$, respectively. By definition,

$$K_i(X) = \pi_{i+1}|BQ\mathscr{P}(X)|$$

and

$$G_i(X) = \pi_{i+1}|BQ\mathscr{M}(X)|,$$

where $|\ |$ denotes the geometric realization of the classifying space B. Many people use K' instead of G, but we feel that there is less chance of typographical error with G.

REMARK 1.1. In the last few years, there have been several improvements on Quillen's definition of K-theory. First, Waldhausen isolated a class of categories which are more general than exact categories to which K-functors may be applied [**Wa**]. These categories, now called Waldhausen categories, were used by Thomason in his classic paper [**T2**] on localization in K-theory. The reader who wants to seriously study algebraic K-theory is strongly encouraged to read these papers. In these notes, the level of K-theory we need is not very high, so we can make do with Quillen's definition.

There is a natural map

$$K_*(X) \to G_*(X),$$

and Quillen's resolution theorem [**Q, Theorem 3**] implies that if X is regular, this map is an isomorphism. The advantage of G-theory is that we can filter the category of coherent sheaves by codimension of support: let $\mathscr{M}^i$ be the subcategory of $\mathscr{M}$ consisting of coherent sheaves the codimension of whose support is $\geq i$. Then we have a decreasing filtration:

$$\mathscr{M} = \mathscr{M}^0 \supset \mathscr{M}^1 \supset \cdots .$$

By Quillen's localization theorem [**Q, Theorem 5**], there is a long exact sequence:

$$\cdots \to G_n(\mathscr{M}^{i+1}) \to G_n(\mathscr{M}^i) \to G_n(\mathscr{M}^i/\mathscr{M}^{i+1}) \to G_{n-1}(\mathscr{M}^{i+1}) \to \cdots .$$

We have [**Q, Corollary 1 to Theorem 4**]:

$$G_n(\mathscr{M}^i/\mathscr{M}^{i+1}) \cong \bigoplus_{x \in X^i} K_n \kappa(x).$$

The composite:

$$G_n(\mathscr{M}^i/\mathscr{M}^{i+1}) \to G_{n-1}(\mathscr{M}^{i+1}) \to G_{n-1}(\mathscr{M}^{i+1}/\mathscr{M}^{i+2})$$

will be denoted by ∂_n. Then $\partial_{n-1}\partial_n$ is zero, and we get a complex:

$$\bigoplus_{x \in X^0} K_n \kappa(x) \to \bigoplus_{x \in X^1} K_{n-1} \kappa(x) \to \cdots \to \bigoplus_{x \in X^j} K_{n-j} \kappa(x) \to \cdots$$

which is called the *Gersten-Quillen complex*. This is the E_1-term of the Brown-Gersten-Quillen (BGQ) spectral sequence:

$$E_1^{r,s}(X) = \bigoplus_{x \in X^r} K_{-r-s} \kappa(x) \Rightarrow K_{-n}(X).$$

Note that this is a fourth quadrant spectral sequence, which accounts for the bizarre looking indexing. As for the E_2-term, we have the following theorem of Quillen [**Q, Proposition 5.6 and Theorem 5.11**]:

THEOREM 1.2. *If X is regular over a field k, then*

$$E_2^{r,s}(X) = H^r(X, \mathscr{K}_{-s}),$$

where $\mathscr{K}_i$ is the Zariski sheaf associated to the presheaf $U \mapsto K_i(\Gamma(U, \mathscr{O}_U))$. In other words, the E_1-term of the BGQ spectral sequence computes the cohomology of the sheaves $\mathscr{K}_s$.

COROLLARY 1.3 (Bloch-Quillen formula). *With hypotheses as in the theorem,*

$$H^r(X, \mathscr{K}_r) = CH^r(X),$$

the Chow group of codimension r-cycles modulo rational equivalence.

In fact, Grayson has shown [Gra1] that for *any* scheme X, we have

$$E_2^{r,-r}(X) = CH^r(X),$$

where $CH^r(X)$ is defined as in the paper of Fulton [F]. This will be of use later.

We note also the following result of Grayson [Gra2].

THEOREM 1.4. *Let X be as above (regular of finite type over a field) and n a positive integer. Then denoting by $E_1^{r,s}(X, \mathbf{Z}/n)$ the Brown-Gersten-Quillen spectral sequence modulo n, we have*

$$E_2^{r,-s}(X, \mathbf{Z}/n) = H^r(X, \mathscr{K}_s/n),$$

where $\mathscr{K}_s$ is the Zariski sheaf associated to the presheaf $U \mapsto K_s(\Gamma(U, \mathscr{O}_U))/n$.

One case in which we shall have occasion to use this formalism is when $\mathscr{X}$ is a regular proper scheme of finite type over a discrete valuation ring A, with generic fibre X and special fibre Y. By using the boundary maps in algebraic K-theory, we can relate the Gersten-Quillen complexes on X, $\mathscr{X}$ and Y as follows ([Sh]). There is an exact sequence of complexes:

$$0 \to E_1(Y)[-1] \to E_1(\mathscr{X}) \to E_1(X) \to 0.$$

This gives the long exact sequence of E_2-terms:

$$\cdots \to E_2^{r,s}(\mathscr{X}) \to E_2^{r,s}(X) \to E_2^{r,s+1}(Y) \to E_2^{r+1,s}(\mathscr{X}) \to E_2^{r+1,s}(X) \cdots.$$

We shall also denote the boundary maps:

$$E_2^{r,s}(X) \to E_2^{r,s+1}(Y)$$

by ∂, and we hope no confusion results because of this. In Chapter 4, we shall study the map:

$$\partial_1 : H^d(X, \mathscr{K}_{d+1}) \to CH^d(Y)$$

in the case where X is of dimension d. Then we often write $CH_0(Y)$ for $CH^d(Y)$ if Y is equidimensional.

1.2. Milnor K-theory. Let k be a field. When $n \geq 1$ we denote by $K_n^M k$ the quotient of

$$k^* \otimes \cdots \otimes k^*$$

(n-factors) by the subgroup generated by

$$a_1 \otimes \cdots \otimes a_n$$

with $a_i + a_j = 1$ for some $i \neq j$. This is called the nth Milnor K-group of k. We define K_0^M to be $\mathbf{Z}$. We denote the image of $a_1 \otimes \cdots \otimes a_n$ in $K_n^M k$ by $\{a_1, \ldots, a_n\}$. When there are several fields involved, we often write $\{a_1, \ldots, a_n\}_k$ to emphasize that this symbol is in the Milnor K-theory of k. There is a natural product $K_i^M k \times K_j^M k \to K_{i+j}^M k$.

Milnor K-theory of discrete valuation rings. If A is a discrete valuation ring with fraction field k and residue field F, then there is a map

$$\partial : K_n^M k \to K_{n-1}^M F$$

which is the valuation v for $n = 1$ and for $n \geq 2$ has the following properties (see [**BT**, Chapter 1, §4]):

(1) if $a_1, \ldots, a_{n-1}$ are units of A and π is a uniformizing parameter, then

$$\partial(\{a_1, \ldots, a_{n-1}, \pi\}) = \{\bar{a}_1, \ldots, \bar{a}_{n-1}\},$$

where bar denotes the residue class.

(2) Let $U_n^0(k)$ be the image of the map

$$A^* \otimes \cdots \otimes A^* \to K_n^M k.$$

Then the kernel of ∂ is $U_n^0(k)$.

(3) For $n = 2$, ∂ is the tame symbol defined by:

$$\partial\{a, b\} = (-1)^{v(a)v(b)} \overline{\left(\frac{a^{v(b)}}{b^{v(a)}} \right)}.$$

(4) Let m be the maximal ideal of A, and let $U_n^i(k)$ be the subgroup of $K_n^M k$ generated by symbols of the form $\{a_1, \ldots, a_n\}$ with $a_1 \in 1 + m^i$. Then we have an exact sequence:

$$0 \to U_n^1(k) \to K_n^M k \to K_n^M F \oplus K_{n-1}^M F \to 0.$$

The last nonzero map is defined by

$$\{a_1, \ldots, a_n\} + \{b_1, \ldots, b_{n-1}, \pi\} \mapsto (\{\bar{a}_1, \ldots, \bar{a}_n\}, \{\bar{b}_1, \ldots, \bar{b}_{n-1}\}),$$

where $a_1, \ldots, a_n, b_1, \ldots, b_{n-1} \in A^*$. This map depends on the choice of π.

The norm. We summarize the basic properties of the norm map in Milnor K-theory. For more details, see the nice treatments in [**Ke**, Kap. 4] and [**Fe/V**, Chapter IX, §3, Theorem 3.8]. Let $k(t)$ be the rational function field in one variable over k. For v an irreducible polynomial of $k[t]$, let $k(v)$ denote the residue field $k[t]/(v)$. Recall [**Mi**, Theorem 2.3] that there is an exact sequence:

$$0 \to K^M_{n+1}k \to K^M_{n+1}k(t) \to \bigoplus_v K^M_n k(v) \to 0.$$

Let v_∞ be the infinite valuation given by $v_\infty(f) = -\deg(f)$. Then there exists a unique map N making the following diagram commute:

$$
\begin{array}{ccc}
K^M_{n+1}k(t) & \xrightarrow{\ \partial\ } & \bigoplus_v K^M_n k(v) \\
\downarrow & & \downarrow N \\
K^M_n k(v_\infty) & \longleftarrow & K^M_n k.
\end{array}
$$

Here the left vertical map is ∂_∞, and the bottom horizontal map is minus the identity. For $v \neq v_\infty$ we denote by N_v the v-component of N and we make N_{v_∞} the identity map. We have (Weil reciprocity):

$$\sum_{\text{all } v} (N_v(\partial_v(x))) = 0$$

for all x in $K^M_n k(t)$. This fact and the definition of N_{v_∞} uniquely characterize N_v. Now let L/k be a finite extension which is generated by one element α. Let M be the normal closure of L over k. Then we can define a norm map:

$$N_\alpha \colon K^M_n L \to K^M_n k$$

as the composite

$$K^M_n L \to K^M_n M \xrightarrow{\ N_{M/k}\ } K^M_n k.$$

In order for this to be a good definition, we should have that if $k \subset L_1 \subset L_2$ are three fields then $N_{L_2/k} = N_{L_1/k} N_{L_2/L_1}$. The norm satisfies the projection formula: if $\alpha \in K^M_n k$, $\beta \in K^M_m L$, then we have:

$$\{\alpha, N(\beta)\}_k = N\{\alpha, \beta\}_L.$$

The norm on K^M_1 is the usual field norm.

The Galois symbol. The Galois symbol is the fundamental relation between Milnor K-theory and Galois cohomology, and it plays a crucial role in class field theory. Let m be a positive integer which is invertible in the field k. Define the Galois symbol on K^M_0 as the projection of $\mathbf{Z}$ onto

$\mathbf{Z}/m = H^0(k, \mathbf{Z}/m(0))$; for K_1^M define it to be the map $h_{m,k}^1: K_1^M k = k^* \to H^1(k, \mathbf{Z}/m(1))$ given by Kummer theory. For K_2^M we define it as follows. Consider the map:

$$h_{m,k}^2: k^*/m \otimes k^*/m \to H^2(k, \mathbf{Z}/m(2))$$

given by $h_{m,k}^2(a \otimes b) = h_{m,k}^1(a) \cup h_{m,k}^1(b)$. We often denote by (a, b) the image of $a \otimes b$ in $H^2(k, \mathbf{Z}/m(2))$. Note that we have a similar projection formula for Galois cohomology [CE, Chapter XII, §8].

LEMMA 1.5. *Let* $a \in k^*$, $a \neq 1$. *Then* $h_{m,k}^2(a \otimes (1-a)) = 0$, *and hence*, $h_{m,k}^2$ *descends to a map*

$$h_{m,k}^2: K_2^M k/m \to H^2(k, \mathbf{Z}/m(2)).$$

PROOF. (Compare [T7, 3.1], where a stronger result is proved.) Let

$$x^m - a = \prod_{i=1}^r f_i(x)$$

be a factorization of $x^m - a$ into a product of monic irreducible polynomials over k. Let $\bar{k}$ be a fixed algebraic closure of k and let L_i be the splitting field of $f_i(x)$. Choose a root b_i of $f_i(x)$ in L_i. Then we have

$$\prod_{i=1}^r N_{L_i/k}(1 - b_i) = 1 - a$$

and by the projection formula in Galois cohomology, we get

$$(a, 1-a)_k = \left(a, \prod_{i=1}^r N_{L_i/k}(1 - b_i) \right)_k = \sum_{i=1}^r N_{L_i/k}(a, 1 - b_i)_{L_i}$$

$$= \sum_{i=1}^r N_{L_i/k}(b_i^m, 1 - b_i)_{L_i} = m \sum_{i=1}^r N_{L_i/k}(b_i, 1 - b_i)_{L_i}$$

$$= 0.$$

In a similar way, one shows that the map

$$h_{m,k}^n: K_n^M k/m \to H^n(k, \mathbf{Z}/m(n))$$

defined by sending $\{a_1, \ldots, a_n\}$ to $h_{m,k}^1(a_1) \cup \cdots \cup h_{m,k}^1(a_n)$ is well defined.

If k is of characteristic $p > 0$, let $W_r \Omega_{k,\log}^n$ be the logarithmic subsheaf of the de Rham-Witt sheaf $W_r \Omega^n$ (see 1.3 below for more details). Then we have the *differential symbol*

$$h_{p^r}^n: K_n^M k/p^r \to H^0(k_{\text{ét}}, W_r \Omega_{k,\log}^n),$$

defined for $r = 1$ by

$$h_p^n(\{a_1, \ldots, a_n\}) = \frac{da_1}{a_1} \wedge \cdots \wedge \frac{da_n}{a_n}.$$

In general this map may be defined in a similar way as the Galois symbol by using the isomorphism

$$k^*/k^{*p^r} \cong H^0(k, W_r\Omega^1_{k,\log})$$

and the product structure on the sheaves $W_r\Omega^i_{k,\log}$. For more details, see [K3] and [BK1]. Kato has conjectured that for any field k and any m prime to the characteristic of k, the map $h^n_{m,k}$ is bijective. The following theorem summarizes what is known about this conjecture.

THEOREM 1.6. *The Galois symbol $h^n_{m,k}$ is an isomorphism at least in the following cases*:

(i) (*classical*) $n = 0, 1$.

(ii) (*Merkur'ev-Suslin* [MS1]). $n = 2$ *and any* m.

(iii) (*Merkur'ev-Suslin* [MS2; Rost]). $n = 3$ *and m is a power of* 2.

(iv) (*Bloch-Kato* [BK]). *k is a complete discretely valued field with residue field of characteristic $p > 0$, m is a power of p and $n \geq 0$*.

(v) (*Bloch-Gabber-Kato* [BK]). *The differential symbol is bijective for any $n \geq 0$ and any field k of characteristic $p > 0$*.

If m is a power of a prime l, we can pass to the inverse limit and get a Galois symbol

$$K^M_i k \to H^i(k, \mathbf{Z}_l(i)),$$

which we can combine with cup-product to get a pairing:

$$K^M_i k \times H^j(k, \mathbf{Q}_l/\mathbf{Z}_l(r)) \to H^{i+j}(k, \mathbf{Q}_l/\mathbf{Z}_l(r+i)).$$

Let $\mathscr{K}^M_i$ be the Zariski sheaf which is the quotient of

$$\mathscr{O}^*_X \otimes \cdots \otimes \mathscr{O}^*_X \qquad (i \text{ times})$$

by the subabelian sheaf generated locally by sections $a_1 \otimes \cdots \otimes a_i$ with $a_j + a_k = 1$ for some $j \neq k$. If I is an ideal of $\mathscr{O}_X$, let $Y = \mathrm{Spec}(\mathscr{O}_X/I)$ with the reduced scheme structure and denote by $i: Y \to X$ the inclusion. By abuse of notation, we write $\mathscr{K}^M_i(\mathscr{O}_X/I)$ instead of $i_*\mathscr{K}^M_i(\mathscr{O}_X/I)$.

Set

$$\mathscr{K}^M_i(\mathscr{O}_X, I) = \mathrm{Ker}[\mathscr{K}^M_i(\mathscr{O}_X) \to \mathscr{K}^M_i(\mathscr{O}_X/I)].$$

1.3. Étale cohomology. For the theory of étale cohomology, we refer the reader to [SGA 4], [SGA 4½], and [M2]. Here we collect together a few facts that we will need. Let X, Y be schemes, and let $f: Y \to X$ be a morphism between them. f is *flat* if for each open set U in X, $\Gamma(f^{-1}(U), \mathscr{O}_{f^{-1}(U)})$ is a flat $\Gamma(U, \mathscr{O}_U)$-module. Let y be a point of Y, $x = f(y)$, A the local ring of X at x, B the local ring of Y at y, and m_x, m_y the maximal ideals of A, B, respectively. Then we say that f is *unramified at y* if m_x generates m_y and the residue field B/m_y is a finite separable field extension of A/m_x. f is *unramified* if it is so at all points y of Y. A morphism $Y \to X$ is *étale* if it is flat and unramified.

We denote by $\mathbf{Et}/X$ the category of schemes locally of finite type and étale over X. The small étale site is this category equipped with the étale topology. The category of schemes finite and étale over X will be denoted by $\mathbf{FEt}/X$. Assume now that X is connected, and let $\bar{x}$ be a geometric point of X. Define a functor $F: \mathbf{FEt}/X \to \mathbf{Groups}$ by sending Y finite étale over X to $\mathrm{Hom}(\bar{x}, Y)$. This functor is prorepresentable by an object $\tilde{X} = \{X_i, \phi_i\}$ of $\mathbf{FEt}/X$, and we define the fundamental group $\pi_1(X, \bar{x})$ with basepoint $\bar{x}$ by:

$$\pi_1(X, \bar{x}) := \mathrm{Aut}_X(\tilde{X}) = \varprojlim_i \mathrm{Aut}_X(X_i).$$

If $\bar{y}$ is another geometric point of X, then we have an isomorphism

$$\pi_1(X, \bar{x}) \to \pi_1(X, \bar{y})$$

which is unique up to inner automorphism of $\pi_1(X, \bar{x})$. We define $\pi_1^{\mathrm{ab}}(X)$ to be the quotient of $\pi_1(X, \bar{x})$ by the closure of its commutator subgroup. This is independent of any choice of basepoint.

Let Y/X be an étale cover. If x is a closed point of X and the covering of $\mathrm{Spec}\,\kappa(x)$ given by $Y \times_X \kappa(x)$ is trivial, we say that x *splits completely in* Y.

The following lemma is very important. It is a weak Čebotarev density theorem.

LEMMA 1.7. *Let X be an irreducible scheme of finite type over $\mathbf{Z}$ such that X_{red} is normal. Suppose Y/X is a finite connected étale cover in which every closed point of X splits completely. Then $Y = X$.*

PROOF. We use the theory of zeta functions [S2]. Recall that the zeta function of a scheme X of finite type over $\mathbf{Z}$ is given by

$$\zeta(X, s) = \prod_{x \in X_0} \frac{1}{1 - N(x)^{-s}}$$

where $N(x)$ is the cardinality of the residue field $\kappa(x)$. Let d be the dimension of X. Then $\zeta(X, s)$ has a simple pole at $s = d$. Since X_{red} is normal, Y_{red} is also normal ([SGA1], Exp. I, Cor. 9.10). Since Y is connected, it is irreducible, and hence $\zeta(Y, s)$ has a simple pole at $s = d$. But if every closed point of X splits completely in Y, then we must have $\zeta(Y, s) = \zeta(X, s)^n$, where $n = [Y:X]$. This implies that $n = 1$.

LEMMA 1.8. *Let $f: Y \to X$ be a finite flat morphism of schemes, with Y normal and X regular. Assume that f is unramified at all points of codimension one of Y. Then f is étale.*

The proof of this lemma may be found in ([SGA2, Exposé X, Théorème 3.4]; see also [Z], [Na1], [Na2]).

We need a similar but easier lemma for Henselian local rings:

LEMMA 1.9. *Let A be a Henselian regular local ring with fraction field K and residue field F. Suppose that L/K is a finite extension in which every prime ideal of height one in A splits completely. Then the extension is trivial.*

PROOF. Let B be the integral closure of A in L. Then B is a Henselian regular local ring, and every codimension one point of $\operatorname{Spec} A$ splits completely in $\operatorname{Spec} B$. Hence the covering B/A is unramified in codimension one, and so by Lemma 1.8, it is étale. Let m be the maximal ideal of A and m' the maximal ideal of B. By using a regular system of parameters of A, we then see that $B/m' \cong A/m$. This means that the map $A \to B$ has a section ([M2], Ch. 1, Theorem 4.2d)), and hence the extension L/K is trivial.

Let $\mathscr{F}$ be a sheaf of abelian groups on X. Then we can find an injective resolution of $\mathscr{F}$ and define the cohomology groups $H^i(X, \mathscr{F})$, which are independent of the chosen injective resolution. In this article, several sheaves will play prominent roles, and we describe some of these. First, let $\mathbf{Z}/n$ denote the constant sheaf with group $\mathbf{Z}/n\mathbf{Z}$. Then the étale cohomology group $H^1(X, \mathbf{Z}/n)$ classifies finite étale Galois covers Y of X together with a homomorphism

$$\operatorname{Aut}(Y/X) \to \mathbf{Z}/n.$$

Then

$$H^1(X, \mathbf{Z}/n) = \operatorname{Hom}(\pi_1^{\mathrm{ab}}(X), \mathbf{Z}/n).$$

Let X be a scheme and n a positive integer invertible on X. We denote by $\mathbf{Z}/n(1)$ the étale sheaf of nth roots of 1 on X. If $i > 0$, $\mathbf{Z}/n(i)$ denotes the tensor product of $\mathbf{Z}/n(1)$ with itself i-times; if $i < 0$, $\mathbf{Z}/n(i) = \operatorname{Hom}(\mathbf{Z}/n(-i), \mathbf{Z}/n)$. We define:

$$H^j(X, \mathbf{Z}_l(i)) := \varprojlim_n H^j(X, \mathbf{Z}/l^n(i))$$

and

$$H^j(X, \mathbf{Q}_l/Z_l(i)) := \varinjlim_n H^j(X, \mathbf{Z}/l^n(i)).$$

If p is a prime number which is *not* invertible on X, then we have the sheaf $\mathbf{Z}/p^m(1)$ on the flat site of X. However, taking tensor powers of this sheaf to get twists does not give anything interesting, so we must do something different. The answer is provided by the logarithmic de Rham-Witt cohomology of Bloch-Illusie-Milne. We give a very brief description, referring the reader to [I] for details.

Let A be a ring of characteristic $p > 0$. We denote by $W_n(A)$ the ring of Witt vectors of A of length n and by

$$W(A) = \varprojlim_n W_n(A)$$

the ring of Witt vectors of A. Given $a \in W(A)$, we denote its components by $(a_0, a_1, \ldots)$. The ring $W(A)$ is equipped with operations F (Frobenius) and V (Verschiebung). We have

$$F(a) \equiv a^p \ (\mathrm{mod}\, p)$$

and

$$V(a_0, a_1, \ldots) = (0, a_1, \ldots).$$

Then $FV = p = VF$ and $W(A)/VW(A) \cong A$. If $x \in A$, we denote by $\underline{x}$ the element $(x, 0, 0, \ldots)$ of $W(A)$, which is called the Teichmüller representative of x in $W(A)$. If k is a perfect field, $W(k)$ is characterized as being the unique (up to isomorphism) complete discrete valuation ring with residue field k in which p is absolutely unramified [**S4**, Chapter II, §5]. If X is a scheme, we define $W_n(\mathscr{O}_X)$ as the Zariski sheaf

$$U \mapsto W_n(\Gamma(U, \mathscr{O}_U)).$$

In the papers [**I**], [**B**], the de Rham-Witt complex $W_n\Omega^\bullet$ is defined and studied closely. The operators F, V defined above on the Witt vectors extend to operators on the de Rham-Witt complex, and we have the differential d, whose relation with F and V may be summarized by several relations, some of which are:

$$FV = p = VF, \qquad F\, dV = d, \qquad F\, d\underline{x} = \underline{x}^{p-1}\, d\underline{x}.$$

Let $W_n\Omega^i_{X,\log}$ denote the subsheaf of $W_n\Omega^i_X$ which is generated étale locally by logarithmic differentials

$$\frac{dx_1}{x_1} \wedge \cdots \wedge \frac{dx_i}{x_i}$$

with $x_j \in \mathscr{O}_X^*$ for $j = 1, 2, \ldots, i$. We will always consider this sheaf in the étale topology, and we often denote it by $\nu_n(i)$ or $\mathbf{Z}/p^n(i)$. It plays the role of $\mathbf{Z}/n(i)$ when n is not invertible on X. For example, we have

$$H^{i+1}_{\mathrm{fl}}(X, \mu_{p^n}) \cong H^i(X, \nu_n(1)),$$

where fl denotes flat cohomology (see [**M1**]).

1.4. Localization in étale cohomology and Bloch-Ogus theory. The Brown-Gersten-Quillen spectral sequence in K-theory has a counterpart in étale cohomology, and it plays an important role here. Let X be any scheme and n a positive integer invertible on X. By splicing together localizations sequences for étale cohomology, we get the *coniveau spectral sequence*:

$$E_1^{p,q} = \bigoplus_{x \in X^p} H^{q-p}(\kappa(x), \mathbf{Z}/n(j-p)) \Rightarrow H^{p+q}(X, \mathbf{Z}/n(j)).$$

Let $\mathscr{H}^r(\mathbf{Z}/n(j))$ be the Zariski sheaf associated to the presheaf

$$U \mapsto H^r(U_{\text{ét}}, \mathbf{Z}/n(j)).$$

Then we have the following theorem of Bloch-Ogus [**BO**]:

THEOREM 1.10. *If X is smooth over a field, then in the coniveau spectral sequence, we have*

$$E_2^{p,q} = H^p(X, \mathscr{H}^q(\mathbf{Z}/n(j))).$$

In other words, the complex:

$$\bigoplus_{x \in X^0} H^r(k(x), \mathbf{Z}/n(j)) \to \bigoplus_{x \in X^1} H^{r-1}(k(x), \mathbf{Z}/n(j-1)) \to \cdots$$

$$\to \bigoplus_{x \in X^{r-1}} H^1(k(x), \mathbf{Z}/n(j-r+1)) \to \bigoplus_{x \in X^r} H^0(k(x), \mathbf{Z}/n(j-r))$$

computes the cohomology of the sheaf $\mathscr{H}^r(\mathbf{Z}/n(j))$.

One situation where we will use this is the following. For X smooth and integral over a field and n invertible on X, we have a commutative diagram (up to sign):

$$
\begin{array}{ccccc}
K_2 k(X)/n & \longrightarrow & \bigoplus_{x \in X^1} k(x)^*/n & \longrightarrow & \bigoplus_{x \in X^2} \mathbf{Z}/n \\
\downarrow & & \downarrow & & \downarrow \\
H^2(k(X), \mathbf{Z}/n(2)) & \longrightarrow & \bigoplus_{x \in X^1} H^1(k(x), \mathbf{Z}/n(1)) & \longrightarrow & \bigoplus_{x \in X^2} \mathbf{Z}/n.
\end{array}
$$

Here the left vertical map is the Galois symbol, the middle is the map obtained from Kummer theory on $k(x)$ and the right vertical map is the obvious map. By the Merker'ev-Suslin theorem and Hilbert's theorem 90, all the vertical maps in the diagram are isomorphisms. From Theorems 1.3 and 1.10, we then get maps

$$H^i(X, \mathscr{K}_2/n) \to H^i(X, \mathscr{H}^2(\mathbf{Z}/n(2))).$$

Using these, we can get the following exact sequence which we will need later:

$$0 \to H^1(X, \mathscr{K}_2)/n \to NH^3(X, \mathbf{Z}/n(2)) \to CH^2(X)_n \to 0,$$

where $NH^3(X, \mathbf{Z}/n(2)) = \mathrm{Ker}(H^3(X, \mathbf{Z}/n(2)) \to H^3(k(X), \mathbf{Z}/n(2)))$.

1.5. Cohomology with compact support. Since we will often be working over number fields, it is crucial that we take the archimedean places into account when doing cohomological calculations. In order to do this, we introduce cohomology with compact support [KS3]. Let U be an open subset of the ring of integers in a number field or a smooth connected curve over a finite field. We denote by k the function field of U. If k is a number field, let X be the full ring of integers in k; if k is a function field in one variable over a finite field $\mathbf{F}$, let X be the unique smooth projective model of k over $\mathbf{F}$. Let Σ be the set of places of k which do not correspond to closed points of U. Thus Σ consists of the archimedean places of k and the complement of U in X. For each $v \in \Sigma$, let k_v be the completion of k with respect to the v-adic valuation. If Y is a scheme of finite type over U, we can form the schemes $Y_v = Y \times_U k_v$. We denote by $Y(\Sigma)$ the disjoint union of these schemes for $v \in \Sigma$ and by $\pi \colon Y(\Sigma) \to Y$ the natural map. If $\mathscr{F}$ is an abelian sheaf on Y, we can pull it back to a sheaf on $Y(\Sigma)$ which we denote by the same letter $\mathscr{F}$. Then we define the cohomology with compact support $H_c^*(Y, \mathscr{F})$ to be:

$$\mathbf{H}^*(Y, \mathrm{Cone}(\mathscr{F} \to R\pi_*\mathscr{F})[-1]).$$

Here $R\pi_*\mathscr{F}$ is the direct image in the derived category and $\mathscr{F} \to R\pi_*\mathscr{F}$ is regarded as a map of complexes in the derived category. Then we get an exact sequence:

$$\cdots \to \bigoplus_{v \in \Sigma} H^{i-1}(X_v, \mathscr{F}) \to H_c^i(X, \mathscr{F}) \to H^i(X, \mathscr{F}) \to \bigoplus_{v \in \Sigma} H^i(X_v, \mathscr{F}) \to \cdots.$$

For another definition, see [**M4**, p. 203].

1.6. Duality. In this subsection we list all the duality theorems we shall require. For proofs of the arithmetic results, we recommend the excellent book of Milne [**M4**], to which we shall give precise references for the statements.

THEOREM 1.11 [**M2**, Chapter VI, §11]. *Poincaré Duality: Let $\overline{X}$ be a smooth, projective, connected variety of dimension d over an algebraically closed field k, let n be a positive integer prime to the characteristic of k. Then*

$$H^{2d}(\overline{X}, \mathbf{Z}/n(d)) \cong \mathbf{Z}/n,$$

and the cup-product pairing

$$H^j(\overline{X}, \mathbf{Z}/n(i)) \times H^{2d-j}(\overline{X}, \mathbf{Z}/n(d-i)) \to \mathbf{Z}/n$$

is a perfect pairing of finite abelian groups. If n is a power of a prime number l, we can pass to the limit over these pairings to get a perfect pairing

$$H^j(\overline{X}, \mathbf{Z}_l(i)) \times H^{2d-j}(\overline{X}, \mathbf{Q}_l/\mathbf{Z}_l(d-i)) \to \mathbf{Q}_l/\mathbf{Z}_l.$$

Let $\mathbf{F}$ be a finite field. Then its absolute Galois group G is isomorphic to $\hat{\mathbf{Z}}$, and $H^1(\mathbf{F}, \mathbf{Q}/\mathbf{Z}) = \mathbf{Q}/\mathbf{Z}$. If M is a finite G-module and $M^* = \mathrm{Hom}(M, \mathbf{Q}/\mathbf{Z})$, we have a perfect pairing (Pontryagin duality):

$$H^1(\mathbf{F}, M) \times H^0(\mathbf{F}, M^*) \to H^1(\mathbf{F}, \mathbf{Q}/\mathbf{Z}) = \mathbf{Q}/\mathbf{Z}.$$

We can combine Poincaré and Pontryagin dualities to get a duality for varieties over finite fields:

THEOREM 1.12. *Let X be a smooth, projective, geometrically connected variety of dimension d over a finite field $\mathbf{F}$. If n is prime to $\mathrm{char}\,\mathbf{F}$, then*

$$H^{2d+1}(X, \mathbf{Z}/n(d)) = \mathbf{Z}/n,$$

and the pairing of finite abelian groups

$$H^j(X, \mathbf{Z}/n(i)) \times H^{2d+1-j}(X, \mathbf{Z}/n(d-i)) \to H^{2d+1}(X, \mathbf{Z}/n(d))$$

is perfect.

When n is a power of char. $\mathbf{F}$, then we have the duality theorem of Milne [**M1**], [**M5**, Cor. 1.12]. To describe this, it will be convenient to use the notation $H^j(X, \mathbf{Z}/p^n(i))$ for $H^{j-i}(X, W_n\Omega^i_{X, \log})$.

THEOREM 1.13. *For X as in Theorem 1.12, the groups $H^j(X, \mathbf{Z}/p^n(i))$ are finite for all i and j and*

$$H^{2d+1}(X, \mathbf{Z}/p^n(d)) = \mathbf{Z}/p^n.$$

The pairing

$$H^j(X, \mathbf{Z}/p^n(i)) \times H^{2d+1-j}(X, \mathbf{Z}/p^n(d-i)) \to H^{2d+1}(X, \mathbf{Z}/p^n(d))$$

is perfect.

It is important to remember that this result does *not* come from combining Pontryagin duality for $\mathbf{F}$ and Poincaré duality for $\overline{X}$. Indeed, the groups $H^j(\overline{X}, \mathbf{Z}/p^n(i))$ are not finite in general. There is a duality theory for $\overline{X}$, but it is a bit more involved and we do not need it here (see [**M1**]).

Local duality.

THEOREM 1.14 [**M4**, Chapter I, §2, Corollary 3]. *Let k be a p-adic field. Then we have:*

$$H^2(k, \mathbf{Q}/\mathbf{Z}(1)) = \mathbf{Q}/\mathbf{Z}.$$

If M is a finite $\mathrm{Gal}(\overline{k}/k)$-module of order prime to char. k, then the pairing:

$$H^i(k, M) \times H^{2-i}(k, M^*(1)) \to H^2(k, \mathbf{Q}/\mathbf{Z}(1)) = \mathbf{Q}/\mathbf{Z}$$

is a perfect pairing of finite groups.

As with the case of a finite field, we can combine Tate and Poincaré duality to get a duality theorem for varieties over local fields.

THEOREM 1.15 [**Sa6**]. *If X is smooth, projective and geometrically connected of dimension d over a local field k and n is prime to char. k, then*

$$H^{2d+2}(X, \mathbf{Z}/n(d+1)) = \mathbf{Z}/n,$$

and the pairing

$$H^j(X, \mathbf{Z}/n(i)) \times H^{2d+2-j}(X, \mathbf{Z}/n(d+1-i)) \to H^{2d+2}(X, \mathbf{Z}/n(d+1))$$

is a perfect pairing of finite groups. Passing to the limit, we also get a perfect pairing with $\mathbf{Z}_l(i)$-coefficients on one side and $\mathbf{Q}_l/\mathbf{Z}_l(d+1-i)$-coefficients on the other side.

Now we discuss Tate's duality theorem for abelian varieties over local fields. Let A be an abelian variety over a local field of characteristic zero, and denote by $\widehat{A}$ the dual abelian variety. Then we have $\widehat{A}(k) = \mathrm{Ext}^1(A, \mathbf{G}_m)$, where the Ext is taken as sheaves in the flat topology over k. This identification gives us a pairing:

$$\widehat{A}(k) \times H^1(k, A) \to H^2(k, \mathbf{G}_m) = \mathbf{Q}/\mathbf{Z}$$

which is called the derived cup-product pairing.

THEOREM 1.16 [**M4**, Chapter I, §3, Corollary 3.4]. *Under the derived cup product pairing, the compact group* $A(k)$ *is Pontryagin dual to the discrete group* $H^1(k, \widehat{A}(\bar{k}))$. *If* $TA = \varprojlim_m A_m$ *is the total Tate module of* A, *then we have the following commutative diagram of exact sequences, with the vertical arrows isomorphisms*:

$$
\begin{array}{ccccccccc}
0 & \to & \widehat{A}(k) & \to & H^1(k, T\widehat{A}) & \to & TH^1(k, \widehat{A}(\bar{k})) & \to & 0 \\
 & & \downarrow & & \downarrow & & \downarrow & & \\
0 & \to & H^1(k, A(\bar{k}))^* & \to & H^1(k, A(\bar{k})_{\mathrm{tors}})^* & \to & (A(k) \otimes \mathbf{Q}/\mathbf{Z})^* & \to & 0
\end{array}
$$

In Chapter 4, we will need the following corollary:

COROLLARY 1.17. *With notation as above, the pairing*

$$\widehat{A}(k) \times (A(k) \otimes \mathbf{Q}/\mathbf{Z}) \to \mathbf{Q}/\mathbf{Z}$$

induced by the pairings above, is trivial. These groups are exact annihilators of one another under the derived cup-product pairing.

Next we review Artin-Verdier duality. For references, see [**Ma**] and [**M4**, II, §2, Proposition 2.6 and §3].

THEOREM 1.18. *Let* X *be the spectrum of the ring of integers in a number field* k, *and let* U *be an open subset of* X. *Then we have*

$$H^3_c(U, \mathbf{G}_m) \cong \mathbf{Q}/\mathbf{Z}.$$

If n *is a positive integer and* $\mathscr{F}$ *is a constructible sheaf of* $\mathbf{Z}/n$-modules, *the Yoneda pairing*

$$H^i_c(U, \mathscr{F}) \times \mathrm{Ext}^{3-i}(\mathscr{F}, \mathbf{G}_m) \to \mathbf{Q}/\mathbf{Z}$$

is a perfect pairing of finite abelian groups.

If n *is invertible on* U, *then the Yoneda pairing reduces to a nondegenerate pairing between cohomology groups*

$$H^i_c(U, \mathscr{F}) \times H^{3-i}(U, \widehat{\mathscr{F}}(1)) \to \mathbf{Q}/\mathbf{Z}.$$

We shall also need the following generalization of Artin-Verdier duality:

THEOREM 1.19 [**M4**, §7, Corollary 7.7]. *Let* $f: X \to V$ *be a smooth morphism of schemes of finite type over* $\mathbf{Z}$ *with* V *regular of dimension one. Assume* X *is connected of pure dimension* d, *and let* n *be a positive integer which is invertible on* X. *Then we have*

$$H^{2d+1}_c(X, \mathbf{Z}/n(d)) = \mathbf{Z}/n,$$

and the cup product pairing

$$H^i(X, \mathbf{Z}/n(j)) \times H_c^{2d+1-i}(X, \mathbf{Z}/n(d-j)) \to H_c^{2d+1}(X, \mathbf{Z}/n(d)) = \mathbf{Z}/n$$

is a nondegenerate pairing of finite groups.

REMARK 1.20. It is important to remember that the proof of Artin-Verdier duality relies on the class field theory of global fields, in particular on the determination of the Brauer group. The higher-dimensional class field theory will depend on this result, so that it does not give another proof of class field theory in the one-dimensional case.

There is a version of Artin-Verdier duality for possibly singular schemes. To state this, let X be a reduced one-dimensional proper scheme of finite type over $\mathbf{Z}$, and consider the following complex of étale sheaves:

$$\mathscr{G} := \bigoplus_{x \in X^0} i_{x_*} \mathbf{G}_{m,x} \to \bigoplus_{x \in X^1} i_{x_*} \mathbf{Z}.$$

Here i_x is the inclusion of the point x into X. We view the left term as being of degree 0 and the right term as being of degree 1. Deninger [De3] considers the hypercohomology groups of $\mathscr{G}$ and proves the following:

THEOREM 1.21.
$$H_c^3(X, \mathscr{G}) \cong \mathbf{Q}/\mathbf{Z},$$

and for any constructible sheaf $\mathscr{F}$ of $\mathbf{Z}/n$-modules on X, the Yoneda pairing

$$H_c^i(X, \mathscr{F}) \times \mathrm{Ext}_X^{3-i}(\mathscr{F}, \mathscr{G}) \to H_c^3(X, \mathscr{G})$$

is a perfect pairing of finite groups.

Finally, we mention Poitou-Tate duality [Po], [T2]. Let k be a number field and M a finite Galois module. Then there is a nine term exact sequence:

$$0 \to H^0(k, M) \to \prod_v H^0(k_v, M) \to H^2(k, M^*(1))^* \to H^1(k, M) \to$$

$$\prod_v{}' H^1(k_v, M) \to H^1(k, M^*(1))^* \to H^2(k, M) \to$$

$$\bigoplus_v H^2(k_v, M) \to H^0(k, M^*(1))^* \to 0.$$

Here $\prod'$ denotes the restricted direct product with respect to the subgroups $H^1(k_v, M)^{nr} = \mathrm{Ker}[H^1(k_v, M) \to H^1(k_v^{nr}, M)]$ and $M^*(1) = \mathrm{Hom}(M, \mathbf{Q}/\mathbf{Z}(1))$.

1.7. The Henselian topology. To prove the reciprocity law for higher dimensional global fields, we shall need a Grothendieck topology which is between the Zariski topology and the étale topology. We will call it the Henselian topology, and it has been developed extensively in a paper of Nisnevich [Ni]. He calls it the completely decomposed topology, and several people call it the Nisnevich topology.

Let X be a scheme and let X_{hen} be the site whose underlying category is the same as $X_{\text{ét}}$ (the small étale site), but with coverings given as follows: a family $f_i : U_i \to U$ is a covering of U if

$$(1) \qquad\qquad\qquad \bigcup_i f_i(U_i) = U.$$

(2) For each $x \in U$, there exist i and a point $x_i \in U_i$ such that $f(x_i) = x$ and $\kappa(x_i) = \kappa(x)$.

It is the second condition which distinguishes this topology from the étale topology. Given $x \in X$, we define the local ring in the usual way as the limit over all Henselian neighborhoods of x. Then the local ring of X at x in the Henselian topology is a Henselian local ring. Note that if X is irreducible, then any covering of X in the Henselian topology contains a Zariski open subset of X, since once of the schemes in the covering must have a point whose residue field is the same as the generic point of X. For more properties of the Henselian topology, see [Ni]. Generally speaking, Henselian local rings have the same cohomological properties as their completions; an advantage of them is that if A is a local domain with fraction field K and A^h is the henselization of A with respect to the maximal ideal, the fraction field of A^h is algebraic over K. A disadvantage of Henselian rings is that they are more difficult to describe explicitly than complete rings.

Let $\mathscr{F}$ be a sheaf of X_{hen}. Then there is a theory of supports and a spectral sequence:

$$E_1^{p,q} = \bigoplus_{x \in X^p} H_x^{p+q}(X, \mathscr{F}) \Rightarrow H^{p+q}(X, \mathscr{F}).$$

Note that this spectral sequence is concentrated in the fourth quadrant. The following technical lemma will be important:

LEMMA 1.22. *Let $\mathscr{F}$ be a sheaf of abelian groups on X_{hen} and assume that for each point x with closure of dimension at least i, the stalk of $\mathscr{F}$ is zero at x. Then $H^i(X, \mathscr{F}) = 0$. In particular, if X is of dimension d, then $H^i(X, \mathscr{F}) = 0$ for all $i > d$.*

PROOF. Consider the spectral sequence above. We claim that for each $r \geq i$, $E_1^{r, i-r}$ is zero. For let $x \in X^r$, and let A_x be the henselization of the local ring of X at x. Then $H_x^i(X, \mathscr{F}) \cong H_x^i(A_x, \mathscr{F})$. Setting $U_x = \text{Spec}(A_x) - x$ and using the localization sequence

$$\cdots H^{i-1}(A_x, \mathscr{F}) \to H^{i-1}(U_x, \mathscr{F}) \to H_x^i(A_x, \mathscr{F}) \to 0,$$

we see that it is enough to show that $H^{i-1}(U_x, \mathscr{F}) = 0$. Note that U_x is a scheme of dimension $r - 1$. Continuing in this way by localizing U_x, we are reduced to showing that $H^0(V, \mathscr{F}) = 0$, where V is a subscheme of X with the property that for any point $x \in V$, the closure of x in X has dimension at least i. By hypothesis, $\mathscr{F}_x = 0$, and since X_{hen} has enough points, this means that $H^0(V, \mathscr{F}) = 0$. This completes the proof of the lemma.

COROLLARY 1.23. *Assume* $\dim X = d$ *and* $f : \mathscr{F} \to \mathscr{G}$ *is a map of sheaves on* X_{hen} *which induces an isomorphism on stalks at points of codimension* ≤ 1. *Then* $H^d(X, \mathscr{F}) \overset{\sim}{\to} H^d(X, \mathscr{G})$.

PROOF. This follows easily by applying Lemma 1.22 to $\ker f$ and $\text{Coker} f$.

REMARK 1.24. The proof of the preceding lemma may seem somewhat obscure. The following example might help clarify it. Let $X = \text{Spec}(k[[x, y]])$ and consider $H^2(X, \mathscr{F})$, where $\mathscr{F}$ is generically zero. We suppress the sheaf $\mathscr{F}$ in what follows. Let P be the closed point and $U = X - P$. Then the map $H^2_P(X) \to H^2(X)$ is surjective. Now the map

$$\bigoplus_{y \in U^1} H^1_y(U) \to H^1(U)$$

is surjective and for each y, the map

$$H^0(\text{Spec}(\mathscr{O}^h_{U_y}) - y) \to H^1_y(U)$$

is surjective. But $\text{Spec}(\mathscr{O}^h_{U_y}) - y$ is the spectrum of a field whose completion is $k((s))((t))$. This field contains the function field of X, and since our sheaf is generically zero, its restriction to these points is zero.

Chapter 2. Class field theory of higher dimensional local fields

2.1. The reciprocity law. In this chapter we study the class field theory of n-dimensional local fields following papers of Kato, Parshin and Fesenko. In defining the reciprocity map and proving the main results, we will follow the cohomological approach of Kato [K1]. For a more extensive survey, see [Fe5]. Parshin's approach in the equicharacteristic case is via topological K-theory. He defines the reciprocity map by viewing the maximal abelian extension F^{ab} of F as a compositum of three extensions obtained from the algebraic closure of the finite constant field, Kummer extensions and the maximal p-extension of F. Fesenko [Fe3], [Fe4] has an approach which employs the group-theoretic technique of Neukirch [N1]. We have followed Kato's approach because it fits in well with the methods of the following chapters. However, Fesenko's method is simpler for higher local class field theory, and we hope to be able to prove the global results some day by a similar method. Those who only want the local theory should follow Fesenko's method.

DEFINITION 2.1. A 0-dimensional local field is a finite field. An n-dimensional local field F is a complete discretely valued field whose residue field k is an $(n-1)$-dimensional local field.

We will use local field for short and denote the characteristic of F by p. Of course, a usual local field (complete, discretely valued with finite residue field) is a 1-dimensional local field. If the characteristics of F and

k are the same, we say that F is an *equicharacteristic* local field. If the characteristics are different, we say that F is a *mixed characteristic* local field.

EXAMPLE 2.2. Let $\mathbf{F}$ be a finite field, and consider the field

$$E = \mathbf{F}((T_1)) \cdots ((T_n))$$

which is defined inductively as the field of formal power series with coefficients in $\mathbf{F}((T_1)) \cdots ((T_{n-1}))$. Then E is an equicharacteristic n-dimensional local field.

EXAMPLE 2.3. Let F be the additive group of series

$$\sum_{i \in \mathbf{Z}} a_i T^i,$$

with coefficients in $\mathbf{Q}_p$ consisting of those with the p-adic valuation of a_i bounded below and

$$\lim_{i \to -\infty} a_i = 0.$$

Then the product of two series can be defined, and one sees easily that F is a mixed characteristic 2-dimensional local field with residue field $\mathbf{F}_p((T))$. The valuation of

$$\sum_{i \in \mathbf{Z}} a_i T^i$$

is given by the infimum of the p-adic valuations of the a_i.

The same construction as in the example can be done with any complete discretely valued field k, and we denote the corresponding field by $k\{\{T\}\}$.

REMARK 2.4. It is well-known (see e.g. [S4]) that a 1-dimensional local field is a finite extension of either $\mathbf{Q}_p$ or of $\mathbf{F}_p((T))$. According to Parshin, any n-dimensional local field is isomorphic to a field of the type $\mathbf{F}((T_1)) \cdots ((T_n))$, $k((T_1)) \cdots ((T_{n-1}))$ or is contained in a field of the type $k\{\{T_1\}\} \cdots \{\{T_{n-k}\}\}((T_{n-k+1})) \cdots ((T_{n-1}))$, where $\mathbf{F}$ is a finite field and k is a p-adic field.

The abelian class field theory of our higher dimensional local fields is a beautiful and elegant generalization of the 0 and 1-dimensional cases. To see this, let us briefly review these cases. First let F be a finite field. Then there is a map

$$\rho \colon K_0 F = \mathbf{Z} \to \mathrm{Gal}(F^{\mathrm{ab}}/F),$$

which sends 1 to the canonical generator Frobenius of $\mathrm{Gal}(F^{\mathrm{ab}}/F) = \mathrm{Gal}(\overline{F}/F)$. The image of ρ is dense in $\mathrm{Gal}(F^{\mathrm{ab}}/F)$ and ρ induces a bijection between the subgroups of finite index of $\mathbf{Z}$ and those of $\mathrm{Gal}(F^{\mathrm{ab}}/F)$. If F is a one dimensional local field, there is a similar map

$$\rho \colon K_1 F = F^* \to \mathrm{Gal}(F^{\mathrm{ab}}/F).$$

The image of ρ is dense and ρ induces a bijection between the open subgroups of finite index of these two groups. The higher-dimensional statement is almost exactly like these cases, except that there is no good topology on the Milnor K_n-groups for $n \geq 3$. This is made up for by a categorical construction due to Kato or by the use of topological K-theory, which is due to Parshin and Fesenko (see §2).

Our first task will be to establish a fundamental isomorphism which generalizes the well-known fact that the Brauer group of a local field is canonically isomorphic to $\mathbf{Q}/\mathbf{Z}$.

THEOREM 2.5. *Let F be an n-dimensional local field. Then the cohomological dimension of F is $n+1$ and*

$$H^{n+1}(F, \mathbf{Q}/\mathbf{Z}(n)) = \mathbf{Q}/\mathbf{Z}.$$

PROOF. We will prove the prime to p part, where p denotes the residue characteristic. Let l be prime number different from p. Let F^{nr} be the maximal unramified extension of F in $\overline{F}$, $\Gamma = \mathrm{Gal}(F^{nr}/F) = \mathrm{Gal}(k^{\mathrm{sep}}/k)$, $I = \mathrm{Gal}(F^{\mathrm{sep}}/F^{nr})$.

We prove the theorem by induction on n. The case $n = 0$ is easy, as $H^1(F, \mathbf{Q}_l/\mathbf{Z}_l) = \mathrm{Hom}(\mathrm{Gal}(\overline{F}/F), \mathbf{Q}_l/\mathbf{Z}_l) = \mathbf{Q}_l/\mathbf{Z}_l$. Assume the conclusion is true for all $m < n$. The extension of groups $(G = \mathrm{Gal}(F^{\mathrm{sep}}/F))$:

$$1 \to I \to G \to \Gamma \to 1$$

gives rise to a spectral sequence

$$H^r(\Gamma, H^s(I, M)) \Rightarrow H^{r+s}(G, M).$$

Since $H^{n+1}(\Gamma, M) = 0$ for any torsion module M and I is of cohomological dimension 1, we get an isomorphism

$$H^{n+1}(G, \mathbf{Q}_l/\mathbf{Z}_l(n)) \to H^n(\Gamma, H^1(I, \mathbf{Q}_l/\mathbf{Z}_l(n))).$$

Now the maximal pro-l quotient of $I = \mathrm{Gal}(\overline{F}/F^{nr})$ is isomorphic to $\mathbf{Z}_l(1)$ as a $\mathrm{Gal}(F^{nr}/F)$-module. All the l-primary roots of unity are contained in F^{nr}, and hence, I acts trivially on $\mathbf{Q}_l/\mathbf{Z}_l(n)$. Then

$$H^1(I, \mathbf{Q}_l/\mathbf{Z}_l(n)) = \mathrm{Hom}(I, \mathbf{Q}_l/\mathbf{Z}_l(n))) = \mathbf{Q}_l/\mathbf{Z}_l(n-1).$$

Hence,

$$H^n(\Gamma, H^1(I, \mathbf{Q}_l/\mathbf{Z}_l(n))) = H^n(\Gamma, \mathbf{Q}_l/\mathbf{Z}_l(n-1)) = \mathbf{Q}_l/\mathbf{Z}_l,$$

by the induction hypothesis. This completes the proof of the prime to p-part of the theorem.

The proof of the p-part is more difficult, and we just give a sketch here. First we recall a few facts from p-adic cohomology theory. Let k be a field of characteristic $p > 0$. Now k contains no nontrivial p^nth roots of unity, so the Galois modules $\mathbf{Q}_p/\mathbf{Z}_p(i)$ will not be of much use. However, in the last fifteen years, the theory of *logarithmic de Rham-Witt cohomology* has provided a satisfactory replacement. Everything is easier when k

is perfect; unfortunately, this will not be the case for the residue field of an n-dimensional local field when $n \geq 2$. For example, if $F = \mathbf{F}_q((T_1))((T_2))$, then $k = \mathbf{F}_q((T))$, which is not perfect. However, k is not "too" imperfect in the sense that $[k : k^p]$ is finite dimensional. In this case de Rham-Witt cohomology works reasonably well. The idea is to relate $H^{n+1}(F, \mathbf{Q}_n/\mathbf{Z}_p(n))$ to a suitable logarithmic de Rham-Witt cohomology group of k. More precisely, we have

$$H^{n+1}(F, \mathbf{Q}_p/\mathbf{Z}_p(n)) \cong \varinjlim_m H^1(k, W_m\Omega^{n-1}_{k, \log}).$$

By induction, this last group is isomorphic to $\varinjlim_m H^0(k_1, W_m\Omega^{n-2}_{k_1, \log})$, where k_1 is the residue field of k. Finally, we see that this group is isomorphic to $\varinjlim_m H^1(\mathbf{F}_q, W_m\Omega^0_{\mathbf{F}_q, \log}) = \mathbf{Q}_p/\mathbf{Z}_p$. For more details, see [K1] and [K3].

From Theorem 2.5, we can define the reciprocity map. We can combine the Galois symbol and cup-product in Galois cohomology to get a pairing:

$$K_n^M F \times H^1(F, \mathbf{Q}/\mathbf{Z}) \to H^{n+1}(F, \mathbf{Q}/\mathbf{Z}(n)) = \mathbf{Q}/\mathbf{Z}.$$

When F is an n-dimensional local field,

$$H^{n+1}(F, \mathbf{Q}/\mathbf{Z}(n)) = \mathbf{Q}/\mathbf{Z},$$

and we get a map

$$\rho_F : K_n^M F \to \mathrm{Hom}(H^1(F, \mathbf{Q}/\mathbf{Z}), \mathbf{Q}/\mathbf{Z}) = \mathrm{Gal}(F^{\mathrm{ab}}/F),$$

which we call the *reciprocity map*. It has the following functorial properties:

(1) The residue field k of F is an $(n-1)$-dimensional local field, so we have a reciprocity map

$$\rho_k : K_{n-1}^M k \to \mathrm{Gal}(k^{\mathrm{ab}}/k).$$

Then the following diagrams commutes:

$$
\begin{array}{ccc}
K_n^M F & \xrightarrow{\rho_F} & \mathrm{Gal}(F^{\mathrm{ab}}/F) \\
\downarrow & & \downarrow \\
K_{n-1}^M k & \xrightarrow{\rho_k} & \mathrm{Gal}(k^{\mathrm{ab}}/k)
\end{array}
$$

where the left vertical map is given by the tame symbol (see Chapter 1, §2) and the right vertical map is given by restriction.

(2) Let E be a finite extension of F. Then the following diagrams commute:

$$
\begin{array}{ccc}
K_n^M F & \xrightarrow{\rho_F} & \mathrm{Gal}(F^{\mathrm{ab}}/F) \\
\uparrow & & \uparrow \\
K_n^M E & \xrightarrow{\rho_E} & \mathrm{Gal}(E^{\mathrm{ab}}/E)
\end{array}
$$

where the left vertical map is given by the norm in Milnor K-theory (Chapter 1, §2) and the right vertical arrow is given by restriction, and

$$
\begin{array}{ccc}
K_n^M F & \xrightarrow{\ \rho_F\ } & \mathrm{Gal}(F^{\mathrm{ab}}/F) \\
\downarrow & & \downarrow \\
K_n^M E & \xrightarrow{\ \rho_E\ } & \mathrm{Gal}(E^{\mathrm{ab}}/E)
\end{array}
$$

where the left vertical arrow is given by restriction and the right vertical arrow is given by the transfer.

THEOREM 2.6 (reciprocity law). *The image of ρ_F is dense in $\mathrm{Gal}(F^{\mathrm{ab}}/F)$. Let E be a finite abelian extension of F. Then ρ induces an isomorphism*:

$$
K_n^M F / N_{E/F} K_n^M E \to \mathrm{Gal}(E/F).
$$

We will not give a complete proof of this theorem. In particular, we will not do the more difficult p analysis, where $p = \mathrm{char.}\ k$. Of course, this is the most important part. However, we hope to convey the idea of the argument. Before doing this, we need some preliminaries. Let π_F be a prime element of F and consider the map

$$
K_n^M F \xrightarrow{\ \partial_{\pi_F}\ } K_n^M k \oplus K_{n-1}^M k
$$

defined by

$$
\partial_{\pi_F}(\{x_1, \ldots, x_n\} + \{y_1, \ldots, y_{n-1}, \pi\}) = (\{\bar{x}_1, \ldots, \bar{x}_n\}, \{\bar{y}_1, \ldots, \bar{y}_{n-1}\}).
$$

This map is well-defined, but depends on the choice of π_F.

Let E/F be a finite extension of fields with residue field extension L/k. Let π_E be a prime element of E and write $\pi_F = u\pi_E^e$, where u is a unit of E.

We define maps

$$
j_{\pi_E, \pi_F} : K_n^M k \oplus K_{n-1}^M k \to K_n^M L \oplus K_{n-1}^M L
$$

and

$$
N_{\pi_E, \pi_F} : K_n^M L \oplus K_{n-1}^M L \to K_n^M k \oplus K_{n-1}^M k
$$

by

$$
j_{\pi_E, \pi_F}(a, b) = (a_L + \{b_L, \bar{u}\}, eb_L),
$$
$$
N_{\pi_E, \pi_F} = (eN_{L/k}(a) - N_{L/k}(\{b, \bar{u}\}), N_{L/k}(b)).
$$

LEMMA 2.7 ([**K1**, II, Lemma 13, p. 667]). *The following diagrams commute*

$$
\begin{array}{ccc}
K_n^M F & \xrightarrow{\ \partial_{\pi_F}\ } & K_n^M k \oplus K_{n-1}^M k \\
\downarrow & & \downarrow{\scriptstyle j_{\pi_E, \pi_F}} \\
K_n^M E & \xrightarrow{\ \partial_{\partial_E}\ } & K_n^M L \oplus K_{n-1}^M L
\end{array}
$$

where the vertical maps are the restriction maps, and

$$K_n^M F \xrightarrow{\partial_{\pi_F}} K_n^M k \oplus K_{n-1}^M k$$

$$\uparrow \qquad\qquad\qquad \uparrow N_{\pi_E, \pi_F}$$

$$K_n^M E \xrightarrow{\partial_{\partial_E}} K_n^M L \oplus K_{n-1}^M L$$

where the vertical maps are norm maps.

In order to prove the main theorem, we need to prove an auxiliary result which is very interesting in its own right.

DEFINITION 2.8. We say that a field F is a B_i-field if for any finite extension E/F, the norm map on Milnor K_i^M-groups

$$K_i^M E \to K_i^M F$$

is surjective.

It is well known that a finite field is B_1, i.e., the norm map is surjective on the multiplicative groups of a finite extension of finite fields. The same is true for K_2 of p-adic fields [**Mo**]. We have the key result:

PROPOSITION 2.9. *Let F be a complete discretely valued field with residue field k and $i \geq 0$ an integer. Then the following are equivalent:*

(i) *F is a B_{i+1}-field.*
(ii) *k is a B_i-field.*

SKETCH OF PROOF. First assume that k is a B_i-field. We show that F is a B_{i+1}-field. By an easy reduction, we may assume E/F is a cyclic extension of prime degree p with residue field extension L/k. We'll do the cases where E/F is unramified or totally tamely ramified.

(A) Unramified case. Let E/F be a finite unramified extension of degree p with residue field extension L/k. Then

$$K_r^M F / N K_r^M E \cong K_r^M k / N K_r^M L \oplus K_{r-1}^M k / N K_{r-1}^M L.$$

To see this, note that since E/F is unramified, L/k is separable, and hence the trace $\mathrm{Tr}_{L/k}: L \to k$ is surjective. This implies that every 1-unit of F is a norm from E [**S**, Chapter V, Proposition 3]. Then $U_r^1(F) \subset N_{E/F} K_r^M E$ for all $r \geq 1$. By Lemma 2.7, we then have

$$K_r^M F / N K_r^M E \cong K_r^M k / N K_r^M L \oplus K_{r-1}^M k / N K_{r-1}^M L.$$

If k is a B_i-field, then the group on the right is zero for $r - 1 \geq i$, and hence the group on the left is zero for $r \geq i + 1$.

(B) Totally tamely ramified case. E/F is totally ramified of degree p and $p \neq \mathrm{char}.\ k$. We claim that

$$K_r^M F / N K_r^M E \cong K_r^M k / p K_r^M k.$$

Indeed every 1-unit of F is a norm from E, and we can use Lemma 2.7 again. Then the conclusion follows from

LEMMA 2.10. *If k is a B_i-field, then it is B_r for all $r \geq i$ and $K_r^M k$ is divisible for all $r > i$.*

PROOF. The first assertion follows immediately from the projection formula: if L/k is a finite extension and $\{a, b\}$ is a symbol in $K_r^M k$ with $a \in K_i^M k$, $b \in K_{r-i}^M k$, we can write $a = N(c)$ for some $c \in K_i^M L$. Then we have

$$N\{c, b\}_L = \{N(c), b\}_k = \{a, b\}_k.$$

For the second assertion, let d be a solution of $x^n = b$, $L = k(d)$, and c as before. Then

$$nN\{c, d\} = N\{c, d^n\}_L = \{N(c), b\}_k = \{a, b\}_k;$$

hence, $\{a, b\}$ is n-divisible.

For the rest of the proof, we will need the following two lemmas ([K1], II, Lemmas 11 and 12).

LEMMA 2.11. *Let F be a field of positive characteristic p and assume $[F : F^p] = p^i < \infty$. Then the norm map $K_i F^{1/p} \to K_i F$ coincides with the isomorphism $\{x_1, \ldots, x_i\} \mapsto \{x_1^p, \ldots, x_i^p\}$.*

LEMMA 2.12. *If k is a B_i-field of characteristic $p > 0$, then $[k : k^p] \leq p^i$.*

We now complete the sketch the proof of Proposition 2.9. First assume that k is a B_i-field. Let E/F be a finite cyclic extension of degree p. By Lemma 2.10, $K_{i+1}^M k / p K_{i+1}^M k = 0$. If char. $F = p > 0$, then since $[k : k^p] \leq p^i$, we have $[F : F^p] \leq p^{i+1}$. Hence, the norm map is surjective on K_{i+1}^M by Lemma 2.11. Conversely, assume that F is a B_{i+1}-field. Then by (A) above, the norm map is surjective on K^M for a cyclic extension l/k. If char. $k = p > 0$, we can take a totally ramified cyclic extension E/F of degree p to deduce that $K_{i+1}^M k = p K_{i+1}^M k$. Hence, $[k : k^p] \leq p^i$, so by Lemma 2.11, the norm map is surjective on K_i^M, for any purely inseparable finite extension l/k. This completes the sketch of the proof of Proposition 2.9.

COROLLARY 2.13. *Let F be an n-dimensional local field. Then F is a B_{n+1}-field.*

PROOF. A finite field is B_1. Then do an induction using Proposition 2.9.

We now sketch the proof of Theorem 2.6 from Proposition 2.9. Using induction and an obvious commutative diagram, we can easily reduce to the case where E/F is a cyclic extension of prime degree p. Let $\chi \in H^1(F, \mathbf{Q}/\mathbf{Z})$ denote the element corresponding to E/F. For any field K and any integer $i \geq 0$, let $H^i(K)$ denote the group $H^i(K, \mathbf{Q}/\mathbf{Z}(i-1))$. To prove the theorem, it will suffice to show that the following sequence is exact:

$$K_n^M E \to K_n^M F \to H^{n+1}(F) \to H^{n+1}(E).$$

Here the first map is the norm on K-theory, the second map is cup-product with χ (see the discussion following Theorem 1.6) and the third is restriction on Galois cohomology. Of course, since χ dies in $H^{n+1}E$, this is equivalent to saying that the map

$$K_n^M F / N K_n^M E \to H^{n+1}(F)_p$$

given by cupping with χ is an isomorphism. We prove this by induction on n. It is easy for $n = 0$, so we assume $n \geq 1$. Again, the proof goes by cases, and we do the unramified and totally tamely ramified ones. In each of the statements, take $r = n$ in the proof of Proposition 2.9.

(A) In this case we may regard χ as a character of k. We have a commutative diagram

$$
\begin{array}{ccc}
K_{n-1}^M k / N K_{i-1}^M L & \longrightarrow & H^n(k)_p \\
\downarrow & & \downarrow \\
K_n^M F / N K_n^M E & \longrightarrow & H^{n+1}(F)_p.
\end{array}
$$

Since k is B_{n-1}, the left vertical map is an isomorphism, and the right vertical map is an isomorphism. By the induction hypothesis, we get the bijectivity of the bottom horizontal map, as desired.

(B) In this case, $\chi = \{\theta, \pi\}$ for some nonzero element θ of $H^0(F)_p = H^0(k)_p$ and some prime element π of F. There is a commutative diagram

$$
\begin{array}{ccc}
K_n(k)/pK_n(k) & \xrightarrow{h_{p,k}^n} & H^n(k, \mathbf{Z}/p(n)) \cong H^n(k)_p \\
\cong \downarrow & & \cong (-1)^n h_F^k \downarrow \\
K_n(F)/N_{E/F}K_n E & \xrightarrow{\{\chi,?\}} & H^{n+1}(F)_p
\end{array}
$$

where the left vertical isomorphism is induced by θ. The top horizontal map is the Galois symbol and the left vertical isomorphism comes from Proposition 2.9. The isomorphism between the two groups at the top right is cup-product with θ. The right vertical arrow is defined by identifying $H^n(k, \mathbf{Z}/p(n-1))$ with $H^n(F^{nr}/F, H^0(F^{nr}, \mathbf{Z}/p(n-1)))$, using the inflation map to get into $H^n(F, \mathbf{Z}/p(n-1))$ and then cupping with the class of $\pi \in H^1(F, \mathbf{Z}/p(1))$ to finally get to $H^{n+1}(F)$. By [K1; II; Lemma 3], this map is independent of the choice of the prime element π. Then the conclusion is a consequence of the following two results.

LEMMA 2.14. *Let F be a complete discretely valued field with residue field k. If m is a positive integer which is invertible in k, the Galois symbol $h_{m,F}^q$ is bijective if and only if $h_{m,k}^q$ and $h_{m,k}^{q-1}$ are bijective.*

As a byproduct of this proof, one can now show:

PROPOSITION 2.15. *Let F be an n-dimensional local field. Then the Galois symbol*

$$h_F^{n+1} \colon K_{n+1}^M / p K_{n+1}^M F \to H^{n+1}(F, \mathbf{Z}/p(n))$$

is bijective for any prime number p which is invertible in F.

2.2. Topological K-theory and the existence theorem. Once the reciprocity law has been proved for higher dimensional local fields, one can ask if there is a generalization of the existence theorem in the class field theory of one-dimensional local fields.

THEOREM 2.16. *Let F be a one-dimensional local field. Then the rule*

$$E/F \to N_{E/F} E^*$$

gives a bijection between the set of finite abelian extensions of F and the set of open subgroups of finite index in F^*. The topology on F^* is induced from the valuation.

In trying to generalize this result to the higher dimensional case, one's first instinct is to just do the same thing. The first problem with this is that the topology on a complete discretely valued field does not in general induce the topology on the residue field (considered as an $n-1$-dimensional local field). It turns out that for $n \geq 3$ there is no good topology on the whole Milnor K_n-group of an n-dimensional local field which will do what is required. There are two ways around this, one due to Kato [K2], and the other to Parshin [P4] and Fesenko [Fe2–Fe5]. We briefly describe them.

Kato's approach. Let $\mathscr{C}_0$ be the category of finite sets and for $n \geq 0$, let $\mathscr{C}_{n+1} = \mathrm{ind}(\mathrm{pro})(\mathscr{C}_n)$. We put

$$\mathscr{C}_\infty = \bigcup_{n=0}^{\infty} \mathscr{C}_n.$$

Here ind and pro are used in the sense of (**SGA4**, Tome 2, Exp. VI, 6.3, 6.10). Then Kato defines a ring object $\mathscr{F}$ of $\mathscr{C}_n$ corresponding to an n-dimensional local field such that there is a natural bijection between F and $[e, \mathscr{F}]$ in $\mathscr{C}_\infty$, where e is the unit object. The multiplicative group is identified with $[e, \mathscr{F}^*]$, where $\mathscr{F}^*$ is the multiplicative group object associated to the ring object $\mathscr{F}$. Then a subgroup N of $K_n F$ is said to be open if the map

$$F^* \times \cdots \times F^* \to K_n^M F/N$$

obtained by composing the natural projection with the symbol map comes from a map in $\mathscr{C}_\infty$:

$$\mathscr{F}^* \times \cdots \times \mathscr{F}^* \to K_n^M F/N.$$

Here $K_n^M F/N$ is considered as an object of $\mathrm{ind}(\mathscr{C}_0) \subset \mathscr{C}_1$. A character

$$\phi \colon K_n^M F \to \mathbf{Q}/\mathbf{Z}$$

is said to be continuous if the induced map:

$$F^* \times \cdots \times F^* \to \mathbf{Q}/\mathbf{Z}$$

comes from a map in $\mathscr{C}_\infty$. With these definitions, Kato proves:

THEOREM 2.17. *Let F be an n-dimensional local field. Then the correspondence $E \to N_{E/F} K_n^M E$ is a bijection between the set of all finite abelian extensions of F and the set of all open subgroups of $K_n^M F$ of finite index.*

This is a beautiful theorem, but it seems difficult to check in practice that a subgroup is open or a character is continuous.

The approach of Parshin and Fesenko. Let τ be the strongest topology on $K_n F$ for which the map

$$F^* \times \cdots \times F^* \to K_n^M F$$

is sequentially continuous with respect to the adic topology on F^* and for which $x_i + y_i \to x + y$, $-x_i \to -x$ if $x_i \to x$ and $y_i \to y$ in $K_n F$. Let Λ be the intersection of all neighborhoods of zero in $K_n^M F$. If F is of positive characteristic, Parshin sets

$$K_n^{M^{\mathrm{top}}} F = K_n^M F / \Lambda ;$$

if F is of characteristic zero, he puts

$$K_n^{M^{\mathrm{top}}} F = K_n^M F / \bigcap_{l \geq 1} l K_n^M F.$$

Note that

$$\lambda = \bigcap_{l \geq 1} l K_n^M (F).$$

Then $K_0^{M^{\mathrm{top}}} F = K_0 F$, $K_1^{M^{\mathrm{top}}} F = K_1 F$, $K_{n+1}^{M^{\mathrm{top}}} F = \mu(F)$, where $\mu(F)$ is the group of roots of unity in F. The following theorem is due to Fesenko [Fe2–Fe5].

THEOREM 2.18. *Let F be an n-dimensional local field. Then there is a reciprocity map*

$$\bar{p} : K_n^{M^{\mathrm{top}}} F \to \mathrm{Gal}(F^{\mathrm{ab}}/F)$$

such that the composite of $\bar{p}$ with the natural map

$$K_n^M F \to K_n^{M^{\mathrm{top}}} F$$

is the reciprocity map defined above. $\bar{p}$ is injective and its image is dense in $\mathrm{Gal}(F^{\mathrm{ab}}/F)$. The correspondence $E \to N_{E/F} K_n^{M^{\mathrm{top}}} E$ is a bijection between the set of finite abelian extensions of F and the set of open subgroups of finite index in $K_n^{M^{\mathrm{top}}} F$.

2.3. Complements in dimension two. Let F be a two-dimensional local field. Then the class field theory of F can be made into a duality by using a good topology on F. This is defined as follows. First assume that the residue field k is of positive characteristic. Set $A = \mathbf{Z}_p[[X]]$ or $\mathbf{F}_p[[X, T]]$ according to whether F is of characteristic zero or p. Let B be the completion of A with respect to the ideal generated by p (resp. T). By ([Na3], 31.12), there is a finite map $\phi: B \to \mathcal{O}_F$ such that
(1) via ϕ, $\mathcal{O}_F$ is a free B-module of finite rank m;
(2) k is finite dimensional over $\mathbf{F}_p((X))$ via the map

$$\mathbf{F}_p((X)) = B/pB \to k$$

(resp., $B/TB \to k$), and the valuation on k is induced by the valuation on $\mathbf{F}_p((X))$. Take a basis $\{e_i\}$ $(i = 1, 2, \ldots, m)$ of $\mathcal{O}_F$ over B and endow $\mathcal{O}_F$ with the topology for which the bijection $B^m \to \mathcal{O}_F$ is a homeomorphism. Then the topology of the unit group U_F is compatible with the multiplicative group structure, and there is a unique topology on F^* which is compatible with the group structure, for which U_F is open and which induces the topology on U_F described above. Finally, we endow $K_2 F$ with the strongest topology for which the symbol map

$$F^* \times F^* \to K_2 F$$

is continuous. When $\operatorname{char} k = 0$, we can define a topology in a similar way, but it depends on some choices (see [K1, II, Remark 3, p. 680]). In the statements below, if char. $k = 0$, we may take the discrete topology on all groups considered.

For a topological group G, let $\operatorname{Hom}_c(G, \mathbf{Q}/\mathbf{Z})$ denote the group of continuous homomorphisms of finite order, where $\mathbf{Q}/\mathbf{Z}$ is given the discrete topology. Then Kato proves [K1, I, II, §3.1, §3.5]:

THEOREM 2.19. *The pairing defined by cup-product and the Galois symbol*:

$$H^1(F, \mathbf{Q}/\mathbf{Z}) \times K_2 F \to \mathbf{Q}/\mathbf{Z}$$

induces an isomorphism

$$H^1(F, \mathbf{Q}/\mathbf{Z}) \to \operatorname{Hom}_c(K_2 F, \mathbf{Q}/\mathbf{Z}).$$

The correspondence $E \mapsto N_{E/F} K_2 E$ *sets up a 1-1 correspondence between the finite abelian extensions of F and the open subgroups of finite index in $K_2 F$ for the topology described above.*

Kato also proves a result on the structure of the Brauer group of a two-dimensional local field. The Kummer map $F^* \to H^1(F, \mathbf{Z}/n(1))$, the identification $\operatorname{Br}(F)_n \cong H^2(F, \mathbf{Z}/n(1))$ and cup-product gives a pairing

$$F^* \times \operatorname{Br}(F) \to \mathbf{Q}/\mathbf{Z}.$$

THEOREM 2.20 [**K1,** I, §8 and II, §3.5]. *With the topology on F^* defined above, the pairing induces an isomorphism*

$$\mathrm{Br}(F) \to \mathrm{Hom}_c(F^*, \mathbf{Q}/\mathbf{Z}).$$

REMARK 2.21. In the two-dimensional case, Koya has an approach which is more similar to the Tate-Nakayama approach in the one-dimensional case [**T2**]. For more on this, see the last part of the next chapter.

Chapter 3. Class field theory of two-dimensional local rings

3.1. The reciprocity law for two-dimensional local rings. Let A be a two-dimensional normal, complete local ring with finite residue field F. We denote the quotient field of A by K and we assume for simplicity that it is of characteristic zero. In this section we describe the class field theory of K following Saito ([**Sa2**], [**Sa7**]). This will be important in the next chapter on the class field theory of varieties over (one-dimensional) local fields.

For each codimension one point x of $X = \mathrm{Spec}(A)$, we denote by A_x the completion of the local ring of A at x and by K_x the quotient field. From the class field theory of the two-dimensional local field K_x, we get a canonical isomorphism (see Chapter 2):

$$H^3(K_x, \mathbf{Q}/\mathbf{Z}(2)) \to \mathbf{Q}/\mathbf{Z}.$$

Let i_x be the restriction map

$$H^3(K, \mathbf{Q}/\mathbf{Z}(2)) \to H^3(K_x, \mathbf{Q}/\mathbf{Z}(2)).$$

THEOREM 3.1. *For any $a \in H^3(K, \mathbf{Q}/\mathbf{Z}(2))$, $i_x(a) = 0$ for almost all x and*

$$\sum_x i_x(a) = 0.$$

For the proof, we refer to [**Sa2**, Theorem 1.1].

Let I_K be the restricted direct product of the $K_2 K_x$ with respect to the subgroups $K_2 A_x$. For each finite subset S of codimension one points of $\mathrm{Spec} A$, put:

$$I_S = \prod_{x \in S} K_2 K_x \times \prod_{x \notin S} K_2 A_x.$$

Then $I_K = \varinjlim_S I_S$ and we endow I_K with the direct limit topology. If $\chi \in H^1(K, \mathbf{Q}/\mathbf{Z})$ is a character, we denote its restriction to $H^1(K_x, \mathbf{Q}/\mathbf{Z})$ by χ_x. There is a pairing

$$\langle\ ,\ \rangle_x \colon H^1(K_x, \mathbf{Q}/\mathbf{Z}) \times K_2 K_x \to \mathbf{Q}/\mathbf{Z}$$

given by cupping with the Galois symbol. Given an element $a \in I_K$ and $\chi \in H^1(K, \mathbf{Q}/\mathbf{Z})$, we have $\langle \chi_x, a_x \rangle_x = 0$ for almost all x, so there is a map

$$\widetilde{\Phi}_K \colon H^1(K, \mathbf{Q}/\mathbf{Z}) \to \mathrm{Hom}_c(I_K, \mathbf{Q}/\mathbf{Z}).$$

If a is in the diagonal image of $K_2 K$, then the reciprocity law tells us that $\langle \chi, a \rangle = \sum_x \langle \chi_x, a_x \rangle_x = 0$. Putting $C_K = I_K / K_2 K$, we get a map

$$\Phi_K : H^1(K, \mathbf{Q}/\mathbf{Z}) \to \operatorname{Hom}(C_K, \mathbf{Q}/\mathbf{Z}).$$

PROPOSITION 3.2. *If A is regular, Φ_K is injective.*

PROOF. Suppose $\chi \in H^1(K, \mathbf{Q}/\mathbf{Z})$ is such that $\Phi_K(\chi) = 0$; let L/K be the cyclic extension corresponding to χ. Then for any codimension one point x of $X = \operatorname{Spec}(A)$, the restriction χ_x of χ to $H^1(K_x, \mathbf{Q}/\mathbf{Z})$ has $\Phi_x(\chi_x) = 0 \in \operatorname{Hom}(K_2 K_x, \mathbf{Q}/\mathbf{Z})$. By the class field theory of K_x (Theorem 2.19), this means that $\operatorname{Spec} A$ splits completely in L. Since A is regular, this means that the extension L/K is trivial (Lemma 1.9).

COROLLARY 3.3. *With hypothesis as in the last Proposition, suppose L/K is a finite abelian extension in which all but finitely many codimension one points of $\operatorname{Spec}(A)$ split completely. Then $L = K$.*

PROOF. By an easy reduction, we may assume that L/K is a cyclic extension and we denote by $\chi \in H^1(K, \mathbf{Q}/\mathbf{Z})$ the corresponding character. Let $\tilde{\Phi}_K$ be as above. By assumption, there is a finite subset S of codimension one points of X such that $\tilde{\psi} = \tilde{\Phi}_K(\chi)$ is trivial on the subgroup $\prod'_{x \notin S} K_2 K_x$ of I_K. Now K_x is the quotient field of a Dedekind domain, and an easy consequence of weak approximation for the multiplicative group of such a field is that the image of the natural map

$$K_2 K \to \prod_{x \in S} K_2 K_x$$

is dense. Hence, $\tilde{\psi}$ is trivial on I_K and all codimension one points split completely. By Proposition 3.2, this implies that the extension is trivial.

We now state without proof Saito's main result.

THEOREM 3.4. *Suppose A is regular. Then the map Φ_K induces an isomorphism between $H^1(K, \mathbf{Q}/\mathbf{Z})$ and the group of continuous homomorphisms of finite order of C_K.*

REMARK 3.5. In the case where the ring A is only assumed to be normal, the result is a bit different. Then there are nontrivial étale covers in which every codimension one point splits completely. To describe these, let $\mathcal{X}$ be a resolution of $\operatorname{Spec}(A)$; that is, a 2-dimensional regular scheme together with a proper birational morphism $f: \mathcal{X} \to \operatorname{Spec} A$. Let $Y = f^{-1}(x)_{\mathrm{red}}$, where x is the closed point and red denotes reduced scheme structure. Then Saito [Sa7] shows that the completely split covers of $X = \operatorname{Spec}(A) - x$ correspond bijectively with the completely split covers of Y. We will not prove this, but in the next chapter we will prove an analogous result in a more global situation, and the proof given will carry over to the situation here. Then the kernel of Φ_K is a sum of r copies of $\mathbf{Q}/\mathbf{Z}$, where r is the rank of the quotient of $\pi_1^{ab}(X)$ which classifies abelian étale covers in which every codimension one point splits completely. An important fact is that this quotient is torsion free.

We now discuss applications of these techniques to duality theorems for curves over p-adic fields. Let X be a smooth, projective, geometrically connected curve over a p-adic field k and denote the function field of X by K. Let P be the set of closed points of X. Then for each $x \in P$, the completion of the local ring of X at x is a two-dimensional local field. Then we have the pairing

$$\langle \; , \; \rangle_x : \mathrm{Br}(K_x) \times K_x^* \to \mathbf{Q}/\mathbf{Z}$$

defined as in Chapter 2, §2. Let $J_K = \prod' K_x^*$ denote the restricted direct product of K_x^* with respect to the subgroups A_x^*. Then for any $b \in K^*$, $b \in A_x^*$ for almost all x, and so K^* may be regarded as a subgroup of J_K; we denote the quotient by $D(K)$. For $a \in \mathrm{Br}(K)$, we denote its image in $\mathrm{Br}(K_x)$ by a_x. By the same type of argument given above, we have:

PROPOSITION 3.6. *For any* $a \in K^*$,

$$\sum_x \langle a, b \rangle_x = 0,$$

so we get a pairing

$$\mathrm{Br}(K) \times D_K \to \mathbf{Q}/\mathbf{Z}.$$

Now let S be a finite subset of closed points of X, and let

$$D_S = \left(\prod_{x \in S} K_x^* \times \bigoplus_{x \notin S} \mathbf{Z} \right) / K^*.$$

Note that D_S is a quotient of $D(K)$ and we may identify it with $\mathrm{Pic}(U_S)$, where $U_S = X - S$. Then it is not difficult to show that the pairing above gives by restriction to $\mathrm{Br}(U_S)$ and passing to the quotient D_S a pairing:

$$\mathrm{Br}(U_S) \times \mathrm{Pic}(U_S) \to \mathbf{Q}/\mathbf{Z}.$$

In particular, when S is empty, we get a pairing

$$\mathrm{Br}(X) \times \mathrm{Pic}(X) \to \mathbf{Q}/\mathbf{Z}.$$

We topologize $\mathrm{Pic}(X)$ as follows: view $\mathrm{Pic}^0(X)$ as a subgroup of $J(k)$, where J denotes the Albanese variety of X over k and give $\mathrm{Pic}^0(X)$ the induced topology. Then there is a unique topology on $\mathrm{Pic}(X)$ which is compatible with its group structure, such that $\mathrm{Pic}^0(X)$ is an open subgroup and the induced topology on $\mathrm{Pic}^0(X)$ is the one just described. Now we have

THEOREM 3.7. *The pairing defined above induces an isomorphism*

$$\mathrm{Br}(X) \to \mathrm{Hom}_c(\mathrm{Pic}(X), \mathbf{Q}/\mathbf{Z}).$$

When k is of characteristic zero, this result is due to Lichtenbaum [Li1], who deduced it starting from Tate's duality theorem for abelian varieties. In [Sa4, Appendix], Saito proves the result also in positive characteristic, and he deduces Tate duality in positive characteristic, where it was not known.

3.2. Koya's approach. Recently, Y. Koya has found an approach to the class field theory of two-dimensional local fields and the quotient fields of two-dimensional local rings which is quite similar to the cohomological approach in the one-dimensional case [T2], [HN], [S4], [Nak1], [Nak2]. We will outline this approach without giving many proofs. For more details, see Koya's papers [Ko1–Ko4].

To give some motivation, we first recall some general facts from the one-dimensional case. Let F be a one-dimensional local field and E a finite Galois extension of F with group G. The group $H^2(G, E^*)$ is cyclic; let a_G be a generator. Denote by $\widehat{H}$ the Tate cohomology groups. Then the theorem of Tate-Nakayama implies that the map:

$$\widehat{H}^{-2}(G, \mathbf{Z}) \to \widehat{H}^0(G, E^*)$$

given by cupping with a_G is an isomorphism (see [T2], [Nak1], and [S4, Chapter IX, §8, Corollary to Theorem 14]). Now the group on the left is G^{ab} and the group on the right is F^*/NE^*; the map is the inverse of the reciprocity map in one dimensional class field theory. Thus, the theorem of Tate-Nakayama "gives" the reciprocity isomorphism of local class field theory. One would like to do the same in higher dimensions. The first thing to try is replacing the multiplicative group in the one-dimensional case with higher Milnor K-theory. Unfortunately, this does not work because Milnor K-theory does not satisfy Galois descent. Koya's idea is to use Lichtenbaum's complexes $\mathbf{Z}(i)$ [Li2] as the satisfactory generalization of the multiplicative group and "modified" hypercohomology groups as the generalization of Tate cohomology. We recall some of the basic expected properties of these complexes on a scheme X.

(i) $\mathbf{Z}(i)$ is a complex of sheaves in the étale and Zariski topologies;

(ii) $\mathbf{Z}(0)$ is the constant sheaf $\mathbf{Z}$ and $\mathbf{Z}(1) = \mathbf{G}_m[-1]$;

(iii) $\mathbf{Z}(i)$ is acyclic outside of $[1, i]$;

(iv) let $\alpha: X_{\mathrm{ét}} \to X_{\mathrm{Zar}}$ be the natural morphism of sites. Then $R^{i+1}\alpha_*\mathbf{Z}(i) = 0$.

(v) let n be a positive integer which is prime to the residue characteristics of X. Then there is a Kummer sequence

$$\mathbf{Z}(i) \xrightarrow{n} \mathbf{Z}(i) \to \mathbf{Z}/n(i) \to \mathbf{Z}(i)[1].$$

If p is a residue characteristic of X, then there are similar sequences

$$\mathbf{Z}(i) \xrightarrow{p^m} \mathbf{Z}(i) \to \mathbf{Z}/p^m(i) \to \mathbf{Z}(i)[1],$$

where $\mathbf{Z}/p^m(i)$ is as in Chapter 1.

For the convenience of the reader, we briefly recall the definition of a complete resolution of a finite group. For more details, see [CE, Chapter XII, §3].

Let G be a finite group and consider an exact sequence

$$\cdots \to B^{-2} \to B^{-1} \to B^0 \xrightarrow{d^0} B^1 \to B^2 \to \cdots ,$$

where each B^j is a finitely generated free $\mathbf{Z}[G]$-module. Such a sequence is called a complete resolution of G if the image of d^0 is the trivial G-module $\mathbf{Z}$ and the truncated sequence:

$$\cdots \to B^{-n} \to B^{-n+1} \to \cdots \to B^{-2} \to B^{-1} \to B^0 \to \mathbf{Z} \to 0$$

is a projective resolution of $\mathbf{Z}$. Given two free resolutions $F_1^\bullet, F_2^\bullet$ of $\mathbf{Z}$, one can get a complete resolution of G by splicing together $F_1^\bullet$ with $\mathrm{Hom}(F_2^\bullet, \mathbf{Z})$ and renumbering. For any G-module M, the cohomology groups of the complex obtained by applying the functor $\mathrm{Hom}(B^j, M)$ give the Tate cohomology groups of G. Now let $C^\bullet$ be a bounded complex of G-modules and consider the double complex $D^{i,j} = \mathrm{Hom}(B^{-i}, C^j)$. The cohomology of the single complex associated to this double complex will be denoted by $\widehat{\mathbf{H}}^n(G, C^\bullet)$, and Koya calls them the *modified cohomology groups* of G. When $C^\bullet$ consists of just one G-module M, these are the same as the Tate cohomology groups of M. If N_G is the norm map on the hypercohomology of groups, then it is not difficult to show that

$$\widehat{\mathbf{H}}^0(G, C^\bullet) = \mathrm{Coker}(\mathscr{H}^0(C^\bullet) \xrightarrow{N_G} \mathbf{H}^0(G, C^\bullet)).$$

Also, for $n \geq 1$, the modified cohomology group coincides with the usual hypercohomology groups of $C^\bullet$. Finally, to a distinguished triangle of complexes is associated a long (infinite in both directions) exact sequence of modified cohomology groups. Then Koya proves the following generalization of the theorem of Tate and Nakayama:

THEOREM 3.8. *Let G be a finite group and $C^\bullet$ a complex of G-modules which is zero outside of degrees 0 and -1. Assume that for each p-Sylow subgroup G_p of G*

(i) $\widehat{\mathbf{H}}^1(G_p, C^\bullet) = 0$.

(ii) $\widehat{\mathbf{H}}^2(G_p, C^\bullet)$ *is generated by an element a_{G_p} whose order is the cardinality of G_p.*
Then for any integer n and any subgroup H of G, cup-product with a_H gives an isomorphism

$$\widehat{\mathbf{H}}^{n-2}(H, \mathbf{Z}) \cong \widehat{\mathbf{H}}^n(H, C^\bullet).$$

As applications of this theorem, if we take E/F an extension of 1-dimensional local fields and $C^\bullet$ to be E^*, we recover local class field theory. If we take E/F an extension of global fields and the complex to be the idèle class group of E, we recover the reciprocity law of global class field theory.

However the novel aspect of Koya's approach is that if we take E/F an extension of two-dimensional local fields and $C^\bullet$ to be $\mathbf{Z}(2)_E$, we recover the reciprocity law (Theorem 2.6 above) (see [Ko1]). Similarly, a suitable idelic version of this complex allows him to prove the reciprocity law (Theorem 3.4 and Remark 3.5 above) for the quotient field of a two-dimensional Henselian normal ring (see [Ko3]). However, it should be noted that in the latter case, Koya must use Saito's main result described above.

Chapter 4. Class field theory for varieties over local fields

4.1. Definition of the reciprocity map and statement of the main theorems. Let k be a field of characteristic zero which is complete with respect to a discrete valuation, with finite residue field $\mathbf{F}$. We shall refer to such a k as a p-adic field, and in this chapter we study the class field theory of smooth projective varieties over k (mainly curves). Our exposition follows papers by Saito [Sa2] and Bloch [B2]. In the recent thesis of Spiess [Sp], new proofs are given for some of the results in this chapter.

Throughout this chapter, let X be a smooth, projective, geometrically connected variety of dimension d over k. We will assume that there exists a regular scheme $\mathscr{X}$ over $\mathscr{O}_k$ whose generic fibre is X. The special fibre of $\mathscr{X}$ will be denoted by Y. Our first task is to define the reciprocity map:

$$\sigma\colon H^d(X, \mathscr{K}_{d+1}) \to \pi_1^{\mathrm{ab}}(X).$$

Actually, we will do this in two ways, which turn out to be the same. First, we use the presentation

$$H^d(X, \mathscr{K}_{d+1}) \cong \mathrm{Coker}\left[\bigoplus_{x \in X^{d-1}} K_2 k(x) \to \bigoplus_{x \in X^d} k(x)^*\right]$$

(see Chapter 1, §1).

For each $x \in X^d$ we have the reciprocity map from the local class field theory of the 1-dimensional local field $k(x)$

$$\sigma_x\colon k(x)^* \to \mathrm{Gal}(k(x)^{\mathrm{ab}}/k(x)) = \pi_1^{\mathrm{ab}}(k(x)).$$

We can compose this with the natural map $\pi_1^{\mathrm{ab}}(k(x)) \to \pi_1^{\mathrm{ab}}(X)$ and expand by linearity to get a map

$$\sigma\colon \bigoplus_{x \in X^d} k(x)^* \to \pi_1^{\mathrm{ab}}(X).$$

LEMMA 4.1. σ *is trivial on the image of*

$$\bigoplus_{x \in X^{d-1}} K_2 k(x)$$

and so gives a map

$$\sigma\colon H^d(X, \mathscr{K}_{d+1}) \to \pi_1^{\mathrm{ab}}(X).$$

It will be convenient to have another description of this map. For this, we recall that there is a commutative diagram (up to sign):

$$
\begin{array}{ccc}
\bigoplus_{x\in X^{d-1}} K_2 k(x) & \longrightarrow & \bigoplus_{x\in X^d} k(x)^* \\
\downarrow & & \downarrow \\
\bigoplus_{x\in X^{d-1}} H^2(k(x),\mathbf{Z}/n(2)) & \longrightarrow & \bigoplus_{x\in X^d} H^1(k(x),\mathbf{Z}/n(1))
\end{array}
$$

The left vertical map is the Galois symbol (see Chapter 1, §1) and the right vertical map is obtained from Kummer theory. Now the cokernel of the bottom horizontal map is $H^d(X,\mathscr{H}^{d+1}(\mathbf{Z}/n(d+1)))$ (Theorem 1.10). This diagram induces a homomorphism

$$
H^d(X,\mathscr{K}_{d+1}) \to H^d(X,\mathscr{H}^{d+1}(\mathbf{Z}/n(d+1))).
$$

Now in the spectral sequence

$$
H^r(X,\mathscr{H}^s(\mathbf{Z}/n(d+1))) \Rightarrow H^{r+s}(X,\mathbf{Z}/n(d+1)),
$$

there is an edge map

$$
H^d(X,\mathscr{H}^{d+1}(\mathbf{Z}/n(d+1))) \to H^{2d+1}(X,\mathbf{Z}/n(d+1)).
$$

On the other hand, by Tate local duality for varieties over local fields (Theorem 1.15), we have

$$
H^{2d+1}(X,\mathbf{Z}/n(d+1)) \cong H^1(X,\mathbf{Z}/n)^* = \pi_1^{ab}(X)/n.
$$

Putting all of these maps together and passing to the inverse limit over n, we get another map

$$
\sigma_1 : H^d(X,\mathscr{K}_{d+1}) \to \pi_1^{ab}(X).
$$

LEMMA 4.2. *We have* $\sigma = \sigma_1$.

PROOF. Let x be a closed point of X with residue field $k(x)$. Then the following diagram commutes:

$$
\begin{array}{ccccc}
k(x)^*/n & \longrightarrow & \pi_1^{ab}(k(x))/n & \longrightarrow & \pi_1^{ab}(X) \\
\downarrow & & \downarrow & & \downarrow \\
H^1(k(x),\mathbf{Z}/n(1)) & \longrightarrow & H^1(k(x),\mathbf{Z}/n)^* & \longrightarrow & H^1(X,\mathbf{Z}/n)^*
\end{array}
$$

Here the left vertical map is given by Kummer theory. The top horizontal map is the reciprocity map in the class field theory of $k(x)$, and the bottom left horizontal map is given by Tate local duality (see Chapter 1, §6). On the other hand, there is a commutative diagram

$$
\begin{array}{ccc}
H^1(k(x),\mathbf{Z}/n(1)) & \longrightarrow & H^1(k(x),\mathbf{Z}/n)^* \\
\downarrow & & \downarrow \\
H^{2d+1}(X,\mathbf{Z}/n(d+1)) & \longrightarrow & H^1(X,\mathbf{Z}/n)^*
\end{array}
$$

where the left vertical map is a Gysin map and the right vertical map is the dual of the map on H^1 induced from the map $x \to X$. Putting these two diagrams together, we get the desired commutativity.

REMARK 4.3. In [Sa2], Saito proves Lemma 4.1 above by using the class field theory of two-dimensional local fields. However, later in his paper, he needs the equality of σ and σ_1, so we thought it was best to prove this first.

The proper map $X \to \operatorname{Spec} k$ induces a trace map

$$H^d(X, \mathscr{K}_{d+1}) \to k^*$$

which may be defined more explicitly in terms of the presentation of $H^d(X, \mathscr{K}_{d+1})$ above by taking the product of the norms to k of elements in $k(x)^*$ for various closed points x of X. We denote the kernel of this map by $V(X)$. The following diagram is commutative

$$
\begin{array}{ccc}
H^d(X, \mathscr{K}_{d+1}) & \longrightarrow & k^* \\
\downarrow & & \downarrow \\
\pi_1^{\mathrm{ab}}(X) & \longrightarrow & \pi_1^{\mathrm{ab}}(k).
\end{array}
$$

Here the right vertical map is the reciprocity map in local class field theory. Denoting the kernel of the bottom horizontal map by $\pi_1^{\mathrm{ab}}(X)^0$, we get the *geometric reciprocity map*

$$\tau \colon V(X) \to \pi_1^{\mathrm{ab}}(X)^0.$$

Now we state the main results in this chapter.

THEOREM 4.4. *The image of τ is finite.*

To state the other main result of the theory, recall from Chapter 1, §1 the map

$$\partial \colon H^d(X, \mathscr{K}_{d+1}) \to CH_0(Y).$$

THEOREM 4.5. *Let $X, \mathscr{X}$ and Y be as above. Then for a character $\chi \in H^1(X, \mathbf{Q}/\mathbf{Z})$ to come from a character of $H^1(\mathscr{X}, \mathbf{Q}/\mathbf{Z})$, it is necessary and sufficient that the corresponding homomorphism*

$$H^d(X, \mathscr{K}_{d+1}) \to \mathbf{Q}/\mathbf{Z}$$

be trivial on $\operatorname{Ker} \partial$.

This is one of the key local results to be used in the reciprocity law for unramified class field theory of schemes of finite type over $\mathbf{Z}$ (see Chapter 5). We will only prove it when X is a curve. For the proof in the general case, see [Sa3].

4.2. Proofs of the theorems. The group $V(X)$ may be described in terms of symbols defined as follows. Let L be a finite extension of k. The product map

$$H^d(X_L, \mathscr{K}_d) \times H^0(X_L, \mathscr{K}_1) \to H^d(X_L, \mathscr{K}_{d+1})$$

and the Bloch-Quillen formula $CH^d(X_L) = H^d(X_L, \mathcal{K}_d)$ gives a map

$$[\ ,\ \}_L : CH^d(X_L) \times L^* \to H^d(X_L, \mathcal{K}_{d+1}).$$

Given $\alpha \in CH^d(X_L)$ and $\beta \in L^*$, we denote by $[\alpha, \beta\}_{L/k}$ the norm down to $H^d(X, \mathcal{K}_{d+1})$ of $[\alpha, \beta\}_L$. Denote by $A_0(X_L)$ the group of zero-cycles of degree zero modulo rational equivalence on X_L. Then it is easy to see that if $\alpha \in A_0(X_L)$, then in fact $[\alpha, \beta\}_{L/k} \in V(X)$. We have the following lemma which will be used constantly:

LEMMA 4.6 [B2, 1.11 and 1.13]. *Suppose X has a k-rational point. As L runs over finite extensions of k, the symbols $[\alpha, \beta\}_{L/k}$ generate $V(X)$.*

This lemma may be proved by using the projection formula and the presentation of $H^d(X, \mathcal{K}_{d+1})$ given in Chapter 1, §1.

We now show how Theorem 4.4 can be reduced to the case of curves.

LEMMA 4.7. *Suppose Theorem 4.4 is true for any curve C_L defined over a finite extension L/k. Then it is true in arbitrary dimension.*

PROOF. By Lemma 4.6, any element η of $V(X)$ may be written as a sum of symbols $[\alpha, \beta\}_{L/k}$ as above, and hence it is enough to consider such symbols. By ([AK] or [C]), there exists a smooth, projective, geometrically connected curve C_L on X_L which passes through the support of α, and we have a commutative diagram

$$
\begin{array}{ccccc}
A_0(C_L) \otimes L^* & \longrightarrow & V(C_L) & \longrightarrow & \pi_1^{\mathrm{ab}}(C_L)^0 \\
\downarrow & & \downarrow & & \downarrow \\
A_0(X_L) \otimes L^* & \longrightarrow & V(X) & \longrightarrow & \pi_1^{\mathrm{ab}}(X)^0
\end{array}
$$

Hence, if the image of $V(C_L)$ in $\pi_1^{\mathrm{ab}}(C_L)^0$ is finite, the image of $V(X)$ in $\pi_1^{\mathrm{ab}}(X)^0$ is torsion. But by Proposition 4.12 below, the torsion subgroup of $\pi_1^{\mathrm{ab}}(X)$ is finite.

The pro-l-part of $\pi_1^{\mathrm{ab}}(X)$ is $H^1(X, \mathbf{Q}_l/\mathbf{Z}_l)^*$. By Tate duality for varieties over local fields (Theorem 1.15), this last group is isomorphic to $H^{2d+1}(X, \mathbf{Z}_l(d+1))$. This group may in turn be computed by using the Hochschild-Serre spectral sequence:

$$E_2^{r,s} = H^r(k, H^s(\overline{X}, \mathbf{Z}_l(d+1))) \Rightarrow H^{r+s}(X, \mathbf{Z}_l(d+1)).$$

The diagram

$$
\begin{array}{ccc}
H^d(X, \mathcal{K}_{d+1}) & \longrightarrow & k^* \\
\downarrow & & \downarrow \\
H^{2d+1}(X, \mathbf{Z}_l(d+1)) & \longrightarrow & H^1(k, H^{2d}(\overline{X}, \mathbf{Z}_l(d+1)))
\end{array}
$$

is commutative, where the bottom horizontal map comes from the spectral sequence and the right vertical map is Kummer theory via the identification $H^{2d}(\overline{X}, \mathbf{Z}_l(d+1)) \cong \mathbf{Z}_l(1)$. Note that $E_2^{0,2d} = H^{2d}(\overline{X}, \mathbf{Z}_l(d+1))^G = \mathbf{Z}_l(1)^G = 0$, so the spectral sequence and the diagram give a map

$$V(X) \to H^2(k, H^{2d-1}(\overline{X}, \mathbf{Z}_l(d+1))).$$

With the identifications we have made, this is the same as the geometric reciprocity map defined above.

Let A be the Albanese variety of X, and let $T_l = \varprojlim_m A(\overline{k})_{l^m}$ be its l-adic Tate module. Then an easy argument with Poincaré duality and the Weil pairing shows that the group $H^{2d-1}(\overline{X}, \mathbf{Z}_l(d+1))$ modulo its torsion subgroup is isomorphic as a G-module to $T_l(1) = T_l \otimes \mathbf{Z}_l(1)$. Note that this torsion subgroup is zero for almost all l. Using Tate local duality for Galois modules, we get that

$$H^2(k, T_l(1))) = T_{lG},$$

where the subscript G denotes the coinvariants by $G = \text{Gal}(\overline{k}/k)$.

PROPOSITION 4.8. *Assume that X has good reduction. Then*

$$H^2(k, H^{2d-1}(\overline{X}, \mathbf{Z}_l(d+1)))$$

is finite for all l and zero for almost all l.

PROOF. By what we have discussed above, it is enough to show that T_{lG} is finite for all l and zero for almost all l. First assume that l is different from the residue characteristic p. Since X has good reduction, the inertia subgroup acts trivially on T_l, and so we have

$$T_{lG} = T_{l\,\text{Gal}(\overline{\mathbf{F}}/\mathbf{F})}.$$

Now $\text{Gal}(\overline{\mathbf{F}}/\mathbf{F})$ is topologically generated by the Frobenius F, and the coinvariants of a module by F is just the cokernel of $1 - F$. Let $V_l = T_l \otimes \mathbf{Q}_l$. Then the $\mathbf{Q}_l$-dimension of the cokernel of $1 - F$ acting on this module is the same as the dimension of the kernel of $1 - F$. But the kernel of $1 - F$ acting on V_l is just $V_l A(\mathbf{F})$. Now $A(\mathbf{F})$ has only finitely many points of l-primary order, so $V_l A(\mathbf{F}) = 0$. This shows that the cokernel of $1 - F$ has $\mathbf{Q}_l$-dimension zero, so the cokernel of $1 - F$ acting on T_l is a finite group. An easy diagram chase shows that this group is the kernel of $1 - F$ acting on $T_l \otimes \mathbf{Q}_l/\mathbf{Z}_l$, and this is just $A(\mathbf{F})(l)$. This group is zero for almost all l since $A(\mathbf{F})$ is a finite group. This completes the proof of the prime to p-part.

For the p-part we need a more sophisticated argument which we just sketch. The complete argument may be found in [B2, 2.4]. If T_{pG} were infinite, there would be a nontrivial homomorphism $T_{pG} \to \mathbf{Z}_p$. By a theorem of Tate [T5], there would be a corresponding map of p-divisible groups over $\mathcal{O}_k$:

$$(A_{p^m})_m \to (\mathbf{Z}/p^m)_m.$$

Since the group on the right is an étale group, this map must factor through the étale quotient of the group on the left. Taking the T_p of the special fibres, we would get a map of $\mathrm{Gal}(\overline{\mathbf{F}}/\mathbf{F})$-modules:

$$T_p A(\overline{\mathbf{F}}) \to \mathbf{Z}_p.$$

But this is ridiculous because Frobenius acts trivially on $\mathbf{Z}_p$ but its action on the group on the left has no fixed points (same argument as for the T_l for $l \neq p$). This shows that T_{pG} is finite.

REMARK 4.9. Note that Weil's theorem on the absolute values of the eigenvalues of Frobenius acting on V_l is not used in the proof of the proposition. The fact that no eigenvalue of Frobenius acting on V_l has absolute value 1 follows immediately from the fact that an abelian variety over a finite field only has finitely many rational points.

We now go about proving Theorem 4.4 for curves. We fix the following notation: X is a smooth, projective, geometrically connected curve over a p-adic field k. We fix a regular proper model $\mathscr{X}$ over the ring of integers $\mathscr{O}_k$ of k.

LEMMA 4.10. *The map* $\tau_n \colon H^1(X, \mathscr{K}_2)/n \to \pi_1^{\mathrm{ab}}(X)/n$ *is injective.*

PROOF. By the theorem of Merkur'ev-Suslin (see Chapter 1, §1) and the work of Bloch, we have an isomorphism of sheaves

$$\mathscr{K}_2/n \cong \mathscr{H}^2(\mathbf{Z}/n(2)),$$

hence an isomorphism

$$H^1(X, \mathscr{K}_2/n) \cong H^1(X, \mathscr{H}^2(\mathbf{Z}/n(2))).$$

From this and the remarks following Theorem 1.10, we get that

$$H^1(X, \mathscr{K}_2)/n \hookrightarrow H^3(X, \mathbf{Z}/n(2)).$$

But $H^3(X, \mathbf{Z}/n(2)) \cong \pi_1^{\mathrm{ab}}(X)/n$.

LEMMA 4.11. *Let* L/k *be a finite extension. If Theorem 4.4 is true for* X_L, *then it is true for* X.

This may be proved easily by a norm argument.

Using Lemma 4.11 and the stable reduction theorem for curves ([DM]), we may assume that $\mathscr{X}$ is semi-stable. Let A be the Jacobian variety of X. Then A has semistable reduction, hence its Néron model $\mathscr{A}$ has special fibre which is an extension of an abelian variety by a torus.

Let I_l be the pro-l part of the image of the map τ. We claim that for l different from $p = \mathrm{char.}\ \mathbf{F}$, I_l is finite. This is because $V(X)$ is generated by symbols coming from $A(L) \otimes L^*$ as L ranges over finite extensions of k. Now $A(L)$ is an extension of a finite group by a pro-p group and L^* is an extension of $\mathbf{Z}$ by a group $(\mathscr{O}_L)$ which is itself an extension of a finite group by a pro-p-group. Summing all of this up, we see that for each extension L

of k, the group $A(L) \otimes L^*$ is an extension of a finite group by a pro-p-group. This implies that

$$V(X) \otimes \mathbf{Q}_l/\mathbf{Z}_l = 0;$$

hence, the same is true of I_l. But I_l is a finitely generated $\mathbf{Z}_l$-module, and it is easy to see that the only such modules with this property are torsion. This proves the claim. To prove finiteness of the pro-p-part and the vanishing of the image of the pro-l-part for almost all l, we use a monodromy argument. For the convenience of the reader, we briefly review this theory. For more details, see [SGA 7, Première Partie, Exposé IX]. We give a different proof from that of Saito [Sa2], and we note that other proofs have been given by Salberger [Sal1] and Spiess [Sp].

Recall that the Tate module $T_l(A)$ of an abelian variety over a local field has a filtration

$$T_l(A) \supset T_l(A)^{\mathrm{ef}} \supset T_l(A)^{\mathrm{et}},$$

where ef stands for the essentially fixed part and et stands for the essentially toric part. The ef part sits in an exact sequence

$$(*) \quad 0 \to T_l(A)^{\mathrm{et}} \to T_l(A)^{\mathrm{ef}} \to T_l B \to 0,$$

where B is an abelian variety with good reduction. For l different from the residue characteristic p, $T_l(A)^{\mathrm{ef}}$ is essentially the part fixed by the inertia group of k. The difference between T_l and the essentially fixed part may be expressed by the exact sequence:

$$(**) \quad 0 \to T_l^{\mathrm{ef}} \to T_l \to \mathrm{Hom}(T_l^{\mathrm{et}}, \mathbf{Z}_l(1)) \to 0.$$

The filtration above coincides with the monodromy filtration for l different from p.

PROPOSITION 4.12. *The torsion subgroup of $T_{l,G}$ is zero for almost all l.*

PROOF. By a norm argument, it suffices to prove the result after a finite extension of k. We take L/k finite such that the essentially toric part of T_l is split. Consider the exact sequence $(*)$. Now $T_{l,G}^{\mathrm{et}}$ is finite and zero for almost all l (it is a sum of copies of the group of l-primary roots of unity in L). As for $(T_l B)_G$, this is finite for all l and zero for almost all l because B has good reduction (this group is isomorphic to the l-primary torsion subgroup of $B(\mathbf{F})$) if $l \neq p$. This proves that T_G^{ef} is finite. Now consider the exact sequence $(**)$. Since the toric part is split, G acts trivially on the right group. Since the coinvariants of the left group is finite and zero for almost all l, we get vanishing of the torsion subgroup of $T_{l,G}$ for almost all l.

For the finiteness of the image of the pro-p-part, we need the notion of the Raynaud extension associated to A. This is a semiabelian scheme $A^\natural$ over $\mathcal{O}_k$ which is an extension of an abelian scheme by a torus. We have

$$T(A_k^\natural) = T^{\mathrm{ef}}(A).$$

It has the same special fibre as the Néron model of A over $\mathcal{O}_k$. For any complete extension L/k, there is a map of rigid analytic spaces

$$A^\sharp(L) \to A(L).$$

For more details on this, see [SGA 7, Part I, Exp. IX §2, 5, 7] and [FC, Chapter II, §1].

Now we can give the proof. Since the group $V(X)$ is generated by symbols $[\alpha, \beta]_{L/k}$ (Lemma 4.6), it will be enough to show that all such symbols are trivial in $H^2(L, V_p(1))$. Let $\widehat{T}^0$ denote the group of cocharacters of T and let C be a set of elements of $\widehat{T}^0$ which forms a basis. For each $v \in C$, there is a map on L-points

$$L^* \to A(L)$$

and an embedding of G-modules.

$$i_v \colon \mathbf{Q}_p(1) \to V_p.$$

Consider the following diagram:

$$
\begin{array}{ccccc}
A(L) \otimes L^* & \longrightarrow & H^1(L, V_p) \otimes H^1(L, \mathbf{Q}_p(1)) & \longrightarrow & H^2(L, V_p(1)) \\
\downarrow & & \downarrow & & \downarrow \\
A(L) \otimes A(L) & \longrightarrow & H^1(L, V_p) \otimes H^1(L, V_p) & \longrightarrow & H^2(L, \mathbf{Q}_p(1))
\end{array}
$$

Here the left and center vertical maps are identity on the left factor and the natural maps induced by the embedding i_v. The right vertical map is given by considering the dual map

$$\mathrm{Hom}(V_p, \mathbf{Q}_p(1)) \to \mathrm{Hom}(\mathbf{Q}_p(1), \mathbf{Q}_p(1))$$

corresponding to i_v, then twisting by 1 (note that V_p is canonically isomorphic to $\mathrm{Hom}(V_p, \mathbf{Q}_p(1))$ because A is the Jacobian variety of a curve, hence is self-dual). The left most horizontal arrows are given by Kummer theory for L^* and for $A(L)$. Finally, the right horizontal arrow on the bottom is given by cup-product using the Weil pairing

$$V_p \otimes V_p \to \mathbf{Q}_p(1)$$

that the composite of the two top horizontal maps is τ is proved by Bloch ([B2], Thm. 1.14).

Assume for the moment the following two lemmas:

LEMMA 4.13. *The diagram above commutes for each* $v \in C$.

LEMMA 4.14. *The map*

$$H^2(L, V_p(1)) \to H^2(L, \mathrm{Hom}(V_p^{\mathrm{et}}, \mathbf{Q}_p(1)) \otimes \mathbf{Q}_p(1)),$$

induced by taking Galois cohomology of the exact sequence (∗∗) *above twisted by* 1, *is injective.*

Then we can complete the proof. Take any symbol in $J(L) \otimes L^*$ and suppose its image in $H^2(L, V_p(1))$ were nonzero. Then by the second lemma above, there would exist some $v \in C$ such that its image under the right vertical map in the diagram would be nonzero. By Corollary 1.17, the composite of the left vertical arrow and the two horizontal arrows following it is zero. This is a contradiction and completes the proof modulo the two lemmas.

PROOF OF LEMMA 4.13. The idea of the proof is easily seen from the case of Tate elliptic curve E with parameter q. This is an elliptic curve over k together with a parameterization of the set $E(k)$ as a quotient of k^* by the subgroup generated by q. For any extension field L/k, we have an exact sequence

$$0 \to q^{\mathbf{Z}} \to L^* \to E(L) \to 0.$$

Doing this for $L = \bar{k}$ and considering the kernel of multiplication by p^n as n varies, we easily get the exact sequence:

$$0 \to \mathbf{Q}_p(1) \xrightarrow{a} V_p(E) \xrightarrow{b} \mathbf{Q}_p \to 0.$$

The Weil pairing e_p may be described explicitly for any $\alpha \in V_p(E)$ by the equation

$$e_p(a(\zeta), \alpha) = \zeta^{b(\alpha)},$$

where ζ is a generator of $\mathbf{Q}_p(1)$. Using similar reasoning, one can prove the commutativity of the diagram used above in the more general situation.

PROOF OF LEMMA 4.14. Consider the exact sequences (∗) and (∗∗) above. Twisting (∗∗) by 1 and taking Galois cohomology, we get the exact sequence

$$\cdots \to H^2(L, V_p^{\mathrm{ef}}(1)) \to H^2(L, V_p(1)) \to H^2(L, \mathrm{Hom}(V_p^{\mathrm{et}}, \mathbf{Q}_p(1))(1)) \to 0.$$

We claim that the group on the left is zero. To see this, we consider part of the Galois cohomology sequence associated with the twist of (∗) by 1

$$\cdots \to H^2(L, V_p^{\mathrm{et}}(1)) \to H^2(L, V^{\mathrm{ef}}(1)) \to H^2(L, V_p B) \to 0.$$

We claim that the end terms of this sequence are zero. For the group on the left, we may make a finite extension of L so that the essentially toric part is a G-direct sum of $\mathbf{Q}_p(1)$, hence the twist by 1 is a direct sum of $\mathbf{Q}_p(2)$. Now by Tate local duality, we have

$$H^2(L, \mathbf{Q}_p(2)) \cong H^0(L, \mathbf{Q}_p(-1))^*.$$

Since L only contains finitely many p-power roots of unity, this last group is trivial. As for the group on the right, it vanishes by the same argument we used in the proof of Proposition 4.8. This completes the proof of the lemma and of Theorem 4.4.

COROLLARY 4.15. *Let X be a smooth projective curve over a p-adic field of characteristic zero. The group $V(X)$ is an extension of a finite group by a divisible group, so $V(X) \otimes \mathbf{Q}/\mathbf{Z} = 0$.*

PROOF. The map $\sigma: H^1(X, \mathscr{K}_2) \to \pi_1^{ab}(X)$ factors through

$$\varprojlim_m H^1(X, \mathscr{K}_2)/m,$$

and by Lemma 4.10, the map:

$$\varprojlim_m (X, \mathscr{K}_2)/m \to \pi_1^{ab}(X)$$

is injective. Hence an element of $W = \operatorname{Ker} \sigma = \operatorname{Ker} \tau$ is divisible in $H^1(X, \mathscr{K}_2)$. Let $W_n = nH^1(X, \mathscr{K}_2) \cap V(X)$, let M be the order of the group of roots of unity in k, M' the order of the torsion subgroup of T_G. Then for any n, $W_{nM} \subset nV(X)$. Put $N = MM'$. Then $W_N \subset M'V(X)$ and $M'V(X) \subset W$. This shows that $W_N = W$ and proves that W is divisible.

DEFINITION 4.16. We say that a covering X'/X is completely split if every closed point of X splits completely in X'. We denote by $\pi_1^{ab}(X)_{cs}$ the quotient of $\pi_1^{ab}(X)$ which classifies the abelian étale covers of X which are completely split.

For the rest of this chapter, let X be a smooth projective, geometrically connected curve over the p-adic field k.

PROPOSITION 4.17. *Let $\mathscr{X}$ be a regular proper model of X over $\mathscr{O}_k$ with special fibre Y. Then there is a canonical isomorphism*

$$\pi_1^{ab}(X)_{cs} \to \pi_1^{ab}(Y)_{cs}.$$

PROOF. The natural surjective map

$$\pi_1^{ab}(X) \to \pi_1^{ab}(\mathscr{X}) = \pi_1^{ab}(Y)$$

induces a surjective map

$$\pi_1^{ab}(X)_{cs} \to \pi_1^{ab}(Y)_{cs}.$$

To show injectivity, we must see that if $Z \to X$ is an abelian completely split covering, $\mathscr{Z}$ is the integral closure of $\mathscr{X}$ in the function field of Z and $\tilde{f}: \mathscr{Z} \to \mathscr{X}$ the extension of f, the restriction of $\tilde{f}$ to the special fibres is étale and completely split. To see this, let y be a closed point of Y, A_y the completion of the local ring of $\mathscr{X}$ at y and K_y the fraction field of A_y. For any closed point x of X, its closure in $\mathscr{X}$ contains a unique closed point y of Y, and x determines a codimension one point of $\operatorname{Spec} A_y$. Now any codimension one point of $\operatorname{Spec} A_y$, except possibly for finitely many which lie above the maximal ideal of $\mathscr{O}_k$, is determined by some closed point of X. Since f is a completely split covering, such codimension one points split completely in $\operatorname{Spec} A_y \times \mathscr{Z}$. By Corollary 3.3, this means that the extension $\operatorname{Spec} A_y \times \mathscr{Z}$ is isomorphic to a disjoint union of copies of $\mathrm{Spec}(A_y)$. This shows that the closed point y is completely split and completes the proof of the proposition.

To determine the structure of $\pi_1^{ab}(X)_{cs}$, we may assume that Y has only ordinary double points (by blowing up $\mathscr{X}$ at closed points, if necessary).

DEFINITION 4.18. The dual graph of Γ of Y is the connected graph with edges corresponding to the singular points of Y and vertices corresponding to the irreducible components of Y. The extremities of an edge corresponding to a singular point P is the set of vertices corresponding to irreducible components on which P lies.

PROPOSITION 4.19. *The cokernel of the map $\tau\colon V(X) \to T_G$ is a free $\widehat{\mathbf{Z}}$-module of finite rank r. It classifies the completely split abelian coverings of X. r is the rank of the torus part of the special fibre of the Néron minimal model of X over k. Let Y be the special fibre of the minimal regular model of X over $\mathscr{O}_k$. We have*

$$r = \mathrm{rank}_{\mathbf{Z}} H^1(\Gamma, \mathbf{Z}).$$

PROOF. By Theorem 4.4, the rank of the cokernel of τ is the rank of $T_{l,G}$ for any prime l. Calculating the rank of this cokernel using the filtration on T_l given in exact sequences $(*)$ and $(**)$ above, we see that the rank of $T_{l,G}$ is exactly the rank of the essentially toric part of T_l, and this is the rank of the torus part of the special fibre of the Néron model of J over k. For the assertion about the reduction Y, let U be the smooth locus of Y, and put $S = Y - U$. For each $y \in S$, let A_y be the henselization of the local ring of Y at y. We have the following commutative diagram:

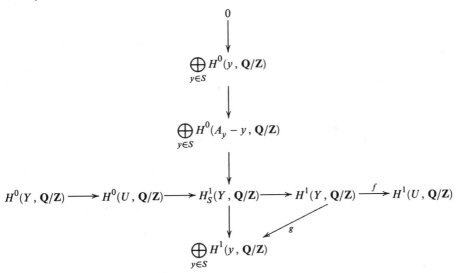

Note that $H^i(A_y, \mathbf{Q}/\mathbf{Z}) = H^i(y, \mathbf{Q}/\mathbf{Z})$ by proper base change. Now U is smooth, so it has no completely split coverings (Lemma 1.7). Hence,

$$H^1(Y, \mathbf{Q}/\mathbf{Z})_{cs} = \mathrm{Ker}\, f \cap \mathrm{Ker}\, g$$

where $H^1(Y, \mathbf{Q}/\mathbf{Z})_{cs}$ is the Pontryagin dual of $\pi_1^{ab}(Y)_{cs}$. Hence, we have the exact sequence

$$H^0(Y, \mathbf{Q}/\mathbf{Z}) \to H^0(U, \mathbf{Q}/\mathbf{Z}) \to \bigoplus_{y \in S} \frac{H^0(A_y - y, \mathbf{Q}/\mathbf{Z})}{H^0(y, \mathbf{Q}/\mathbf{Z})}$$

$$\to H^1(Y, \mathbf{Q}/\mathbf{Z})_{cs} \to 0.$$

Now $H^0(Y, \mathbf{Q}/\mathbf{Z}) = \mathbf{Q}/\mathbf{Z}$ and $H^0(U, \mathbf{Q}/\mathbf{Z}) = \bigoplus_v \mathbf{Q}/\mathbf{Z}$, where the sum runs over the irreducible components v of Y. For any $y \in S$, we have

$$H^0(A_y - y, \mathbf{Q}/\mathbf{Z}) = \mathbf{Q}/\mathbf{Z} \oplus \mathbf{Q}/\mathbf{Z}$$

because Y is an ordinary double point. Hence, for $y \in S$, the cokernel of the map

$$H^0(y, \mathbf{Q}/\mathbf{Z}) \to H^0(A_y - y, \mathbf{Q}/\mathbf{Z})$$

is isomorphic to $\mathbf{Q}/\mathbf{Z}$. We thus have an exact sequence

$$\bigoplus_v \mathbf{Q}/\mathbf{Z} \to \bigoplus_{y \in S} \mathbf{Q}/\mathbf{Z} \to H^1(Y, \mathbf{Q}/\mathbf{Z})_{cs} \to 0.$$

Each v corresponds to a vertex of the dual graph Γ and each y corresponds to an edge. Hence, we have $H^1(Y, \mathbf{Q}/\mathbf{Z})_{cs} = H^1(\Gamma, \mathbf{Q}/\mathbf{Z})$. This group is cotorsion free and hence $\pi_1^{ab}(Y)_{cs}$ is torsion free. By Proposition 4.17, $\pi_1^{ab}(X)_{cs}$ is also torsion free. This completes the proof of the proposition.

For a curve Y over a finite field $\mathbf{F}$, we have the reciprocity map

$$CH_0(Y) \to \pi_1^{ab}(Y)$$

(see Chapter 5 below for the definition). Let $CH_0(Y)^0$ denote the kernel of the degree map which sends a closed point y to $[\mathbf{F}(y) : \mathbf{F}]$.

PROPOSITION 4.20. *Let Y be a proper connected curve over a finite field. Then $CH^0(Y)^0$ is a finite group, and the reciprocity map induces an isomorphism:*

$$CH_0(Y)^0 \cong \pi_1^{ab}(Y)_{tors}.$$

PROOF. We use Deninger's duality for Y (Theorem 1.21). It gives an isomorphism:

$$\pi_1^{ab}(X)/n = H^1(X, \mathbf{Z}/n)^* \cong \mathrm{Ext}_Y^2(\mathbf{Z}/n, \mathscr{G}),$$

where $\mathscr{G}$ is the complex in Deninger's theorem. To compute this last group, consider the exact sequence of sheaves

$$0 \to \mathbf{Z} \xrightarrow{n} \mathbf{Z} \to \mathbf{Z}/n \to 0.$$

Taking the long exact Ext sequence with values in $\mathscr{G}$, we see that the map

$$\mathrm{Ext}_Y^1(\mathbf{Z}, \mathscr{G})/n \to \mathrm{Ext}_Y^2(\mathbf{Z}/n, \mathscr{G})$$

is injective. But $\text{Ext}_Y^1(\mathbf{Z}, \mathcal{G}) = H^1(Y, \mathcal{G})$, and this last group may be calculated by using the spectral sequence

$$E_1^{r,s} = H^s(Y, \mathcal{G}^r) \Rightarrow H^{r+s}(Y, \mathcal{G}).$$

We have

$$E_1^{0,1} = H^1(Y, \mathcal{G}^0) = H^1\left(Y, \bigoplus_{y \in Y^0} i_{y_*} \mathbf{G}_{m,y}\right) = 0$$

by Hilbert's theorem 90. Also,

$$E_1^{1,0} = \text{Coker}\left[H^0\left(Y, \bigoplus_{y \in Y^0} i_{y_*} \mathbf{G}_{m,y}\right) \to H^0\left(Y, \bigoplus_{y \in Y^1} i_{y_*} \mathbf{Z}\right)\right],$$

and this group is easily seen to be $CH_0(Y)$. We have thus shown that $CH_0(Y)/n$ injects into $\pi_1^{\text{ab}}(Y)/n$. Taking the inverse limit over all n, we get that the profinite completion of $CH_0(Y)$ injects into $\pi_1^{\text{ab}}(Y)$. But $CH_0(Y)$ is finitely generated, as can be seen easily by comparing it with $CH_0(Y')^0$, where Y' is the normalization. Hence, we get the injectivity of $CH_0(Y)$ into $\pi_1^{\text{ab}}(Y)$. Surjectivity follows from Proposition 4.19. To see the finiteness of $CH_0(Y)^0$, note that the composite

$$CH_0(Y)^0 \to CH_0(Y')^0 \to CH_0(Y)^0$$

is the identity. Since $CH_0(Y')^0$ is a finite group, so is $CH_0(Y)^0$.

Now we can summarize much of the discussion above by the following theorem, which is crucial for unramified class field theory of surfaces.

THEOREM 4.21. *Let* X *be a smooth, projective, geometrically connected curve over a p-adic field of characteristic zero. Let* $\mathcal{X}$ *be a regular proper model of* X *over* $\mathcal{O}_k$ *with special fibre* Y. *Then we have the following diagram of exact sequences*:

$$
\begin{array}{ccccccccc}
0 & \longrightarrow & E_Y & \longrightarrow & H^1(Y, \mathbf{Q}/\mathbf{Z}) & \longrightarrow & CH_0(Y)^* & \longrightarrow & 0 \\
 & & \| & & \downarrow & & \downarrow & & \\
0 & \longrightarrow & E_Y & \longrightarrow & H^1(X, \mathbf{Q}/\mathbf{Z}) & \xrightarrow{\sigma^*} & H^1(X, \mathcal{K}_2)^* & &
\end{array}
$$

Here E_Y *is the dual of* $\pi_1^{\text{ab}}(Y)_{\text{cs}}$.

PROOF. This follows immediately from Propositions 4.19 and 4.20.

COROLLARY 4.22. *For a character* $\chi \in H^1(X, \mathbf{Q}/\mathbf{Z})$ *to come from* $H^1(\mathcal{X}, \mathbf{Q}/\mathbf{Z})$, *it is necessary and sufficient that the corresponding homomorphism induced by* σ

$$\chi_\sigma : H^1(X, \mathcal{K}_2) \to \mathbf{Q}/\mathbf{Z}$$

be trivial on the kernel of the boundary map $H^1(X, \mathscr{K}_2) \xrightarrow{\partial} CH_0(Y)$.

COROLLARY 4.23. *The map* $\partial: H^1(X, \mathscr{K}_2) \to CH_0(Y)$ *is surjective.*

PROOF. This follows from the diagram in the theorem and the fact that $CH_0(Y)$ is finitely generated.

COROLLARY 4.24. *Suppose X has good reduction. Then the image of the map*

$$\sigma: H^1(X, \mathscr{K}_2) \to \pi_1^{ab}(X)$$

is dense. In particular, $\tau: V(X) \to T_G$ *is surjective.*

PROOF. The first statement follows from the fact that when X has good reduction, the group E_Y in the diagram in Theorem 4.21 is zero. The second statement follows from the finiteness of T_G when X has good reduction (Proposition 4.8).

The following result is crucial for the global theory.

PROPOSITION 4.25. *Let X be a smooth, projective, geometrically connected curve over a p-adic field k. Assume that X has a good reduction with special fibre Y and that the absolute ramification index of k is 1. Denote by T_l the Tate module of the Jacobian of X and by $\widetilde{T}_l$ the Tate module of the Jacobian of Y. Then the natural map*

$$T_{lG} \to \widetilde{T}_{l, G_{\mathbf{F}}}$$

is an isomorphism for all primes l.

PROOF. For l different from the residue characteristic of k, this follows easily from the proof of Proposition 4.8. We take $l = p$. It suffices to prove the following.

CLAIM. Let $T_p^{\text{ét}}$ be the maximal étale quotient of T_p. Then the natural map

$$T_{pG} \to T_{pG}^{\text{ét}}$$

is an isomorphism.

To see this, let T_p^0 be the connected part of T_p. We claim that $T_{pG}^0 = 0$. If this were not the case, there would exist a nontrivial map of Galois modules:

$$T_p^0 \to \mathbf{Z}/p$$

which induces an isomorphism of Galois modules from some quotient H of T_p^0 to $\mathbf{Z}/p$. Since the inertia subgroup of G acts trivially on H, it extends to a finite flat group scheme over $\mathscr{O}_k$. By a theorem of Fontaine [**Fo**] (using the hypothesis that k is absolutely unramified), this induces an isomorphism of group schemes, and hence T_p^0 has $\mathbf{Z}/p$ as a quotient. But this is ridiculous since the geometric special fibre of T_p^0 is a sum of α_p's and μ_p's [**Mu**, §14, p. 136 and §15, p. 147].

REMARK 4.26. In the paper of Kato-Saito [**KS2**], this result is proved in the case where the absolute ramification index is less than $p - 1$. The proof given here was suggested by Kato, and it may also be used to show this using the Fontaine-Lafaille theory [**FL**].

Chapter 5. Unramified class field theory

In this chapter we discuss the unramified class field theory of a scheme X of finite type over $\mathbf{Z}$. We define the reciprocity map from the Chow group of 0-cycles to the abelianized fundamental group and prove the reciprocity law for unramified class field theory. Note that the reciprocity law will follow from the reciprocity law for the function field of X to be proved in Chapter 6. However, unramified class field theory by itself has a nice geometric flavor which helps one gain an understanding of the subject. That is why we do it separately here. Readers who insist on economy of exposition are invited to go directly to Chapter 6.

5.1. Definition and basic properties of the reciprocity map. Let X be a regular connected projective scheme of finite type over $\mathbf{Z}$. There are two possibilities for X. First, X could be a smooth projective variety over a finite field. This is the geometric case, and will be denoted by (G). Second, X could be a regular projective model over $\mathbf{Z}$ of a smooth projective variety over $\mathbf{Q}$. This is the arithmetic case, and we denote this by (A). The unramified class field theory of X will be slightly different depending on which case X falls into.

Denote the abelianized fundamental group of X by $\pi_1^{\mathrm{ab}}(X)$. Let $Z_0(X)$ denote the group of 0-cycles on X. If x is a closed point of X, the residue field $\kappa(x)$ is a finite field because X is of finite type over $\mathbf{Z}$. The absolute Galois group $G_x = \mathrm{Gal}(\overline{\kappa(x)}/\kappa(x))$ is isomorphic to $\hat{\mathbf{Z}}$ with a canonical generator, the *Frobenius* which we shall denote by ϕ_x. We have $\phi_x(a) = a^{N(x)}$ for any $a \in \overline{\kappa(x)}$, where $N(x)$ is the cardinality of $\kappa(x)$. We can consider G_x as a subgroup of $\pi_1^{\mathrm{ab}}(X)$ by viewing it as the decomposition group of x (well defined since any two decomposition groups are conjugate), and hence, we can view ϕ_x as an element of $\pi_1^{\mathrm{ab}}(X)$. Then we define the *reciprocity map*

$$\theta_X \colon Z_0(X) \to \pi_1^{\mathrm{ab}}(X)$$

by $\theta(x) = \phi_x$. A quicker and clearly equivalent way to define the reciprocity map is to send a closed point x to the image of the Frobenius under the map

$$i_{x_*} \colon \pi_1^{\mathrm{ab}}(\kappa(x)) \to \pi_1^{\mathrm{ab}}(X),$$

where $i_x \colon x \to X$ is the inclusion.

Now we split up the discussion.

CASE (G). X is a smooth, projective, geometrically connected variety over a finite field $\mathbf{F}$. Let $\overline{\mathbf{F}}$ be an algebraic closure of $\mathbf{F}$, and put $G = \mathrm{Gal}(\overline{\mathbf{F}}/\mathbf{F})$.

If $f: Y \to X$ is any morphism of schemes, there is an induced map

$$f_* : \pi_1^{ab}(Y) \to \pi_1^{ab}(X)$$

obtained by pulling back étale coverings from X to Y. If X and Y are smooth and projective of the same dimension over a finite field, we can get a map f^* defined as follows. Let n be a positive integer. Then $\pi_1^{ab}(X)/n = H^1(X, \mathbf{Z}/n)^*$, and by Poincaré duality (Theorem 1.12), this group is isomorphic to $H^{2d}(X, \mathbf{Z}/n(d))$, where d is the dimension of X (if n is divisible by char. $\mathbf{F}$, we mean these groups in the sense of logarithmic de Rham-Witt cohomology (Chapter 1, §2)); similarly for Y. Now there is a natural map

$$H^{2d}(X, \mathbf{Z}/n(d)) \to H^{2d}(Y, \mathbf{Z}/n(d))$$

and this gives the desired pullback map. Note that the same argument works over any field over which one has Poincaré duality, e.g., an algebraically closed field or a p-adic field.

We need the following technical lemma:

LEMMA 5.1. (1) *Let* $f: Y \to X$ *be a finite map of varieties over* $\mathbf{F}$. *Then the diagram*

$$
\begin{array}{ccc}
CH_0(Y) & \xrightarrow{\theta_Y} & \pi_1^{ab}(Y) \\
\downarrow & & \downarrow \\
CH_0(X) & \xrightarrow{\theta_X} & \pi_1^{ab}(X)
\end{array}
$$

is commutative, where the vertical arrows are the push-forward maps.

(2) *Assume that* Y *and* X *are smooth and projective of the same dimension. Then the diagram*

$$
\begin{array}{ccc}
CH_0(Y) & \xrightarrow{\theta_Y} & \pi_1^{ab}(Y) \\
\uparrow & & \uparrow \\
CH_0(X) & \xrightarrow{\theta_X} & \pi_1^{ab}(X)
\end{array}
$$

is commutative, where the vertical maps are pull-back maps on π_1^{ab} *defined above.*

Now set $\pi_1^{ab}(X)^0 = \mathrm{Ker}(\pi_1^{ab}(X) \to \pi_1^{ab}(\mathbf{F}))$. Then it is easy to see that the reciprocity map restricted to cycles of degree 0 has image in $\pi_1^{ab}(X)^0$.

PROPOSITION 5.2. *The reciprocity map factors through rational equivalence of cycles to give a map*

$$CH_0(X) \to \pi_1^{ab}(X).$$

PROOF. We will give a proof based on functoriality properties of θ. The reader should note that some of these functoriality properties are actually quite deep as they rely on Poincaré duality. First suppose X is the projective line $\mathbf{P}_\mathbf{F}^1$. Then any 0-cycle of degree zero is the divisor of a function, and $\pi_1^{ab}(\mathbf{P}^1) = \mathrm{Gal}(\overline{\mathbf{F}}/\mathbf{F})$. Then if two 0-cycles ξ, η are rationally equivalent, $\xi - \eta$ is of degree zero, and so maps to $\pi_1^{ab}(\mathbf{P}_\mathbf{F}^1)^0$, which is trivial. Hence, the lemma is true for $\mathbf{P}_\mathbf{F}^1$. Next, let C be any smooth, projective curve over $\mathbf{F}$ and let f be a rational function on C. We can view f as a morphism

$$f: C \to \mathbf{P}_\mathbf{F}^1.$$

There is a commutative diagram

$$
\begin{array}{ccc}
Z_0(\mathbf{P}_\mathbf{F}^1) & \longrightarrow & Z_0(C) \\
\downarrow{\scriptstyle\theta} & & \downarrow{\scriptstyle\theta} \\
\pi_1^{ab}(\mathbf{P}_\mathbf{F}^1) & \longrightarrow & \pi_1^{ab}(C)
\end{array}
$$

Now (f) is simply the image in $Z_0(C)$ of the divisor $(0)-(\infty)$ of $\mathbf{P}_\mathbf{F}^1$. By what we have just proved for the projective line and a simple diagram chase, we see that $\theta((f)) = 0$ in $\pi_1^{ab}(C)$. Now take an arbitrary smooth, projective variety X. The subgroup of $Z_0(X)$ of cycles rationally equivalent to zero is generated by the divisors of functions on curves on X. Let D be such a curve, C its normalization. Then we can define the divisor of a function f on D by taking $g_*(f)$, where we consider f as a function on C and $g : C \to D$ is the natural map. There is a commutative diagram:

$$
\begin{array}{ccc}
Z_0(C) & \longrightarrow & Z_0(X) \\
\downarrow & & \downarrow \\
\pi_1^{ab}(C) & \longrightarrow & \pi_1^{ab}(X)
\end{array}
$$

By what we proved for curves and this diagram, we see that $\theta_X((f)) = 0$ in $\pi_1^{ab}(X)$. This completes the proof of the proposition.

REMARK 5.3. In the work of Kato-Saito, the previous proposition is proved by reduction to the case of curves and then appealing to the reciprocity law on curves. The purpose of our proof is to show how using more cohomological information can actually reduce it to the projective line, in which case it is trivial. But the fact that the abelianized fundamental group is contravariant functorial for finite morphisms is highly nontrivial, and depends on duality theory for l-adic étale and logarithmic de Rham-Witt cohomology.

By the proposition, we get a map

$$\theta: CH_0(X) \to \pi_1^{ab}(X).$$

Let $A_0(X)$ be the subgroup of $CH_0(X)$ of cycles of degree zero. Consider the commutative diagram:

$$
\begin{array}{ccccccccc}
0 & \longrightarrow & A_0(X) & \longrightarrow & CH_0(X) & \longrightarrow & \mathbf{Z} \\
& & \downarrow & & \downarrow{\scriptstyle\theta} & & \cap \\
0 & \longrightarrow & \pi_1^{\mathrm{ab}}(X)^0 & \longrightarrow & \pi_1^{\mathrm{ab}}(X) & \longrightarrow & \mathrm{Gal}(\overline{\mathbf{F}}/\mathbf{F}) & \longrightarrow & 0
\end{array}
$$

We get a map

$$
\theta_0 \colon A_0(X) \to \pi_1^{\mathrm{ab}}(X)^0.
$$

Now we can state the reciprocity law in the geometric case.

THEOREM 5.4 (Kato-Saito [KS2]). θ_0 *is an isomorphism of finite groups.*

CASE (A). Let X be a regular, connected scheme which is proper and flat over $\mathbf{Z}$. Let X_η be the generic fibre of X, which is a smooth, projective variety over $\mathbf{Q}$. Denote by k the algebraic closure of $\mathbf{Q}$ in the function field of X_η; then X_η is geometrically connected over k. We denote the ring of integers of k by $\mathscr{O}_k$.

We would like to show that the reciprocity map θ defined above factors through rational equivalence of cycles, but this is not quite the case. We must restrict ourselves to coverings which are unramified at infinity, in the following sense.

DEFINITION 5.5. Let $Y \to X$ be a finite flat morphism of schemes of finite type over $\mathbf{Z}$. We say that f is unramified at infinity if every real point of X splits completely into real points of Y.

Let $\widetilde{\pi_1^{\mathrm{ab}}}(X)$ denote the quotient of $\pi_1^{\mathrm{ab}}(X)$ which classifies coverings which are unramified at infinity. We shall denote by $\tilde{\theta}$ the composite of θ and the projection onto $\widetilde{\pi_1^{\mathrm{ab}}}(X)$.

PROPOSITION 5.6. $\tilde{\theta}$ *factors through rational equivalence of cycles to give a map*

$$
\tilde{\theta} \colon CH_0(X) \to \widetilde{\pi_1^{\mathrm{ab}}}(X).
$$

PROOF. We must show that if Y is an irreducible, one dimensional subscheme of X and f is a function on Y, then $\tilde{\theta}((f)) = 0$ in $\widetilde{\pi_1^{\mathrm{ab}}}(X)$. Let Y' be the normalization of Y. Then the following diagram commutes:

$$
\begin{array}{ccccc}
Z_0(Y') & \longrightarrow & Z_0(Y) & \longrightarrow & Z_0(X) \\
\downarrow & & \downarrow & & \downarrow \\
\widetilde{\pi_1^{\mathrm{ab}}}(Y') & \longrightarrow & \widetilde{\pi_1^{\mathrm{ab}}}(Y) & \longrightarrow & \widetilde{\pi_1^{\mathrm{ab}}}(X)
\end{array}
$$

Thus, we can reduce the proposition to the case of a 1-dimensional regular scheme of finite type over $\mathbf{Z}$. This is the classical case and can be found in [T6, Theorem A, p. 188]. This completes the proof of the proposition.

THEOREM 5.7 (Bloch; Kato-Saito; reciprocity law in the arithmetic case). *Let X be a regular, projective, connected scheme of finite type over $\mathbf{Z}$. Then the reciprocity map $\tilde{\theta}$ defined above is an isomorphism of finite groups.*

5.2. Proof of the reciprocity law in the geometric case. It might be useful to list the main steps.

STEP 1 (Katz-Lang [KL]). Prove that $\pi_1^{ab}(X)^0$ is finite.

STEP 2. (Lang [L2]; see also Lemma 1.7). Prove that θ_0 is surjective.

STEP 3. (Colliot-Thélène/Sansuc/Soulé [CTSS]; Milne [M3]). When X is a surface, define another map λ between $A_0(X)$ and $\pi_1^{ab}(X)^0$, and prove that it is injective. *A priori*, λ may not be the same as θ_0, but Colliot-Thélène, Sansuc and Soulé show ([CTSS], Proposition 1) that these maps are the same. This will not be needed.

STEP 4. (Colliot-Thélène/Raskind [CTR2]). Use a counting argument to show that both θ_0 and λ are isomorphisms when X is a surface.

STEP 5. (Colliot-Thélène; see [KS2]). Show that θ_0 is an isomorphism for X of any dimension by using a hyperplane section argument.

PROOF OF STEP 1. First, we show that $\pi_1^{ab}(X)^0$ is finite by using the duality theory outlined in Chapter 1, §3 above. We have $\pi_1^{ab}(X) = H^1(X, \mathbf{Q}/\mathbf{Z})^*$, where $*$ denotes Pontryagin dual. Let l be a prime number, which may equal char. $\mathbf{F}$. Then we have

$$(*) \qquad H^1(X, \mathbf{Q}_l/\mathbf{Z}_l)^* \cong H^{2d}(X, \mathbf{Z}_l(d)).$$

The group on the bottom right may be computed by using the Hochschild-Serre spectral sequence $(G = \mathrm{Gal}(\overline{\mathbf{F}}/\mathbf{F}))$:

$$H^r(G, H^s(\overline{X}, \mathbf{Z}_l(d))) \Rightarrow H^{r+s}(X, \mathbf{Z}_l(d)).$$

Since G is of cohomological dimension one for profinite modules (because it is for torsion modules), we get the short exact sequence:

$$0 \to H^1(G, H^{2d-1}(\overline{X}, \mathbf{Z}_l(d))) \to H^{2d}(X, \mathbf{Z}_l(d)) \to H^{2d}(\overline{X}, \mathbf{Z}_l(d)) \to 0.$$

The following diagram is commutative:

$$
\begin{array}{ccccccc}
0 \to & \pi_1^{ab}(X)^0 & \to & \pi_1^{ab}(X) & \to & \mathrm{Gal}(\overline{\mathbf{F}}/\mathbf{F}) & \to 0 \\
& \downarrow & & \downarrow & & \downarrow & \\
0 \to & H^1(G, H^{2d-1}(\overline{X}, \mathbf{Z}_l(d))) & \to & H^{2d}(X, \mathbf{Z}_l(d)) & \to & H^{2d}(\overline{X}, \mathbf{Z}_l(d)) & \to 0
\end{array}
$$

The middle verticle arrow is the projection of $\pi_1^{ab}(X)$ onto its pro-l component followed by the identification $(*)$ above. The left vertical arrow is defined to make the diagram commutative. The group on the bottom right is isomorphic to $\mathbf{Z}_l$, and the right vertical arrow is bijective on the pro-l-components. Hence, the left vertical arrow is an isomorphism on pro-l components. Thus it will suffice to show that the bottom left group in this diagram is finite for all l and zero for almost all l. Let A be the Albanese variety of X and put $T_l = \varprojlim_m A(\overline{\mathbf{F}}))$. Then $H^{2d-1}(\overline{X}, \mathbf{Z}_l(d))$ is an extension of a finite group by T_l. Now for any torsion or profinite G-module M, $H^1(G, M)$ is just the coinvariants of G acting on M, and hence it suffices to show that the coinvariants of G acting on $T_l(A)$ is finite and zero for almost all l. This is the same argument used in the proof of Proposition 4.8. This concludes the proof of Step 1.

STEP 2. The image of θ is dense in $\pi_1^{ab}(X)$ and θ_0 is surjective.

PROOF. Suppose that the image of θ were not dense in $\pi_1^{ab}(X)$. Consider the map obtained from θ by taking Pontryagin duals

$$\theta^*: H^1(X, \mathbf{Q}/\mathbf{Z}) \to \prod_{x \in X_0} \mathbf{Q}/\mathbf{Z}.$$

Then θ^* is not injective. Now θ^* is given by taking an abelian covering Y of X and restricting it to closed points. Then to say that θ^* is not injective is to say that there exists a finite abelian, connected, étale covering Y of X whose restriction to all closed points of X is trivial. But this contradicts Lemma 1.7. Hence, the image of θ is dense in $\pi_1^{ab}(X)$, and one sees easily that the same is true for the image of θ_0. But θ_0 has target a finite group by Step 1, so its image must be all of $\pi_1^{ab}(X)^0$. This completes the proof of Step 2.

STEP 3. Assume X is a surface. There exists an injective map

$$\lambda: A_0(X) \to \pi_1^{ab}(X)^0.$$

By Step 1, this proves that $A_0(X)$ is finite.

PROOF ([PA], [CTR2]). Let l be a prime number, which we will take to be different from the characteristic of $\mathbf{F}$ for the moment. We define λ_l as a composite of four maps:

$$\lambda_l: A_0(X)\{l\} \to NH^3(X, \mathbf{Q}/\mathbf{Z}_l(2)) \to H^3(X, \mathbf{Q}_l/\mathbf{Z}_l(2))$$
$$\to H^4(X, \mathbf{Z}_l(2))\{l\} \to \pi_1^{ab}(X)^0\{l\}.$$

The first map

$$A_0(X)\{l\} \to NH^3(X, \mathbf{Q}_l/\mathbf{Z}_l(2))$$

is defined as follows. By the Merkur'ev-Suslin Theorem [MS] and the work of Bloch, there is an exact sequence (see Chapter 1, §4):

$$(*): 0 \to H^1(X, \mathscr{K}_2) \otimes \mathbf{Q}_l/\mathbf{Z}_l \to NH^3(X, \mathbf{Q}_l/\mathbf{Z}_l(2)) \to CH^2(X)\{l\} \to 0,$$

where

$$NH^3(X, \mathbf{Q}/\mathbf{Z}_l(2)) = \mathrm{Ker}[H^3(X, \mathbf{Q}_l/\mathbf{Z}_l(2)) \to H^3(k(X), \mathbf{Q}_l/\mathbf{Z}_l(2))].$$

By an argument involving the Weil conjectures, $H^3(X, \mathbf{Q}_l/\mathbf{Z}_l(2))$ is finite and zero for almost all l. The first nonzero group in the exact sequence $(*)$ is therefore a finite l-divisible group, hence it is zero. Hence we have a bijection between the last two nonzero groups. The second map is the natural inclusion. The third map is induced from the Bockstein map:

$$H^3(X, \mathbf{Q}_l/\mathbf{Z}_l(2)) \to H^4(X, \mathbf{Z}_l(2)),$$

and since the group on the left is finite, it is an isomorphism onto the l-primary part of the group on the right. Finally, the fourth map is induced from Poincaré duality at the level of X and the identification (definition) $\pi_1^{ab}(X)^* = H^1(X, \mathbf{Q}/\mathbf{Z})$. Since λ_l is the composite of injective maps, it is injective and $A_0(X)\{l\}$ is finite. For the p-part ($p = $ char. $\mathbf{F}$), see [CTR2] and [Gro].

STEP 4. θ_0 is an isomorphism if X is a surface.

PROOF. Consider the l-primary component θ_l of θ. By Step 1, we know that this map is surjective. By Step 3, we know that $A_0(X)\{l\}$ is finite. Hence, the order of $A_0(X)\{l\}$ is less than or equal to the order of $\pi_1^{ab}(X)^0\{l\}$. On the other hand, λ_l is injective, hence the order of these two groups must be the same and both are isomorphisms.

STEP 5. Show that θ_0 is an isomorphism for X of any dimension.

PROOF. Assume that X is of dimension $d \geq 3$ and that the result is true for varieties of dimension less than d. Take some $\alpha \in \mathrm{Ker}\,\theta$. Let l_1, l_2 be distinct prime numbers. By the Bertini theorems [AK], [C], there exist smooth hyperplane sections Y_i defined over finite extensions $\mathbf{F}_i/\mathbf{F}$ of degree a power of l_i and passing through the support of α. Then we have a commutative diagram

$$
\begin{array}{ccc}
A_0(Y_i) & \longrightarrow & \pi_1^{ab}(Y_i)^0 \\
\downarrow & & \downarrow \\
A_0(X) & \longrightarrow & \pi_1^{ab}(X)^0
\end{array}
$$

By the weak Lefschetz theorem or by ([SGA 2] Exp. XII, Cor. 3.5), the right vertical map is an isomorphism. This, the diagram above, and the induction hypothesis show that α is killed by the degree of $\mathbf{F}_i/\mathbf{F}$ in $A_0(X)$ for $i = 1, 2$. Hence, α is zero. This completes the proof of the reciprocity law in the geometric case.

5.3. Proof of the reciprocity law in the arithmetic case. We outline the steps used to prove that $\tilde{\theta}$ is an isomorphism in the case where X is 2-dimensional. Unfortunately, we have not succeeded in reducing the general case to the 2-dimensional case as we did in the geometric case. The general case can be found in a paper of Saito [Sa3], and the technique is quite similar to the 2-dimensional case.

In the case where X is a smooth arithmetic surface, Bloch [B2] proved Theorem 5.7, and the case of a general arithmetic surface was proved by Kato and Saito. Our proof of the 2-dimensional case is due to Kato and is different from his proof in the paper with Saito [KS2]. The big difference is the use of a result of Jannsen to simplify the proof. As in case (G), we list the main steps.

STEP 1 (Katz-Lang). Prove the finiteness of $\pi_1^{ab}(X)$.

STEP 2 (Jannsen). Prove a "Hasse principle", which implies that $\tilde{\theta}$ is an isomorphism modulo n, for any positive integer n.

STEP 3 (Bloch). Prove the finiteness of $CH_0(X)$ and conclude.

PROOF OF STEP 1. Put $S = \operatorname{Spec} \mathcal{O}_k$. It is well known that $\pi_1^{ab}(S)$ is finite, and we will show that $\pi_1^{ab}(X/S) := \operatorname{Ker}(\pi_1^{ab}(X) \to \pi_1^{ab}(S))$ is finite. To do this, consider the following commutative diagram:

$$
\begin{array}{ccc}
H^1(\mathcal{O}_k, \mathbf{Q}/\mathbf{Z}) & \longrightarrow & H^1(k, \mathbf{Q}/\mathbf{Z}) \\
\downarrow & & \downarrow \\
H^1(X, \mathbf{Q}/\mathbf{Z}) & \longrightarrow & H^1(X_\eta, \mathbf{Q}/\mathbf{Z}).
\end{array}
$$

Since X and $\mathcal{O}_k$ are regular schemes, the horizontal maps are injective. Taking Pontryagin duals, we get the commutative diagram with surjective horizontal maps

$$
\begin{array}{ccc}
\pi_1^{ab}(X_\eta) & \longrightarrow & \pi_1^{ab}(X) \\
\downarrow & & \downarrow \\
\pi_1^{ab}(k) & \longrightarrow & \pi_1^{ab}(\mathcal{O}_k).
\end{array}
$$

Assume for a moment that X has a section over S. Then we have splittings

$$
\pi_1^{ab}(X) = \pi_1^{ab}(\mathcal{O}_k) \oplus \pi_1^{ab}(X/S)
$$
$$
\pi_1^{ab}(X_\eta) = \pi_1^{ab}(k) \oplus \pi_1^{ab}(X_\eta/k).
$$

From these splittings and the diagram above, it is enough to show that $\pi_1^{ab}(X_\eta/k)$ is finite. Now do not assume that $f: X \to S$ has a section, and let T be a finite extension of S over which there is a section. Then using what we have just done, we can descend this finiteness from X_T to X, because the cokernel of the natural map $\pi_1^{ab}(X_T) \to \pi_1^{ab}(X)$ is finite. Hence, we are reduced to showing that $\pi_1^{ab}(X_\eta/k)$ is finite. By taking Pontryagin duals of the groups in the exact sequence:

$$
0 \to H^1(k, \mathbf{Q}/\mathbf{Z}) \to H^1(X_\eta, \mathbf{Q}/\mathbf{Z}) \to H^1(\overline{X}_\eta, \mathbf{Q}/\mathbf{Z})^{\operatorname{Gal}(\bar{k}/k)},
$$

we see that it is enough to show that $H^1(\overline{X}_\eta, \mathbf{Q}/\mathbf{Z})^{\operatorname{Gal}(\bar{k}/k)}$ is finite. Let $\mathfrak{p}$ be a prime ideal of $\mathcal{O}_k$ modulo which X has good reduction, and let $\bar{\mathfrak{p}}$ be a geometric point of $\mathcal{O}_k$ lying over $\mathfrak{p}$. By the smooth and proper base change theorem, we have

$$
H^1(\overline{X}_\eta, \mathbf{Q}/\mathbf{Z})^G \cong H^1(\overline{Y}, \mathbf{Q}/\mathbf{Z}.)^{\operatorname{Gal}(\overline{\mathbf{F}}/\mathbf{F})},
$$

modulo p-torsion, where p denotes the residue characteristic of $\mathbf{F} = \mathcal{O}/\mathfrak{p}$ and Y is a good reduction modulo $\mathfrak{p}$. By what we proved in the geometric case (see Step 1 of case (G) above), this latter group is finite. Specializing at another prime not lying above p modulo which X has good reduction, we get the finiteness of the p-primary part. This completes the proof of Step 1.

STEP 2. A "Hasse Principle". We need the following theorem of Jannsen:

THEOREM 5.8 (Jannsen [J1].)*Let X be a smooth, projective variety over an algebraic number field k. Let l be a prime number and S a finite set of places of k including the archimedean places, the places of bad reduction for X and places lying above l. Denote by M the maximal divisible subgroup of the $\mathrm{Gal}(\bar{k}/k)$-module $H^i(\overline{X}, \mathbf{Q}_l/\mathbf{Z}_l(j))$. Then if $i \neq 2(j-1)$, the natural map*

$$H^2(k, M) \to \bigoplus_{v \in S} H^2(k_v, M)$$

is an isomorphism.

REMARK 5.9. We could have stated the theorem as saying that $H^2(k, M)$ is isomorphic to the direct sum of all of the $H^2(k_v, M)$. However, if v is a place not in S, then one can show that $H^2(k_v, M) = 0$.

We now describe some of the main technical points of the proof. In the following, X denotes an arithmetic surface, that is, a regular, connected, proper scheme of finite type over $\mathbf{Z}$, whose function field is of transcendence degree one over $\mathbf{Q}$. The generic fibre of X will be denoted by X_η, which we regard as a smooth, projective, geometrically connected curve over an algebraic number field k.

Let $D_n(X_\eta)$ be the image of $H^3(X, \mathbf{Z}/n(2))$ in $H^3(k(X), \mathbf{Z}/n(2))$ and put

$$D(X_\eta) = \varinjlim_n D_n(X_\eta).$$

LEMMA 5.10. *The n-torsion subgroup of $D(X_\eta)$ is exactly $D_n(X_\eta)$.*

PROOF. This follows easily from the divisibility of $H^2(k(X), \mathbf{Q}/\mathbf{Z}(2))$, which follows in turn from the surjectivity of the Galois symbol

$$K_2 k(X) \to H^2(k(X), \mathbf{Q}/\mathbf{Z}(2)).$$

By the Merkur'ev-Suslin theorem and the work of Bloch, we have an exact sequence:

$$(**)_k \quad 0 \to H^1(X_\eta, \mathscr{K}_2) \otimes \mathbf{Q}/\mathbf{Z} \to H^3(X_\eta, \mathbf{Q}/\mathbf{Z}(2)) \to D(X_\eta) \to 0.$$

Let v be a place of k, k_v the completion and $X_{\eta_v} = X x_k k_v$. Defining the relevant groups for X_{η_v}, we get exact sequences which we shall refer to as $(**)_v$. The following lemma is crucial .

LEMMA 5.11. *Let S be as above. For all $v \notin S$, $D_n(X_{\eta_v}) = 0$, and the natural map*

$$D_n(X_\eta) \to \bigoplus_{v \in S} D_n(X_{\eta_v})$$

is an isomorphism.

We will just give an idea of the proof. It is a bit technical, but the idea is reasonably simple. The "important part" of $D(X_\eta)$ will be the contribution from $H^2(k, H^1(\overline{X}, \mathbf{Q}/\mathbf{Z}(2)))$, and by Jannsen's Hasse principle, this group is isomorphic to the sum of the local groups. The technical parts of the lemma are just a reduction to this key point.

Once we have the lemma, we can make headway towards the proof of the main theorem. For v a finite place of K, let $Y_v = X x_{\mathscr{O}_k} \mathbf{F}_v$. Recall [Sh] that there is an exact sequence:

$$\text{(S)} \quad H^1(X_\eta, \mathscr{K}_2) \to \bigoplus_v CH_0(Y_v) \to CH^2(X) \to 0.$$

Let S be a finite set of places of k including the bad reduction places of X_η, and let $V = \operatorname{Spec}\mathscr{O}_k - S$, $U = f^{-1}(V)$ where $f: X \to \operatorname{Spec}\mathscr{O}_k$ is the structure morphism. Then U is smooth over V. Define $A_v = H^1(X_{\eta_v}, \mathscr{K}_2)/n$, $A'_v = CH_0(Y_v)/n$, $B_v = H^3(X_{\eta_v}, \mathbf{Z}/n(2))$, $B'_v = H^2(Y_v, \mathbf{Z}/n(1))$. Consider the following diagram with exact rows:

$$
\begin{array}{ccccccccc}
0 & \longrightarrow & H^1(X_\eta, \mathscr{K}_2)/n & \longrightarrow & H^3(X, \mathbf{Z}/n(2)) & \longrightarrow & D_n(X_\eta) & \longrightarrow & 0 \\
& & \downarrow a_n & & \downarrow b_n & & \downarrow c_n & & \\
0 & \longrightarrow & \left(\bigoplus_{v \in S} A_v\right) \oplus \left(\bigoplus_{v \notin S} A'_v\right) & \longrightarrow & \left(\bigoplus_{v \in S} B_v\right) \oplus \left(\bigoplus_{v \notin S} B'_v\right) & \longrightarrow & \bigoplus_v D_n(X_{\eta_v}) & \longrightarrow & 0
\end{array}
$$

By the lemma, we have $\operatorname{Coker} a_n \cong \operatorname{Coker} b_n$.
Using duality theory, one shows:

LEMMA 5.12. $\operatorname{Coker} b_n \cong \pi_1^{\mathrm{ab}}(U)/n$.

Assume for the moment that $CH_0(X)$ is finitely generated. Then we can prove the reciprocity law for X. By step 1 and Lemma 1.7, we know that the reciprocity map is surjective. Thus, it suffices to show that the dual map

$$\widetilde{H}^1(X, \mathbf{Q}/\mathbf{Z}) \to \operatorname{Hom}(CH_0(X), \mathbf{Q}/\mathbf{Z}),$$

where $\widetilde{H}^1(X, \mathbf{Q}/\mathbf{Z})$ is the Pontryagin dual of $\widetilde{\pi_1^{\mathrm{ab}}}(X)$ (see just before Proposition 5.6), is surjective. So let α be an element of $\operatorname{Hom}(CH_0(X), \mathbf{Q}/\mathbf{Z})$. By the localization sequence, α comes from a homomorphism

$$\bigoplus_v CH_0(Y_v) \to \mathbf{Q}/\mathbf{Z},$$

which is trivial on the image of $H^1(X_\eta, \mathscr{K}_2)$. Composing with the map

$$H^1(X_{\eta_v}, \mathscr{K}_2) \to CH_0(Y_v),$$

we get a homomorphism

$$\left(\bigoplus_{v \in S} H^1(X_{\eta_v}, \mathscr{K}_2) \oplus \bigoplus_{v \notin S} CH_0(Y_v) \right) \to \mathbf{Q}/\mathbf{Z}$$

which is trivial on the image of $H^1(X_\eta, \mathscr{K}_2)$. By the diagram above and Lemma 5.12, α comes from an element β of $H^1(U, \mathbf{Q}/\mathbf{Z})$. If $v \in S$, the image of this element in $H^1(X_{\eta_v}, \mathbf{Q}/\mathbf{Z})$ corresponds to a homomorphism $H^1(X_{\eta_v}, \mathscr{K}_2) \to \mathbf{Q}/\mathbf{Z}$ which is trivial on

$$\operatorname{Ker}(H^1(X_{\eta_v}, \mathscr{K}_2) \to CH_0(Y_v)),$$

hence, by Corollary 4.22, β extends to $H^1(X_v, \mathbf{Q}/\mathbf{Z})$. But an element of $H^1(U, \mathbf{Q}/\mathbf{Z})$ which comes from $H^1(X_v, \mathbf{Q}/\mathbf{Z})$ for $v \in S$ comes from $H^1(X, \mathbf{Q}/\mathbf{Z})$. We leave it to the reader to show that β comes from $\tilde{H}^1(X, \mathbf{Q}/\mathbf{Z})$.

STEP 3. Show that $CH_0(X)$ is finite.

First we state and prove a technical result which will be important to the proof. This is not simply a result, but an important technique in the subject. It goes back to C. Moore [Mo], was refined and reinterpreted by Chase-Waterhouse [CW], generalized by Bloch [B2] and reinterpreted again by Somekawa [So]. It might be worth it to take some time to explain it in detail. Let k be a number field. Then for each place v of k, we have the Galois (Hilbert) symbol (see Chapter 1, §1 for more details):

$$\alpha_v \colon K_2 k_v \to \mu(k_v).$$

Consider the sequence

(M) $K_2 k \to \bigoplus_v \mu(k_v) \to \mu(k) \to 0$.

Here the first map is the sum over all places v of the local Galois symbols and the second map is given by taking a family of roots of unity (ζ_v) and sending it to $\prod \zeta_v^{n_v/n}$, where n_v is the number of roots of unity in k_v and n the number of roots of unity in k. A beautiful theorem of C. Moore [Mo] states that sequence (M) is exact. Another elementary proof of this result was given by Chase-Waterhouse [CW]. Inspired by this latter proof, Bloch [B2] found a generalization of this result to the case of a smooth projective curve over a number field and the roots of unity are replaced by fundamental groups.

We now explain Bloch's exact sequence in the case of a curve. The exposition here is a compositum of arguments in [B2] and [KS2]. Let S be a finite set of places of k containing all the places for which $T_{G_{k_v}}$ are infinite groups and let n be any positive integer which annihilates the finite group T_{G_k}.

THEOREM 5.13. *There is an exact sequence*

(B) $\quad V(X_\eta) \xrightarrow{\partial} \bigoplus_{v \in S} V(X_{\eta_v})/n \oplus \bigoplus_{v \notin S} T_{G_{k_v}} \to T_{G_k} \to 0$.

Before giving the proof, we note that this result implies that $CH_0(X)$ is finite. This can be seen from the localization sequence (S) following Lemma 5.11 and Lemma 5.14 below.

PROOF OF THEOREM 5.13. First we show that the image of ∂ lands in the direct sum (it goes *a priori* into the direct product). We need the following lemma:

LEMMA 5.14. *Let U be an nonempty open subscheme of $\operatorname{Spec} \mathscr{O}_k$ such that for all $v \in U$, $e_v = 1$, where e_v is the absolute ramification index of k_v. Then we have*

$$(T)_{G_{k_v}} = (T)_{G_{\mathbf{F}_v}}$$

for all $v \in U$.

PROOF. This follows immediately from Proposition 4.25.

Now class field theory for smooth projective curves over finite fields and the isomorphism given by the lemma imply that for $v \in U$ as in the lemma,

$$T_{G_{k_v}} = CH_0(Y_v)_{\text{tors}}.$$

But the image of the natural map

$$V(X) \to \prod_v CH_0(Y_v)$$

lies in the direct sum, and this proves that the image of ∂ lies in the direct sum as well.

Let n be the order of the finite group T_{G_k} and let S be as above. We take a modulus $\mathscr{M} = \sum n_i v_i$, where the sum runs over v in S and n divides n_i for all i. Consider the map

$$f: V(X_\eta) \to \bigoplus_{v \in S} V(X_{\eta_v})/n;$$

we denote by $V(X_\eta, \mathscr{M})$ its kernel. Using approximation (see Lemma 5.20 below), one shows that

LEMMA 5.15. *The map f is surjective.*

By this lemma, it suffices to prove exactness of the sequence:

(B′) $\quad V(X_\eta, \mathscr{M}) \to \bigoplus_{v \notin S} T_{G_{k_v}} \to T_{G_k} \to 0$.

It also follows from this lemma that we can extend S, if necessary.

LEMMA 5.16. *The sequence (B′) is a complex.*

PROOF. Let n be a positive integer which kills the finite group T_{G_k} and set $M = H^1(X_{\bar{\eta}}, \mathbf{Z}/n(2))$. By Poitou-Tate duality (see Chapter 1, §6), there is an exact sequence

$$H^2(k, M) \to \bigoplus_{\text{all } v} H^2(k_v, M) \to H^0(k, \operatorname{Hom}(M, \mathbf{Q}/\mathbf{Z}(1)))^* \to 0.$$

Just as in the discussion preceding Proposition 4.8, we can define a map:

$$\tau\colon V(X) \to H^2(k, M).$$

Given an element a of $V(X_\eta, \mathscr{M})$, it goes to zero in $H^2(k_v, M)$ for all $v \in S$ by definition, and $\tau(a)$ goes to zero in $H^0(k, \operatorname{Hom}(M, \mathbf{Q}/\mathbf{Z}(1)))^* = T_{G_k}/n$), since n kills T_{G_k}. This shows that the sequence (B') is a complex.

LEMMA 5.17. *Let L be a finite extension of k. Then exactness of (B') for X_{η_L} implies exactness for X_η.*

PROOF. Put $G_L = \operatorname{Gal}(\bar{k}/L)$. Then we can find a finite set of primes S' of L and a modulus $\mathscr{M}'$ with support in S' such that the following diagram commutes:

$$
\begin{array}{ccccccc}
V(X_{\eta_L}, \mathscr{M}') & \longrightarrow & \displaystyle\bigoplus_{w \notin S'} T_{G_{L_w}} & \longrightarrow & T_{G_L} & \longrightarrow & 0 \\
\downarrow & & \downarrow{\scriptstyle N_2} & & \downarrow{\scriptstyle N_3} & & \\
V(X_\eta, \mathscr{M}) & \longrightarrow & \displaystyle\bigoplus_{v \notin S} T_{G_{k_v}} & \longrightarrow & T_{G_k} & \longrightarrow & 0 \\
& & \downarrow & & \downarrow & & \\
& & 0 & & 0 & &
\end{array}
$$

To prove the lemma, it will suffice to show that the map $\operatorname{Ker} N_2 \to \operatorname{Ker} N_3$ is surjective. Now $\operatorname{Ker} N_3$ is generated by elements of the form $(1 - g)t$ for $t \in T$. Let G' be the intersection of the conjugates of G_L in G_k. By Čebotarev density, there is a place $v \notin S$ and a v' on $\bar{k}^{G'}$ such that the Frobenius associated to v'/v is equal to g. Let w be the place of L lying under v'. Then $(1 - g)t \bmod (1 - G_{L_w})$ lies in $\operatorname{Ker}(T_{G_{L_w}} \to T_{G_{k_v}})$. This completes the proof of the lemma.

DEFINITION 5.18. Let U be an open subscheme of $\operatorname{Spec} \mathscr{O}_k$. We call a point v of U a Bloch point (resp., an l-Bloch point) if the natural map $T_{G_{k_v}} \to T_{G_k}$ is an isomorphism (resp., an isomorphism on l-primary components).

LEMMA 5.19. *For any nonempty U, there exists U'/U unramified such that U' has a Bloch point.*

PROOF. Take any $v \in U$ at which X has good reduction and consider the group $M = H^1(\overline{X}, \mathbf{Q}/\mathbf{Z})$. Then T is the Pontryagin dual of M so $T_H = (M^H)^*$ for any group H which acts on these modules. Since the group

T_{G_k} is finite, there exists a finite extension L/k contained in k_v such that $M^{G_{k_v}} = M^{G_L}$. Let w be the place of L defined by the embedding of L into k_v. There exists a connected étale scheme U'/U with function field L such that $w \in U'_0$. Then w is a Bloch point of U'.

Let l be a prime number. To prove exactness of (B'), it will suffice to prove exactness of the l-primary components for each prime l. Using the last lemma to replace k by a finite extension if necessary, we may assume that the Galois representation $\rho: G_k \to \mathrm{Aut}_{\mathbf{Z}_l} T_l$ has image in $1 + l\,\mathrm{End}_{\mathbf{Z}_l}(T_l)$, k is totally imaginary, and there exists a Bloch point of $\mathcal{O}_L$. Let l^{m-1} annihilate the l-primary part of T_{G_k} and let $H = \rho^{-1}(1 + l^m\,\mathrm{End}\,T_l)$, $\mathscr{G} = G/H$. Then $\mathscr{G}$ is a finite l-group and there exists a nonempty subset W of $\mathscr{G}$ such that for any $a \in W$ and any $g \in Ha$, we have $T_{l,\langle g \rangle} = T_{l,G_k}$. W is a union of conjugacy classes.

To prove exactness, we will need the following approximation lemma:

LEMMA 5.20. *Let k be a number field, let X be a smooth projective variety over k and let S be a finite set of places of k including the places of bad reduction of X. Given a finite set of places $R = \{v_1, \ldots, v_s\}$ of k which is disjoint from S, positive integers $n_1, \ldots, n_s$ and points $b_i \in X(k_{v_i})$ $(i = 1, 2, \ldots, s)$, there exists a finite extension L/k, places $w_1, \ldots, w_s$ of L with $L_{w_i} \cong k_{v_i}$ and a point $b \in X(L)$ such that*

$$b \equiv b_i \bmod m_i^{n_i}$$

for all i. Here m_i is the maximal ideal of $\mathcal{O}_{k_{v_i}}$.

This is proved by Bloch ([B2], Lemma 3.3) as a consequence of Hilbert's irreducibility theorem.

Now we prove the exactness of (B'). By Corollary 4.24, the group $\bigoplus_{v \notin S} T_{G_{k_v}}$ is generated by elements of the form $\tau_v[b_v, c_v\}$, where $[b_v, c_v\}$ is defined just before Lemma 4.6. We will show by using approximation that any such element is congruent modulo the image of θ to an element of $T_{G_{k_v}}$ for some $v \notin S$ such that $T_{G_{k_v}} = T_{G_k}$. This will prove exactness. Let J be the Jacobian of X. We take an element $\gamma = \sum_i [b_{v_i}, c_{v_i}\}$ with $b_i(J(k_{v_i})$ and $v_i(c_{v_i}) = 1$. Choose places u_j of k, not in S or among the v_i such that the Frobenius conjugacy classes fill out the conjugacy classes of $\mathscr{G}$. By the approximation lemma, we find a finite extension L of k, places $w_1, \ldots, w_{r+s}$ of L and an element $b' \in J(L)$ such that

$$L_{w_i} = k_{v_i}, \qquad i = 1, \ldots, r;$$
$$L_{w_i} = k_{u_{i-r}}, \qquad i = r+1, \ldots, s;$$
$$b'_{u_i} \equiv b_i \bmod \pi_i, \qquad i = 1, \ldots, r;$$
$$b'_{u_i} \equiv 0 \bmod \pi_i, \qquad i = r+1, \ldots, r+s.$$

We need to choose an appropriate c' in L^* so that $[b', c'\}$ approximates γ. Towards this end, fix a modulus of L

$$\mathcal{N} = \sum_{\substack{z \mid v \\ v \in S}} t_z z$$

such that if $c' \cong 1 \mod \mathcal{N}$, then $c'_z \in L_z^{*n_z}$, where $\mathcal{M} = \sum n_v v$ is the original modulus. For such a c', we have $[b', c'\} \in V(X_\eta, \mathcal{M})$. We write (c') for the divisor of c' and we denote by R the set of places of k lying under the primes u_i $(i = 1, \ldots, r+s)$ of k'.

LEMMA 5.21. *Given integers* a_i $(i = r+1, \ldots, r+s)$, *there exists* $c' \in L^*$ *such that*

$$(c') = \sum_{i=1}^{r}(u_i) + \sum_{i=r+1}^{r+s} a_i(u_i) + (\bar{u}),$$

for some place $\bar{u}$ *of* L *dividing a* $\bar{v}$ *which is not in* S *or* R.

PROOF. Consider the divisor

$$D = \sum_{i=1}^{r}(u_i) + \sum_{i=r+1}^{r+s} a_i(u_i).$$

Let $L_{\mathcal{N}}$ be the ray class field modulo $\mathcal{N}$. This is a finite abelian extension of L unramified outside the primes occuring in $\mathcal{N}$, and by global class field theory for number fields, $\mathrm{Gal}(L_{\mathcal{N}}/L)$ is isomorphic to

$$\left(\bigoplus_{w \notin \mathrm{supp}\,\mathcal{N}} \mathbf{Z} \right) / P_{\mathcal{N}},$$

where $P_{\mathcal{N}}$ is the subgroup of L^* consisting of units congruent to 1 modulo $\mathcal{N}$. By Čebotarev density, there exists a prime $(\bar{u})$ of L which does not lie over a prime in S or R whose image under the Artin map is equal to the image of $-D$ in $\mathrm{Gal}(L_{\mathcal{N}}/L)$. Hence, there exists $(c') \in L^*$ such that $c' \equiv 1 \pmod{\mathcal{N}}$ and

$$(c') = \sum_{i=1}^{r}(u_i) + \sum_{i=r+1}^{r+s} a_i(u_i) + (\bar{u}).$$

Our task is to choose the integers a_i in the lemma so that the Frobenius conjugacy class of $(\bar{u})$ lies in W, hence $(\bar{u})$ is a Bloch point of $\mathrm{Spec}\,\mathcal{O}_L$. This will prove the exactness of the sequence (B').

LEMMA 5.22. *Let* g_j, $j = 1, \ldots, s$ *be elements of* $\mathcal{G}$ *whose conjugacy classes* $[g_j]$ *fill out the conjugacy classes of* $\mathcal{G}$. *Then the* g_j *generate* $\mathcal{G}$.

PROOF. This is true for any finite group $\mathcal{G}$. Let $\mathcal{H}$ be the subgroup generated by the g_j. If $\mathcal{H}$ is normal, it must be all of $\mathcal{G}$, so we assume it is not normal. Then we have

$$\mathscr{G} = \bigcup_{g \in \mathscr{G}/\mathscr{H}} g\mathscr{H}g^{-1}.$$

But this is ridiculous because the order of the right-hand side is less than the order of $\mathscr{G}$ because the $g\mathscr{H}g^{-1}$ are not disjoint.

Let $G_L = \mathrm{Gal}(\bar{k}/L)$, and let $G_{\mathscr{N}}$ be the ray class group modulo $\mathscr{N}$ (see the proof of Lemma 5.21). Put

$$P = \mathrm{Ker}(G_L \to G_{\mathscr{N}}), \quad Q = \mathrm{Ker}(G_L \subset G_k \to \mathscr{G}).$$

Then there is an exact sequence of groups

$$1 \to P \cap Q \to G_L \to G_{\mathscr{N}} \times \mathscr{G}.$$

Fix $\omega \in W$, and let $\theta \in G_L$ map to $\sum_{i=1}^{r}(u_i) \in G_{\mathscr{N}}$. Let $\bar{\theta}$ be the image of θ in $\mathscr{G}$. Let g_j be Frobenius conjugacy classes mapping to $u_{j+r} \in G_{\mathscr{N}}$ for $j = 1, \ldots, s$. Since $L_{u_{j+r}} \cong k_{w_j}$ and the w_j fill out the Frobenius conjugacy classes of $\mathscr{G}$, by Lemma 5.22, there exists a word α in the g_j mapping to $\bar{\theta}^{-1}\omega \in \mathscr{G}$. Let $a = \theta\alpha \in G_L$. Then a maps to $\sum_{i=1}^{r}(u_i) + \sum_{i=r+1}^{r+s} a_i(u_i)$, for some $a_i \in \mathbf{Z}$, and to $\omega \in \mathscr{G}$. By Cebotarev density, there exists a place $(\bar{u})$ of L not lying over S such that the Frobenius conjugacy class of $(\bar{u})$ meets $a^{-1}(P \cap Q)$. Hence, there exists $c' \in L^*$, $c' \cong 1 \pmod{\mathscr{N}}$ such that

$$(c') = \sum_{i=1}^{r}(u_i) + \sum_{i=r+1}^{r+s} a_i(u_i) + (\bar{u}).$$

The Frobenius conjugacy class of $(\bar{u})$ maps to ω, so $(\bar{u})$ is a Bloch point for $\mathrm{Spec}\,\mathscr{O}_L$. This completes the proof of Theorem 5.13.

REMARKS 5.23. (i) In [KS3, Theorem 5.4], a higher dimensional generalization of Bloch's exact sequence is proved. The proof uses some approximation lemmas to reduce to the 1-dimensional case. We will not give the details here.

(ii) In the recent thesis of M. Spiess [Sp], a generalization of Artin-Verdier duality is proved for arithmetic surfaces. This together with the finiteness theorem for $CH_0(X)$ just proved, gives another proof of the reciprocity law.

Chapter 6. Global class field theory of arithmetic schemes

In this section we state and prove the global reciprocity law for fields which are finitely generated over the prime subfield. For ease of exposition, we shall exclude the p-primary part in characteristic $p > 0$. We begin by stating the general reciprocity law in a geometric way. To many readers, this may look nothing like what they know in the one-dimensional case. To motivate it better, we briefly review the presentation of the one-dimensional case via idèles (which is the way most people learn it) and then by ideals (the original presentation). Then we unravel the geometric statement of the reciprocity law in the one-dimensional case so that we can compare it to the other presentations. After that, we hope that the reader will be better prepared for what lies ahead. Those who are quite happy with the geometric definition right away are invited to go directly to the proof. Sections 6.4–6.7 are entirely based on unpublished notes of Kato-Saito. Any errors are probably due to the author of the present paper.

6.1. Statement of the reciprocity law. Let X be a proper integral scheme of finite type over $\mathbf{Z}$ with function field K. We take a base scheme T together with a finite map $f: X \to T$. For example, we may take $T = X$ (see Remark 6.10 below for more on the role of T). We denote by $\mathbf{F}$ the algebraic closure of the prime field in K. If K is of characteristic zero, let Σ be the set of archimedean places of $\mathbf{F}$. Define

$$D_I(X/T) = H_\Sigma^d(T, f_* \mathscr{K}_d^M(\mathscr{O}_X, I)),$$

where the subscript Σ denotes "cohomology with compact support" as defined in Chapter 1. If K is of positive characteristic, then Σ plays no role.

In section 6.4 below, we define the reciprocity map, and in Sections 6.4–6.7 we prove:

THEOREM 6.1. *Assume* char. $K = 0$. *Then*

(1) *For any nonzero ideal I of $\mathscr{O}_X$, the group $D_I(X/T)$ is finite;*

(2) *The reciprocity map $\varprojlim_{I \neq 0} D_I(X/T) \to \pi_1^{\mathrm{ab}}(K)$ is an isomorphism;*

(3) *For any regular open subscheme U of X, the reciprocity map induces an isomorphism*

$$\varprojlim_{I_U = \mathscr{O}_U} D_I(X/T) \cong \pi_1^{\mathrm{ab}}(U).$$

THEOREM 6.2. *Let X be a projective geometrically connected variety over a finite field $\mathbf{F}$. Let K be the function field of X. Denote by $\pi_1^{\mathrm{ab}}(K)^0$ the kernel of the natural map $\pi_1^{\mathrm{ab}}(K) \to \pi_1^{\mathrm{ab}}(\mathbf{F})$ and by $D_I(X/T)^0$ the kernel of the norm map $D_I(X/T) \to \mathbf{Z}$. Then*

(1) *For any nonzero I, the group $D_I(X/T)^0$ is finite;*

(2) *The reciprocity map, $\varprojlim_{I \neq 0} D_I(X/T)^0 \to \pi_1^{\mathrm{ab}}(K)^0$ is an isomorphism;*

(3) *Let U be a regular open subscheme of X. The reciprocity map induces an isomorphism*

$$\varprojlim_{I_U = \mathscr{O}_U} D_I(X/T)^0 \cong \pi_1^{\mathrm{ab}}(U)^0.$$

If X is only assumed to be proper, we have the same statements except we must replace the groups $D_I(X/T)^0$ by their prime-to-p subgroups and the Galois group by the quotient by its maximal pro-p-subgroup.

REMARK 6.3. There is a big difference between the structure of the Galois group of the maximal abelian extension of a number field and a function field over a number field. Recall that the Leopoldt conjecture predicts that a number field with r_2 nonconjugate complex places has $1 + r_2$ independent $\mathbf{Z}_l$-extensions for any prime l. The situation for a function field is totally different, as can be best summed up by the following:

PROPOSITION 6.4. *Let K be a field which is finitely generated over the prime field and l a prime number different from* char. K. *Let k be the algebraic closure of the prime field in K. Then for any $\mathbf{Z}_l$-extension L/K, there exists a $\mathbf{Z}_l$-extension k'/k such that $L = Kk'$.*

This is well-known for a function field in one variable over a finite field, and the general result was pointed out by Fein, Saltman and Schacher [FSS] and independently by the author. It is probably known to many people. The proof is an easy induction using the fact, due to Katz-Lang [KL], that there are no unramified $\mathbf{Z}_l$-extensions. Partly for this reason, we feel that doing analysis on higher-dimensional idèle groups is probably not the best way to try to prove the analytic continuation of L-functions for higher-dimensional schemes. Also, the topology on higher-dimensional idèle groups is very difficult to work with because it is not separated. Similar advice goes to those who seek to formulate a nonabelian generalization of the reciprocity law. For example, for a function field in one variable over a finite field, étale cohomological techniques are very powerful for proving properties of L-functions, and Laumon has a purely geometric formulation of the Langlands correspondence [La].

6.2. Idelic presentation of class field theory. Before entering into the higher-dimensional case, it might be helpful to review the one-dimensional case from several points of view. First, we shall present the usual way of describing the class field theory of global fields by using idèles. The higher-dimensional case may also be presented in this way (at least in dimension two; see [KS1] and [FP]), but the presentation is extremely complicated. Also, there is a very good reason to believe that there will be no corresponding good analytic theory as in the one-dimensional case (see Remark 6.3 above).

Let K be a global field, i.e., an algebraic number field of finite degree over $\mathbf{Q}$ or a function field in one variable over a finite field. We denote places of K by v and the completion of K at v by K_v. For v nonarchimedean, let $\mathscr{O}_v$ be the ring of integers in K_v. If v is an archimedean place, let $\mathscr{O}_v = K_v$. If S is a finite set of places of K containing the archimedean places, let K_S denote the ring of functions which are units outside S. Set

$$\mathbf{A}_S = \prod_{v \in S} K_v \times \prod_{v \notin S} \mathscr{O}_v$$

equipped with the product topology. $\mathbf{A}_S$ is a locally compact topological ring containing K_S embedded diagonally. This is the ring of S-adèles of K; $\mathbf{A}_K$ denotes the ring of adèles of K which is the direct limit of $\mathbf{A}_S$ as S runs over finite sets of places of K containing the archimedean places. The idèle group of K is the group of units of $\mathbf{A}_K$ and is denoted by $\mathbf{I}_K$. Its topology is not induced from that of A_K, but rather by viewing it as a subspace of $A_K \times A_K$ via the embedding $I_K \to A_K \times A_K$ which sends x to (x, x^{-1}). Then K^* embedded diagonally in I_K is a discrete subgroup; the quotient I_K/K^* is called the idèle class group of K and is denoted by C_K.

Let K^{sep} be a separable closure of K and let G be the absolute Galois group. For each place v, let G_v be a decomposition group. Denoting by ab the maximal abelian quotient, for each place v, we can view G_v^{ab} in a canonical way as a subgroup of G^{ab}. Then the local reciprocity map $K_v^* \to G_v^{ab}$ can be expanded into a global reciprocity map $I_K \to G^{ab}$. One form of the reciprocity law states that the image of K^* under this map is trivial, so we get a map $\rho: C_K \to G^{ab}$.

THEOREM 6.5. (i) *If K is a function field in one variable over a finite field* **F**, *then ρ is injective and its image is the subgroup of G consisting of all elements whose image in* $\mathrm{Gal}(\overline{\mathbf{F}}/\mathbf{F})$ *is an integral power of the Frobenius.*

(ii) *If K is a number field then ρ is surjective, and its kernel D_K is the connected component of identity of C_K.*

(iii) *Let L/K be a finite abelian extension. Then the reciprocity map induces an isomorphism*:

$$C_K/N_{L/K}C_L.$$

(iv) (*Existence Theorem*) *Given any open subgroup H of finite index in C_K, there exists a finite abelian extension L/K such that $N_{L/K}C_L = H$.*

REMARK 6.6. The connected component of identity of C_K is known to have the following structure ([AT], Chapter 9, Theorem 3): let r_1 be the number of real places of K and r_2 the number of nonconjugate complex places. Then D_K is the product of the real numbers, r_2 circles and $r_1 + r_2 - 1$ solenoids.

We now outline the theory of conductors. For each place v of $\mathscr{O}_K$, let m_v be the maximal ideal of the ring of integers $\mathscr{O}_v$. Let $M = (M_v)$ be a modulus, i.e. a family of nonnegative integers which are zero for almost all v. We denote the subgroup

$$\prod \mathscr{O}_v^{*M_v}$$

by I_K^M, and we put $C_K^m = K^* I_K^M/K^*$. For complex places of K, we use the convention that $\mathscr{O}_v^{*M_v} = \mathscr{O}_v^*$. For real places we take $M_v = 0$ or 1, and $\mathscr{O}_v^{*M_v} = \mathbf{R}^*$ if $M_v = 0$ and $\mathbf{R}^{*2}$ if $M_v = 1$. Then by the existence theorem in class field theory, there is a finite abelian extension K_M/K such that the reciprocity map

$$C_K \to \mathrm{Gal}(K^{ab}/K)$$

induces an isomorphism:

$$C_K/C_K^M \cong \mathrm{Gal}(K_M/K).$$

The field K_M is called the *ray class field modulo M*. Every finite abelian extension of K is contained in such a field, and the union of these fields over all moduli is K^{ab}. For the zero modulus, we get the Hilbert class field of K, which is the maximal abelian extension of K which is unramified at all places (including infinity). There is an obvious ordering on moduli given by $M \leq N$ if $M_v \leq N_v$ for all v. The *conductor* of a finite abelian extension L/K is the smallest modulus M such that $L \subset K_M$. One often writes the conductor as a formal product $I = \prod_v m_v^{M_v}$. When there are no infinite places involved, we often speak of the conductor as an ideal of $\mathscr{O}_K$.

REMARK 6.7. It is not true in general that the conductor of K_M is M.

6.3. The ideal theoretic description of class field theory. We briefly describe the main results of class field theory as described ideal-theoretically. The exposition is little more than a summary of the excellent treatment in [N1].

Let $\mathscr{O}_K$ be the ring of integers in a number field K. Consider a formal product $I = \prod_v m_v^{M_v}$, where $M_v = 0$ or 1 if v is archimedean. Let J_I be the group of fractional ideals which are prime to I, and let P_I be the subgroup of J_I generated by the divisors of elements $f \in K^*$ with

$$f \equiv 1 \pmod{m_v^{M_v}}$$

for each v. For archimedean places, this is an empty condition unless v is real and $M_v = 1$, in which case it means that f should be positive. Then with the notation as above, we have

PROPOSITION 6.8. *For each finite place v of K and $\alpha \in \mathbf{I}_K$, let $v(\alpha)$ denote the normalized v-adic valuation of α. Then the map*

$$\mathbf{I}_K \to J,$$
$$\alpha \mapsto \prod m_v^{v(\alpha)}$$

induces an isomorphism $\mathbf{C}_K / \mathbf{C}_K^M \cong J/J_M$.

Our treatment of the higher-dimensional case below will resemble the ideal theoretic formulation more than the idèle theoretic formulation.

Now we will reinterpret the reciprocity law for global fields in a form which is suitable for easy generalization. For the moment we will assume that K has no real place. Let X be the spectrum of the ring of integers of K if we are in the number field case and a smooth projective model of K if we are in the function field case. Let $\mathscr{K}_1(\mathscr{O}_X)$ be the Zariski sheaf of units on X. For a coherent ideal I of $\mathscr{O}_X$, set

$$\mathscr{K}_1(\mathscr{O}_X, I) = \mathrm{Ker}(\mathscr{K}_1(\mathscr{O}_X) \to \mathscr{K}_1(\mathscr{O}_X/I)).$$

Then $H^1(X, \mathscr{K}_1(\mathscr{O}_X, I))$ is an idèle class group with modulus I. In order to relate it to the idèle class group as defined above, we calculate this group explicitly. The theory of supports gives an exact sequence

$$\cdots \to H^0(K, \mathscr{K}_1(\mathscr{O}_X, I)) \to \bigoplus_{x \in X^1} H^1_x(X, \mathscr{K}_1(\mathscr{O}_X, I)) \to H^1(X, \mathscr{K}_1(\mathscr{O}_X, I))$$

$$\to H^1(K, \mathscr{K}_1(\mathscr{O}_X, I)).$$

The last group is zero because the sheaf $\mathscr{K}_1(\mathscr{O}_X, I)$ restricted to the generic point $\operatorname{Spec} K$ is just the sheaf of units on K, and the cohomology is the Picard group of K, which is trivial. The group on the left is just K^*. Thus the important part of the calculation is of the groups $H^1_x(X, \mathscr{K}_1(\mathscr{O}_X, I))$ for x a closed point of X. We have $H^1_x(X, \mathscr{K}_1(\mathscr{O}_X, I)) \cong H^1_x(A_x, \mathscr{K}_1(\mathscr{O}_X, I))$, where A_x is the local ring of X at x. Let K_x be the fraction field of A_x (this is just K). This last group may be calculated by using the localization sequence

$$\cdots H^0(A_x, \mathscr{K}_1(\mathscr{O}_X, I)) \to H^0(K_x, \mathscr{K}_1(\mathscr{O}_X, I)) \to H^1_x(A_x, \mathscr{K}_1(\mathscr{O}_X, I))$$

$$\to H^1(A_x, \mathscr{K}_1(\mathscr{O}_X, I)).$$

Now the group on the right is zero (local rings have trivial Zariski cohomology). Let S be the support of $\mathscr{O}_X/I$. Then the group on the left is isomorphic to A^*_x if x does not belong to S, since in that case the sheaf $\mathscr{K}_1(\mathscr{O}_X, I)$ restricted to A_x is just the sheaf of units on $\operatorname{Spec} A_x$. Then we have

$$H^1_x(X, \mathscr{K}_1(\mathscr{O}_X, I)) \cong K^*/A^*_x \cong \mathbf{Z}.$$

If x does belong to S, then the group on the left is $\operatorname{Ker}(A^*_x \to (A_x/IA_x)^*)$, and this kernel is isomorphic to $1 + IA_x$ as a multiplicative group. We conclude that $H^1_x(X, \mathscr{K}_1(\mathscr{O}_X, I)) = K^*/1 + IA_x$. Summing everything up, we have

$$H^1(X, \mathscr{K}_1(\mathscr{O}_X, I)) = \operatorname{Coker}\left(K^* \to \bigoplus_{x \in S}(K^*/1 + A_x) \oplus \bigoplus_{v \notin S} \mathbf{Z}\right).$$

It is not difficult to see that this is precisely the group we called J/J_I above. If $\widehat{A}_x$ denotes the henselization (or completion) of A_x at x and $\widehat{K}_x$ its fraction field, then since the group $\widehat{K}_x/1 + I\widehat{A}_x$ is discrete, we have an isomorphism

$$K_x/1 + IA_x \cong \widehat{K}_x/1 + I\widehat{A}_x.$$

Thus, we have

$$H^1(X, \mathscr{K}_1(\mathscr{O}_X, I)) \cong \operatorname{Coker}\left(K^* \to \left(\bigoplus_{x \in S}\widehat{K}^*_x/1 + I\widehat{A}_x \oplus \bigoplus_{x \notin S} \mathbf{Z}\right)\right).$$

Note that by weak approximation, K^* is dense in $\prod_{v \in S} K^*_v$, and hence the map

$$\bigoplus_{v \notin S} \mathbf{Z} \to H^1(X, \mathscr{K}_1(\mathscr{O}_X, I))$$

is surjective. Let U be the complement of S in X. When we define the reciprocity map below, we will see that the composite

$$\bigoplus_{v \notin S} \mathbf{Z} \to H^1(X, \mathscr{K}_1(\mathscr{O}_X, I)) \to \pi_1^{\mathrm{ab}}(U)$$

sends an element 1_v to the Frobenius substitution at v.

We want to show how to define the reciprocity map in a geometric way, assuming Artin-Verdier duality. As mentioned in the discussion of Artin-Verdier duality in Chapter 1, its proof involves class field theory. *Thus our definition of the reciprocity map in this way for dimension 1 is circular.* However, it will serve as motivation for the definition in higher dimensions. We use the following notation: I is a coherent ideal of $\mathscr{O}_X$ with support S, $i\colon S \to X$ is the inclusion, U is the complement of S in X and $j\colon U \to X$ is the inclusion. Let n be an integer which is invertible on U. We will define a map of Zariski sheaves $\mathscr{K}_1(\mathscr{O}_X, I) \to \mathscr{H}^1(j_! \mu_n)$, where $\mathscr{H}^1(j_! \mu_n)$ is the sheaf associated to the presheaf $V \to H^1(V_{\text{ét}}, j_! \mu_n)$.

First we show that the natural map:

$$\mathscr{H}^1(j_! \mu_n) \xrightarrow{g} \mathscr{H}^1(\mu_n)$$

is injective. To see this, it is enough to show that the map on stalks is injective. At the generic point η, the stalks are isomorphic. Let x be a point of codimension one and A the local ring of X at x. The exact sequence of sheaves:

$$0 \to j_! j^* \mu_n \to \mu_n \to i_* i^* \mu_n \to 0$$

gives a long exact sequence of cohomology:

$$H^0(A, \mu_n) \xrightarrow{f} H^0(A/I_x, \mu_n) \to H^1(A, j_! \mu_n) \xrightarrow{g} H^1(A, \mu_n) \to H^1(A/I_x, \mu_n),$$

where $I_x = I \otimes_{\mathscr{O}_X} \mathscr{O}_{X,x}$. But the map f is surjective: any n-th root of unity in A/I_x may be lifted to A, since A is local. This proves the injectivity of g.

Now consider the commutative diagram of exact sequences of sheaves:

$$
\begin{array}{ccccc}
0 \longrightarrow & \mathscr{K}_1^M(\mathscr{O}_X, I) & \longrightarrow & \mathscr{K}_1^M(\mathscr{O}_X) & \longrightarrow & \mathscr{K}_1^M(\mathscr{O}_X/I) \\
 & \big\downarrow & & \big\downarrow & & \big\downarrow \\
0 \longrightarrow & \mathscr{H}^1(j_! \mu_n) & \longrightarrow & \mathscr{H}^1(\mu_n) & \longrightarrow & \mathscr{H}^1(i^* \mu_n)
\end{array}
$$

Here the middle and right vertical maps are given by Kummer theory. This induces the left vertical map.

Now the map of sheaves we have just defined induces a map on cohomology $H^1(X, \mathscr{K}_1(\mathscr{O}_X, I)) \to H^1(X, \mathscr{H}^1(j_! \mu_n))$. The morphism change of sites $\pi\colon X_{\text{ét}} \to X_{\text{Zar}}$ gives rise to a spectral sequence

$$H^r(X_{\text{Zar}}, R^s \pi_* j_! \mu_n) \Rightarrow H^{r+s}(X_{\text{ét}}, j_! \mu_n),$$

and the sheaf $\mathscr{H}^1(j_!\mu_n)$ is just $R^1\pi_*j_!\mu_n$. The spectral sequence gives an edge map $H^1(X, \mathscr{H}^1(j_!\mu_n)) \to H^2(X, j_!\mu_n)$. By Artin-Verdier duality (Chapter 1), $H^2(X, j_!\mu_n)$ (which we call $H^2_c(U, \mu_n)$ in Chapter 1) is canonically isomorphic to $H^1(U, \mathbf{Z}/n)^*$, and this last group is just $\pi_1^{\mathrm{ab}}(U)/n$. Taking the inverse limit over all n and all U, we get the reciprocity map

$$\varprojlim_{I \neq 0} H^1(X, \mathscr{K}_1(\mathscr{O}_X, I)) \to \pi_1^{\mathrm{ab}}(K).$$

A reinterpretation of Theorem 6.5 via Proposition 6.8 is that this map is an isomorphism if K is of characteristic zero and an isomorphism on geometric parts if K is of positive characteristic.

Now suppose $\dim X = d$. We would like to define the reciprocity map for X in a similar way by replacing the K_1^M-groups with K_d^M-groups and the μ_n-sheaves with twists $\mu_n^{\otimes d}$. This should be possible, but at the moment we must use the Henselian topology to do it.

6.4. Definition of the reciprocity map in higher dimension. Let X be a d-dimensional proper integral scheme of finite type over $\mathbf{Z}$ with function field K. We take a base scheme T together with a finite map $f: X \to T$. For example, we may take $T = X$ (see Remark 6.10 below for more on the role of T). We define the reciprocity map for extensions of degree prime to the characteristic of K. Define as before

$$D_I(X/T) = H^d_\Sigma(T, f_*\mathscr{K}_d^M(\mathscr{O}_X, I)).$$
$$D_I^h(X/T) = H^d_\Sigma(T_{\mathrm{hen}}, f_*\mathscr{K}_d^M(\mathscr{O}_X, I)),$$

where the subscript hen denotes the Henselian topology defined in Chapter 1. There is a natural map $D_I(X/T) \to D_I^h(X/T)$ given by the change of sites morphism $\pi: X_{\mathrm{hen}} \to X_{\mathrm{Zar}}$ and the map

$$\mathscr{K}_d^M(\mathscr{O}_X, I) \to \pi_*\mathscr{K}_d^M(\mathscr{O}_{X_{\mathrm{hen}}}, I),$$

so it will suffice to define the reciprocity map in the Henselian topology. In fact, this is the only point in this paper where we need the Henselian topology, and we hope someone will be able to do what we do below purely in the Zariski topology.

Let U be a (Zariski) open set of X. We will define a map

$$\varprojlim_{I \neq 0} D_I^h(X/T) \to \pi_1^{\mathrm{ab}}(U),$$

where I runs over coherent ideals of X such that $I_U = \mathscr{O}_U$.

Let n be a positive integer, and let U' be a nonempty open subscheme of U such that
  (1) U' is smooth over its image in $\operatorname{Spec} \mathbf{Z}$.
  (2) n is invertible on U'.

We denote by $j: U' \to X$ the inclusion. Of course, (2) implies that n is not divisible by char. K.

Let S be the finite set of generic points of codimension one subschemes of $X - U$. We choose a coherent ideal I such that $I_U = \mathcal{O}_U$ and for every $x \in S$, $1 + I A_x^h \subset A_x^{h*n}$, where A_x^h is the henselization of the local ring of X at x. We want to define a Galois symbol

$$\mathcal{K}_d^M(\mathcal{O}_X, I) \to \mathcal{H}^d(j_!\mathbf{Z}/n(d))$$

on X_{hen}. This should be possible, but with current technology, we can only do something slightly weaker which will suffice for our purposes. For any sheaf on X_{Zar} or X_{hen} and any point $x \in X$, let $\mathcal{H}_x^i \mathcal{F}$ be the sheaf on X_{hen} associated to the presheaf $U \mapsto H_x^i(U, \mathcal{F})$. We set

$$\widetilde{\mathcal{F}} = \text{Ker}\left[\mathcal{H}_x^0(\mathcal{F}) \to \bigoplus_{x \in X^1} \mathcal{H}_x^1(\mathcal{F})\right].$$

Let $\widetilde{\mathcal{K}}_d^M(\mathcal{O}_X, I)$ be the sheaf obtained from $\mathcal{K}_d^M(\mathcal{O}_X, I)$ by doing this procedure, and let $\widetilde{\mathcal{H}}^d(j_!\mathbf{Z}/n(d))$ be the analogous sheaf for $\mathcal{H}^d(j_!\mathbf{Z}/n(d))$. We want to define a map

$$\widetilde{\mathcal{K}}_d^M(\mathcal{O}_X, I) \to \widetilde{\mathcal{H}}^d(j_!\mathbf{Z}/n(d)).$$

If the ideal I satisfies $I_\eta = \mathcal{O}_{X,\eta}$ ($\eta = $ generic point of X), then $\mathcal{K}_d^M(\mathcal{O}_X, I)$ is the maximal subsheaf of $i_{\eta*} K_d^M K$ whose stalk at any codimension one point x of any scheme étale over X is the same as the image of the stalk of $\widetilde{\mathcal{K}}_d^M(\mathcal{O}_X, I)$ at x.

Let x be a codimension one point of X and A_x^h the Henselization of X at x. We denote generic points of $\text{Spec}\, A_x^h$ by η_x and their residue fields by K_x. By the remark we have just made, it suffices to define maps making the following square commutative:

$$
\begin{array}{ccc}
\mathcal{K}_d^M(\mathcal{O}_X, I)_x & \longrightarrow & \mathcal{K}_d^M(\mathcal{O}_X, I)_{\eta_x} \\
\downarrow & & \downarrow \\
\mathcal{H}^d(j_!\mathbf{Z}/n(d))_x & \longrightarrow & \mathcal{H}^d(j_!\mathbf{Z}/n(d))_{\eta_x}
\end{array}
$$

The right vertical map is the Galois symbol in Chapter 1, §2. To define the left vertical maps, first assume that $x \in U'$. Then the stalk of the sheaf $\widetilde{\mathcal{K}}_d^M(\mathcal{O}_X, I)$ at x is just the Milnor K-group $K_d^M(A_x^h)$ and the stalk of $\widetilde{\mathcal{H}}^d(j_!\mathbf{Z}/n(d))$ is the étale cohomology group $H^d(A_x^h, \mathbf{Z}/n(d))$. The map is the Galois symbol defined in Chapter 1. Now assume that $x \in S$. Then the stalk of $\widetilde{\mathcal{K}}_d^M(\mathcal{O}_X, I)$ at x is the group $K_d^M(A_x^h, I)$. Since A_x^h is a local ring, we know that this group is generated by symbols $\{a_1, \ldots, a_d\}$, where one of the a_i belongs to $1 + I A_x^h$ [KS3, Lemma 1.3.1]. As for the étale cohomology side, the stalk of $\mathscr{H}^d(j_!\mathbf{Z}/n(d))$ at x is $H^d(A_x^h, j_!\mathbf{Z}/n(d))$. Let $\mathbf{F}_x$ be the residue field of A_x. By the proper base change theorem, this group is the same as $H^d(\mathbf{F}_x, i^* j_!\mathbf{Z}/n(d))$, and this last group is clearly zero. Thus, to define our Galois symbol, all we need to do is see the commutativity of the diagram

$$
\begin{array}{ccc}
K_d^M(A_x^h, I_x) & \longrightarrow & K_d^M(K_x) \\
\downarrow & & \downarrow \\
H^d(A_x^h, j_!\mathbf{Z}/n(d)) & \longrightarrow & H^d(K_x, \mathbf{Z}/n(d))
\end{array}
$$

Here the left vertical arrow is the zero map, the right vertical arrow is the Galois symbol for K and the horizontal arrows are the inclusions. By our assumption that $1 + IA_x^h \subset A_x^{h*n}$, the composite of the top horizontal and right vertical arrow is also zero. This gives the desired commutativity and defines the map. Taking the top cohomology groups, we get a map

$$
H^d(X, \widetilde{\mathscr{K}}_d^M(\mathscr{O}_X, I)) \to H^d(X, \mathscr{H}^d(j_!\mathbf{Z}/n(d))).
$$

From Corollary 1.23 and the construction of the Galois symbol above, we get a map $H_\Sigma^d(X, \mathscr{K}_d^M(\mathscr{O}_X, I)) \to H^d(X, \mathscr{H}^d(j_!\mathbf{Z}/n(d)))$. From the spectral sequence

$$
H^r(X, \mathscr{H}^s(j_!\mathbf{Z}/n(d))) \Rightarrow H^{r+s}(X, j; \mathbf{Z}/n(d)),
$$

we get an edge map

$$
H^d(X, \mathscr{H}^d(j_!\mathbf{Z}/n(d)) \to H^{2d+1}(X, j; \mathbf{Z}/n(d)).
$$

By the extension of Artin-Verdier duality (Theorem 1.19), we have

$$
H^{2d+1}(X, j_!\mathbf{Z}/n(d)) \cong H^1(U^1, \mathbf{Z}/n)^* = \pi_1^{\text{ab}}(U')/n.
$$

Finally, composing with the natural map $\pi_1^{\text{ab}}(U') \to \pi_1^{\text{ab}}(U)$, we get our desired map

$$
H_\Sigma^d(X, \mathscr{K}_d^M(\mathscr{O}_X, I)) \to \pi_1^{\text{ab}}(U)/n.
$$

Passing to the inverse limit over all I with $I_U = \mathscr{O}_U$, we get the reciprocity map

$$
\varprojlim_{I_U=\mathscr{O}_U} H_\Sigma^d(X, \mathscr{K}_d^M(\mathscr{O}_X, I)) \to \pi_1^{\text{ab}}(U).
$$

Now taking the limit over all open U in X, we get the reciprocity map for the function field K of X:

$$
\varprojlim_{I \neq 0} H_\Sigma^d(X, \mathscr{K}_d^M(\mathscr{O}_X, I)) \to \pi_1^{\text{ab}}(K).
$$

REMARK 6.9. In the original papers of Kato-Saito [**KS1, KS3**], the reciprocity map was defined by passage to the higher local fields of X using the theory of Parshin chains. The definition we have just given is simpler conceptually, but it has the defect that it does not work at the moment for

p-extensions in characteristic $p > 0$. It is quite possible that recent developments in the theory of the de Rham-Witt complex for open varieties will be enough to accomplish this.

REMARK 6.10. The role of the base scheme T in the reciprocity maps may seem funny. The reason for it is that we only know how to define the norm for this relative version of the class group. More precisely, let $X' \to X$ be a finite flat map. As shown below, there is a norm map

$$D_{I'}(X'/T) \to D_I(X/T)$$

for an appropriately chosen ideal I' of $\mathcal{O}_{X'}$, and this plays a crucial role in the proof of the reciprocity law. We would like to define a norm map

$$H_\Sigma^d(X', \mathcal{K}_d(\mathcal{O}_{X'}, I')) \to H_\Sigma^d(X, \mathcal{K}_d^M(\mathcal{O}_X, I)),$$

but we do not know how to do this at the moment. In the original paper of Kato-Saito [KS3], the Henselian topology was used to get around this problem. This topology has the advantage that for a finite map f and a sheaf $\mathcal{F}$ on X', the sheaves $R^i f_* \mathcal{F}$ are zero (the functor f_* is exact). This is not true in the Zariski topology.

6.5. Some properties of the reciprocity map.

LEMMA 6.11. *Let L/K be a finite abelian extension and U a nonempty normal open subscheme of X whose integral closure in L is étale over U. Then there exists an ideal I of X with $I_U = \mathcal{O}_U$ such that the reciprocity map induces:*

$$D_I(X/T) \to \operatorname{Gal}(L/K).$$

PROOF. Let $n = [L:K]$ and let S be the finite set of generic points of codimension one subschemes of $X - U$. For $x \in S$, let A_x be the henselization of the local ring of X at x. Then choose an ideal I of X such that $I_U = \mathcal{O}_U$ and for any $x \in S$, we have:

$$1 + I A_x \subset A_x^{*n}.$$

Since U is normal, $\operatorname{Gal}(L/K)$ is a quotient of $\pi_1^{ab}(U)/n$, and an examination of the definition of the reciprocity map in Section 6.4 shows that I will do the job.

The set of ideals such that the reciprocity map induces

$$D_I(X) \to \operatorname{Gal}(L/K)$$

is ordered by inclusion. Since $\mathcal{O}_X$ is noetherian, there is a largest such ideal.

DEFINITION 6.12. With notation as in the lemma, the largest ideal I such that the reciprocity map

$$D_I(X/T) \to \operatorname{Gal}(L/K)$$

is defined is called the conductor of L/K.

We give some more properties of the reciprocity map with only indications of proofs. Recall that Kato [**K5**] calls a ring A *nice* if there is a smooth ring A' of finite type over a field or an excellent Dedekind domain such that A is ind-étale over A'. Nice implies regular, and it is hoped that the converse is true. A scheme is called nice if all of its local rings are nice.

LEMMA 6.13. *Let X, T be integral schemes of finite type over $\mathbf{Z}$ such that X is finite over T, and let I be a nonzero ideal of $\mathscr{O}_X$. Then for any closed integral subscheme S of X, there exists a nonzero ideal J of $\mathscr{O}_S$ and a map*

$$\psi_J : D_J(S/T) \to D_I(X/T).$$

Let J' be the largest ideal in $\mathscr{O}_S$ such that this map is defined. Then ψ_J factors through $\psi_{J'}$. If x is a point of S which is contained in a nice open subscheme of X and such that $I_x = \mathscr{O}_{X,x}$, then $J'_x = \mathscr{O}_{S,x}$. The image of ψ_J is independent of J.

For the groups $D_I^h(X/T)$, this lemma is proved in [**KS3**, Proposition 2.9], where the theory of Parshin chains is used. This lemma is crucial for the proof of the reciprocity law given below, and it would be nice to have a more global proof.

Let x be a closed point of X, $i_x : x \to T$ the natural map and I a nonzero coherent ideal of $\mathscr{O}_X$. We denote by $D_{\mathscr{O}_{X,x}}(X/T)$ the hypercohomology group $H^d(X, Ri_x^! f_* \mathscr{K}_d(\mathscr{O}_X))$ and by $D_{I,x}(X/T)$ the hypercohomology group $H^d(X, Ri_x^! f_* \mathscr{K}_d^M(\mathscr{O}_X, I))$. If $x \notin \operatorname{Supp} I$, these are isomorphic. Then there is a map $\partial_x : D_{\mathscr{O}_{X,x}} \to \mathbf{Z}$. Composing with $\mathscr{K}_d^M(\mathscr{O}_X, I) \to \mathscr{K}_d(\mathscr{O}_X)$, we get a map $D_{I,x}(X/T) \to \mathbf{Z}$.

LEMMA 6.14. *Let $X \to T$ be as above, and let x be a closed point of X. Let K_x be the henselization of K at x. Then there is a natural map*

$$\varprojlim_{I \neq 0} D_{I,x}(X/T) \to \operatorname{Gal}(K_x^{\mathrm{ab}}/K_x)$$

making the following diagram commute:

$$
\begin{array}{ccc}
\varprojlim\limits_{I \neq 0} D_{I,x}(X,T) & \longrightarrow & \operatorname{Gal}(K_x^{\mathrm{ab}}/K_x) \\
\downarrow & & \downarrow \\
\mathbf{Z} & \longrightarrow & \operatorname{Gal}(\overline{\kappa(x)}/\kappa(x))
\end{array}
$$

The bottom horizontal map sends 1 to the Frobenius element.

LEMMA 6.15. *Let L/K be a finite abelian extension, and let I be a nonzero ideal of $\mathscr{O}_X$ such that the reciprocity map induces*

$$D_I(X/T) \to \operatorname{Gal}(L/K).$$

If x is contained in a nice open subscheme U of X which does not meet the support of $\mathscr{O}_X/I$, then ∂_x is an isomorphism, and the composite

$$\mathbf{Z} \xrightarrow{\partial_x^{-1}} D_{\mathscr{O}_{X,x}}(X/T) \to D_I(X/T) \to \mathrm{Gal}(L/K)$$

sends 1 to the Frobenius element over x.

LEMMA 6.16. *With notation as in the last lemma, there is a natural surjective map*

$$\gamma \colon D_I(X/T) \to CH_0(U).$$

PROOF. From Lemma 6.13, one can show that there is a natural map

$$D_{\mathscr{O}_X}(X) \to CH_0(X),$$

and the composite

$$\bigoplus_{x \in U_0} \mathbf{Z} \to D_I(X/T) \to D_{\mathscr{O}_X}(X/T) \to CH_0(X) \to CH_0(U)$$

is the natural projection. This gives the surjectivity of γ.

6.6. Proof of the reciprocity law, part I: approximation and the norm theorem. The aim of this section is to prove the following theorem which, while being interesting in its own right, is a key point in the proof of the main theorem. We keep the notation as above. In particular, X is an integral proper scheme of finite type over $\mathbf{Z}$ with function field K and $f \colon X \to T$ is a finite map with T integral. Let I be an ideal of X. Suppose X' is an integral scheme which is finite and flat over X, and L is the function field of X'. We shall see below that there is an ideal I' of $\mathscr{O}_{X'}$ and a norm map $N_{X'/X} \colon D_{I'}(X'/T) \to D_I(X/T)$ such that the composite of the natural map

$$\varprojlim_{I \neq 0} D_I(X/T) \to \varprojlim_{I' \neq 0} D_{I'}(X'/T)$$

with $N_{X'/X}$ is multiplication by $[L : K]$. In Proposition 6.11 above, we showed that there exists a nonzero ideal I of $\mathscr{O}_X$ such that the reciprocity map induces

$$D_I(X/T) \to \mathrm{Gal}(L/K).$$

Then we have

THEOREM 6.17. *With notation as above, suppose L is abelian over K. Then there is an exact sequence*

$$D_I(X'/T) \xrightarrow{N_{X'/X}} D_I(X/T) \to \mathrm{Gal}(L/K) \to 0.$$

Before giving the proof, we prove the existence of the norm map. Recall the sheaves $\widetilde{\mathscr{K}}_d^M(\mathscr{O}_X, I)$ defined in §6.4.

Proposition 6.18 [KS3, Proposition 4.2]. *For a finite flat map* $X' \overset{g}{\to} X$ *as above, there exists an ideal* I' *of* X' *such that there is a norm map*

$$g_* \widetilde{\mathscr{K}}_d^M(\mathscr{O}_X, I') \to \widetilde{\mathscr{K}}_d^M(\mathscr{O}_X, I).$$

Proof. It will clearly suffice to prove the following lemma.

Lemma 6.19. *Let* A *be a discrete valuation ring with fraction field* K. *Let* L *be a finite extension of* K *and* B *the integral closure of* A *in* L. *Let* I *be an ideal of* A. *Then the norm map on Milnor* K*-theory* $K_d^M(L) \to K_d^M(K)$ *induces a map of relative Milnor* K*-groups*

$$\widetilde{K}_d^M(B, IB) \to \widetilde{K}_d^M(A, I).$$

Here $\widetilde{K}_d^M(A, I)$ *denotes the image of* $K_d^M(A, I)$ *in* $K_d^M K$ *and similarly for* $\widetilde{K}_d^M(B, IB)$.

Proof [KS3, Lemma 4.3]. If $d \leq 2$ the result follows from the exactness of the sequence

$$K_d^M(A, I) \to K_d(A) \to K_d(A/I)$$

(see [**DS**]) and the existence of a norm map $K_d(B/IB) \to K_d(A/I)$ in Quillen K-theory. So we assume $d \geq 2$. First we show that the norm maps $\widetilde{K}_d^M(B)$ into $\widetilde{K}_d^M(A)$ and $\widetilde{K}_d^M(B, mB)$ into $\widetilde{K}_d^M(A, m)$, where m is the maximal ideal of A. Consider the commutative diagram

$$
\begin{array}{ccc}
K_d^M(K) & \overset{\partial_1}{\longrightarrow} & K_{d-1}^M(A/m) \\
\uparrow & & \uparrow \\
K_d^M(K) & \overset{\partial_2}{\longrightarrow} & \underset{i}{\bigoplus} K_{d-1}^M(B/m_i)
\end{array}
$$

where $m_1, \ldots, m_s$ are the maximal ideal of B lying over m, the vertical maps are given by the norm and ∂_j is the tame symbol. Now $\mathrm{Ker}\,\partial_1 = \widetilde{K}_d^M(A)$ and $\mathrm{Ker}\,\partial_2 = \widetilde{K}_d^M(B)$. This proves the assertion for $\widetilde{K}_d^M(B)$. As for the relative groups, this follows by a similar argument using Lemma 2.7 and property (4) of the tame symbol given in Chapter 1, §1.

Now let $U(I)$ be the subgroup of $\widetilde{K}_d^M(A)$ generated by symbols of the form $\{a_1, \ldots, a_d\}$, where $a_1 \in 1 + I$ and $a_2, \ldots, a_d \in K^*$, and define $U(IB)$ in a similar way. Then an argument similar to the one in [**K6**, Proposition 2] shows that the norm maps $U(IB)$ to $U(A)$. Now $\widetilde{K}_d^M(A, I) \subset U(I)$ and $U(m^{i+1}) \subset \widetilde{K}_d^M(A, m^i)$ for all $i \geq 0$. By [**K5**, §2, Proposition 2], there exists a nonzero subideal J of IB such that the norm maps $\widetilde{K}_d^M(B, J)$ to $\widetilde{K}_d^M(A, I)$. If the extension of residue fields $B/m_i \to A/m$ is separable, then the symbol map

$$K_2^M(B, IB) \otimes K_{d-2}^M(A) \to \widetilde{K}_d^M(B, IB)/\widetilde{K}_d^M(B, JB)$$

is surjective (see [**K1,** II, §1.3, Lemma 6]) and we are done by the remark above for $d = 2$. This takes care of the case where $[L : K]$ is invertible in the residue field A/m. If the residue field extension is not separable, then $\widetilde{K}_d^M(A, m)/\widetilde{K}_d^M(A, I)$ is killed by a power of p ($p = $ char. A/m), and we may replace L/K by LK'/K', where $[K' : K]$ is prime to p [**BT,** Chapter 1, §5.9, diagram (15) of norm maps]. Thus, we are reduced to the case where $[L : K] = p$. Then by the above argument, the symbol map is surjective, and we are reduced to the case $d = 2$ again. This completes the proof of the lemma.

For a nonzero ideal I' of $\mathscr{O}_{X'}$ there is a sufficiently small ideal I of $\mathscr{O}_X$ such that the map

$$g^* \mathscr{K}_d^M(\mathscr{O}_X, I) \to \mathscr{K}_d^M(\mathscr{O}_{X'}, I')$$

is defined. Thus, we get a map $g^* : D_{I'}(X'/T) \to D_I(X/T)$.

LEMMA 6.20. (1) *The composite*

$$\varprojlim_{I \neq 0} D_I(X/T) \xrightarrow{g^*} \varprojlim_{I' \neq 0} D_{I'}(X'/T) \xrightarrow{N_{X'/X}} \varprojlim_{I \neq 0} D_I(X/T)$$

is multiplication by $[L : K]$.

(2) *The following diagrams commute:*

$$
\begin{array}{ccc}
\varprojlim_{I \neq 0} D_I(X/T) & \longrightarrow & \mathrm{Gal}(K^{\mathrm{ab}}/K) \\
{\scriptstyle g^*}\downarrow & & \downarrow \\
\varprojlim_{I' \neq 0} D_{I'}(X'/T) & \longrightarrow & \mathrm{Gal}(L^{\mathrm{ab}}/L)
\end{array}
$$

where the right vertical map is transfer, and

$$
\begin{array}{ccc}
\varprojlim_{I \neq 0} D_I(X/T) & \longrightarrow & \mathrm{Gal}(K^{\mathrm{ab}}/K) \\
{\scriptstyle N_{X'/X}}\uparrow & & \uparrow \\
\varprojlim_{I' \neq 0} D_{I'}(X'/T) & \longrightarrow & \mathrm{Gal}(L^{\mathrm{ab}}/L)
\end{array}
$$

where the right vertical map is restriction.

PROOF. This follows easily from the definition of the reciprocity map and the corresponding properties of the Galois symbol (see Chapter I, §1).

Before proving the theorem we need one more lemma.

LEMMA 6.21 (approximation lemma). *Let C be a one-dimensional scheme which is separated and of finite type over $\mathbf{Z}$, and let X be a quasiprojective scheme which is smooth over C. Let x_i ($1 \leq i \leq n$) be closed points of X*

which have distinct images in C. *Then there exists a one-dimensional integral closed subscheme* D *of* X *on which each* x_i *lies as a regular point.*

PROOF. Let $f\colon X \to C$ be the structure map and put $v_i = f(x_i)$. If the residue fields of x_i and v_i are the same, this is proved by Bloch [**B2**, Lemma 3.3]. We may assume that X is affine and after taking the integral closure of C in X, we may assume that f has geometrically connected fibres. By one dimensional class field theory, there is a finite abelian cover $r\colon C' \to C$ which is étale over any v_i such that for each i and each closed point w of C' lying over v_i, w and x_i have the same residue field. Let $X' = X \times_C C'$ and $f'\colon X' \to C'$ the natural map. Let $Q = \{v_1, \dots, v_n\}$, and let Q' be the set of points of C' lying over points of Q. For $w \in Q'$, let P_w be the set of points of X' which lie over w and over $r(w)$. Now

$$\#P_w = [\kappa(w) : \kappa(v)] \le \#\kappa(w),$$

where $v = r(w)$. We can then apply Bloch's approximation result to the semilocalizations A of X' at $\bigcup_{w \in Q'} P_w$ and R of C' at Q'. Thus, after making a base change by an open neighborhood U of Q' in C, we can find an étale morphism $s\colon X' \to \mathbf{A}^t_{C'}$ which is injective on the subset $\bigcup_{w \in Q'} P_w$. Let k' be the function field of C'. For a morphism $\operatorname{Spec}(R) \to C'$, let $T(R)$ denote the set $\operatorname{Hom}_{C'}(\operatorname{Spec}(R), \mathbf{A}^t_{C'})$, which is isomorphic (as set) to t-copies of R. Let T be the set of all $a \in T(k')$ such that $s^{-1}(a)$ is irreducible. Then by Hilbert's irreducibility theorem, the image of $T \to \prod_{w \in Q'} T(k'_w)$ is dense in the adic topology. For $w \in Q'$ and $x \in P_w$, let $T_{w,x}$ be the subset of $T(\mathscr{O}_{k'_w})$ consisting of all elements whose image in $T(\kappa(w))$ is the same as the image of x under

$$s \times_{C'} w\colon X' \times_{C'} w \to \mathbf{A}_{C'} \times_{C'} w.$$

Then $T_{w,x}$ is an open and closed subset of $T(k'_w)$ and for distinct $x, y \in P_w$, these sets are disjoint. For each v_i choose one w_i lying over it and let Q'' be the set containing the w_i for $i = 1, 2, \dots, n$. Then there is an $a \in T$ such that its image in $T(k'_w)$ is contained in one of the $T_{w,x}$ if $w \in Q''$ and is contained in

$$T(k'_w) - \bigcup_{x \in P_w} T_{w,x}$$

if $w \notin Q''$. After making a base change over C as before, the closure D'' of the image of a is regular, and hence $D' = s^{-1}(D'')$ is a one-dimensional regular closed subscheme of X' on which just one of the P_w lies if $w \in Q''$ and none of the P_w lies if $w \notin Q''$. Let D be the image of D' in X. Then D is a one-dimensional, closed, irreducible subscheme of X on which all of the x_i lie and D' contains just one point lying over each x_i for all i. To see that each x_i is regular on D, we use the following sublemma:

SUBLEMMA 6.22. *Let* $\psi \colon X' \to X$ *be a finite étale Galois covering with Galois group* G, *and let* D' *be a closed integral subscheme of* X' *with image* D *in* X. *Let* $x' \in D'$ *and assume that as sets,*

$$\psi^{-1}(\psi(x')) \cap D' = x'.$$

Then we have an isomorphism of henselizations:

$$\mathscr{O}^h_{D,\psi(x')} \cong \mathscr{O}^h_{D',x'}.$$

We leave the proof of this to the reader.

PROOF OF THEOREM 6.17. The proof is by induction on the dimension of X. The assertion in dimension zero is trivial and in dimension one, it is a well-known fact from class field theory (see Theorem 6.5, (iii) and Proposition 6.8). We thus assume that $\dim X \geq 2$. That the sequence is a complex follows from Lemma 6.20 (2). The surjectivity of the reciprocity map follows from Lemma 1.7. Let U be an open subscheme of X which is such that the map $X' \to X$ is étale over U and $I_U = \mathscr{O}_U$. Let $t \in U$ and let $t' \in X'$ lying over t. Let Y be the closure of t in X and Y' the closure of t' in X'. Then we may identify $\mathrm{Gal}(\kappa(t')/\kappa(t))$ with a subgroup of $\mathrm{Gal}(L/K)$ and by Lemma 6.13, we have a commutative diagram

$$
\begin{array}{ccccccc}
D_{J'}(Y'/T) & \xrightarrow{N_{Y'/Y}} & D_J(Y/T) & \longrightarrow & \mathrm{Gal}(\kappa(t')/\kappa(t)) & \longrightarrow & 0 \\
\downarrow & & \downarrow & & \downarrow & & \\
D_{I'}(X'/T) & \xrightarrow{N_{X'/X}} & D_I(X/T) & \longrightarrow & \mathrm{Gal}(L/K) & \longrightarrow & 0
\end{array}
$$

where J, J' are appropriate ideals of $\mathscr{O}_Y$, $\mathscr{O}_{Y'}$ such that the maps are defined. Taking U sufficiently small, we may assume that there is a one-dimensional scheme C of finite type over $\mathbf{Z}$ and a smooth morphism $h \colon U \to C$ with geometrically connected fibres. Fix $c \in D_I(X/T)$ whose image in $\mathrm{Gal}(L/K)$ is trivial. We can write c as the image of a zero-cycle $\tilde{c} = \sum_{i=1}^n n_i x_i$ with $n_i \geq 0$.

CLAIM. We may assume that the points $v_i = h(x_i)$ are distinct.

To see this, for a fixed $v = v_i$, let Y be the closure of $h^{-1}(v)$ in X, and let t be its generic point. Consider the diagram above for Y and X. Let M be the set of indices $1 \leq j \leq n$ such that $h(x_j) = v$. Then by Čebotarev density, there exists a closed point $x \in U \cap Y$ such that the images of the zero-cycles x and $\sum_{j \in M} n_j x_j$ in $D_J(Y/T)$ have the same image in $\mathrm{Gal}(\kappa(t')/\kappa(t))$. By induction and the diagram, the image of the cycle $x - \sum_{j \in M} n_j x_j$ in $D_I(X/T)$ lies in the image of $N_{X'/X}$. Hence, we may replace $\sum_{j \in M} n_j x_j$ by x. This proves the claim.

By the claim and the approximation lemma, we can find a one-dimensional closed integral subscheme D of X on which each x_i lies as a regular point. Applying the diagram above to the case $Y = D$ and using the one-dimensional case, we see that c is in the image of $N_{X'/X}$. This completes the proof of the theorem.

COROLLARY 6.23. *With notation as above, let L be any finite extension of K and X' the integral closure of X in L. Let I' be any ideal such that the norm map*

$$N_{X'/X} \colon D_{I'}(X'/T) \to D_I(X/T)$$

is defined. Then the cokernel of $N_{X'/X}$ is finite.

PROOF. This can be reduced to the abelian case by a standard argument using p-Sylow subgroups.

The norm theorem has the following important application.

THEOREM 6.24 (Bloch [**B2**, Theorem 4.2]; Kato and Saito [**KS3**, Theorem 6.1]). *Let X be any scheme of finite type over $\mathbf{Z}$. Then $CH_0(X)$ is a finitely generated abelian group. If X is separated and connected, we have two possibilities*:

(1) *If X is nonempty and proper and some prime number p is nilpotent on X, then $CH_0(X) = \mathbf{Z} \oplus T$, where T is a finite group.*

(2) *If X does not satisfy (1) then $CH_0(X)$ is finite.*

PROOF. First we prove the theorem in the case of an affine scheme U. If U' is a dense open subscheme of U with complement Y, then we have an exact sequence

$$CH_0(Y) \to CH_0(U) \to CH_0(U') \to 0.$$

By induction, if $CH_0(U')$ is finite then so is $CH_0(U)$. We may then assume that U is a dense open subscheme of X with the following properties. There is a regular connected affine scheme S of finite type over $\mathbf{Z}$ and a smooth projective morphism $f \colon X \to S$ with fibres of dimension one such that $f_* \mathscr{O}_X = \mathscr{O}_S$. First assume that f has a section $i \colon S \to X$. Let I be the image of the induced map $i_* \colon CH_0(S) \to CH_0(X)$. Then $CH_0(X)$ is generated by I and the image of

$$\bigoplus_{s \in S} CH_0(X_s)^0.$$

By induction, $CH_0(S)$ is finite, and hence so is I, so the finiteness of $CH_0(X)$ follows from Bloch's exact sequence (Theorem 5.13, Remark 5.23).

Now do not assume f has a section. By replacing X by a nonempty open set, we may assume that there is a finite étale covering S'/S such that, denoting by $'$ basechange by S', there is a section of the morphism $X' \to S'$. Embed U into a projective scheme V of finite type over $\mathbf{Z}$ [**Na4**], and let V' be the integral closure of V in the function field of $U' = U \times_S S'$. Then there exists an ideal I' of $\mathscr{O}_{V'}$ such that we have a commutative diagram:

$$
\begin{array}{ccc}
D_{I'}(V'/V) & \xrightarrow{N_{V'/V}} & D_{\mathscr{O}_X}(V) \\
\downarrow & & \downarrow \\
CH_0(U') & \xrightarrow{\quad N \quad} & CH_0(U).
\end{array}
$$

Here the vertical arrows are defined by Lemma 6.16, and they are surjective. By Corollary 6.23, the cokernel of the top horizontal map is finite. Hence the same is true of the bottom map. This shows that $CH_0(U)$ is finite.

For the general case, we first note that the case where X is of dimension one is classical (may be reduced to finiteness of the Picard group of a one-dimensional regular scheme of finite type over $\mathbf{Z}$). Assume that X is of dimension at least two and the result is true for schemes of dimension less than that of X. If X is integral and proper over $\mathbf{Z}$ then there exists a dense open affine scheme U such that $Y = X - U$ is connected. If the characteristic of the function field of X is zero then we can choose U so that no prime number is nilpotent on Y, and then we will get the finiteness of $CH_0(Y)$. From the exact sequence:

$$CH_0(Y) \to CH_0(X) \to CH_0(U) \to 0,$$

what we have just proved for affine schemes and induction, we get the result. If the characteristic of K is nonzero, then some prime number is nilpotent on X and by induction, $CH_0(Y)$ is isomorphic to $\mathbf{Z} \oplus T$ for some finite group T. Since there is a nontrivial degree map

$$CH_0(X) \to \mathbf{Z},$$

$CH_0(X)$ is not finite. Next, we do the case where X is separated and irreducible. By a theorem of Nagata [Na4], we may then embed X as a dense open subscheme of an integral proper scheme over $\overline{X}$ over $\mathbf{Z}$. By what we have just proved, we get the result for X except possibly if $\overline{X}$ satisfies (1) of the theorem and $X \neq \overline{X}$. In this case, let x be a closed point of $\overline{X} - X$. Then the map $K_0 k(x) \to CH_0(\overline{X})$ is nonzero, but the map $K_0 k(x) \to CH_0(X)$ is trivial. This implies that $CH_0(X)$ is a finite group. A straightforward induction on the number of irreducible components of X completes the proof in the general case. This completes the proof of Theorem 6.24.

6.7. Completion of the proof of the reciprocity law. The proof is by induction on the dimension of X. The case of dimension zero is trivial and dimension one is "classical" class field theory (see §2 above). We thus assume that $\dim X \geq 2$ and that the assertion is true for all proper schemes of finite type over $\mathbf{Z}$ of dimension less than that of X.

STEP 1. Assume that we are given the following data: U is an open subset of X, S is an integral scheme of dimension greater than zero, and $f : U \to S$ is a smooth morphism with geometrically connected fibres and having a section $i : S \to X$. Let $\widetilde{S}$ be the closure of $i(S)$ in X. If I is an ideal of $\mathscr{O}_X$ with $I_U = \mathscr{O}_U$, there exists an ideal J of $\mathscr{O}_S$ such that we have a natural map (see Lemma 6.13 above)

$$D_J(\widetilde{S}/T) \to D_I(X/T).$$

Let Γ_I denote the cokernel of this map; by Lemma 6.13, it is independent of the choice of J. For any morphism $R \to S$, we set $U_R = U \times_S R$ and $\pi_1^{\mathrm{ab}}(U/R) := \mathrm{Ker}(\pi_1^{\mathrm{ab}}(U_R) \to \pi_1^{\mathrm{ab}}(R))$. The section i gives splittings

$$\pi_1^{\mathrm{ab}}(U) \cong \pi_1^{\mathrm{ab}}(S) \times \pi_1^{\mathrm{ab}}(U/S);$$
$$\pi_1^{\mathrm{ab}}(U/k) \cong \pi_1^{\mathrm{ab}}(k) \times \pi_1^{\mathrm{ab}}(U/k),$$

where k is the function field of S. By the theorem of Katz-Lang [**KL**], $\pi_1^{\mathrm{ab}}(U/k)$ is finite, and so replacing S by a nonempty open subscheme S', we may assume that the natural map $\pi_1^{\mathrm{ab}}(U/k) \to \pi_1^{\mathrm{ab}}(U/S)$ is an isomorphism if k is of characteristic zero and an isomorphism on prime-to-p parts if char. $k = p > 0$. Now for sufficiently small I, the reciprocity map

$$\widetilde{\Phi}_I : D_I(X/T) \to \pi_1^{\mathrm{ab}}(U/k)$$

is defined and by Lemma 6.13, it kills the image of $D_J(\widetilde{S})$. Thus we get a map:

$$\Phi_I : \Gamma_I \to \pi_1^{\mathrm{ab}}(U/k).$$

By Lemmas 1.7 and 6.15, this map is surjective.

PROPOSITION 6.25. Γ_I *is a torsion group and* Φ_I *induces an isomorphism*:

$$\Gamma_I \to \pi_1^{\mathrm{ab}}(U/k)$$

if k is of characteristic zero and

$$\Gamma_I' \to \pi_1^{\mathrm{ab}}(U/k)',$$

where $'$ denotes prime to p-part, if k is of characteristic $p > 0$.

COROLLARY 6.26. *Suppose U satisfies the assumptions above. Then the reciprocity map*

$$\varprojlim_{I_U = \mathscr{O}_U} D_I(X/T) \to \pi_1^{\mathrm{ab}}(U)$$

is an isomorphism if char. $k = 0$ *and an isomorphism on pro-prime-to-p geometric parts if* char. $K = p > 0$. *If U is proper over k, the map is an isomorphism on p-parts as well.*

The corollary follows immediately from the proposition by induction on the dimension of $\widetilde{S}$.

PROOF OF PROPOSITION 6.25. The proof will continue until the end of Step 2 below. First we show that Γ_I is a torsion group. For each closed point s of S, let X_s be the closure in X of the fibre U_s of the map f over s and fix an ideal I_{X_s} of $\mathscr{O}_{X_s}$ such that there is a map

$$h_s : D_{I_s}(X_s/T) \to D_I(X/T).$$

Let $D_{I_s}(X_s/T)^0$ denote the kernel of the natural map $D_{I_s}(X_s/T) \to K_0(\kappa(s))$ $= \mathbf{Z}$. By induction, this group is torsion. By Lemmas 1.7, 6.13 and 6.16, the map:

$$\bigoplus_{s \in S_0} D_{I_s}(X_s/T) \to D_I(X/T)$$

is surjective. Since this map factors as

$$\bigoplus_{s \in S_0} D_{I_s}(X_s/T)^0 \oplus \bigoplus_{s \in S_0} \mathbf{Z} \to D_{I_s}(X_s/T)^0 \oplus D_J(\tilde{S}/T) \to D_I(X/T),$$

we get a surjective map

$$\bigoplus_{s \in S_0} D_{I_s}(X_s/T)^0 \to \Gamma_I.$$

This proves that Γ_I is torsion.

STEP 2. We prove that Φ_I is an isomorphism if char. $k = 0$ and an isomorphism on prime-to-p parts if char. $k = p > 0$. First we prove:

LEMMA 6.27. *Let S' be an integral scheme which is finite over S with function field k'. Let X' be the integral closure of X in $U' = U \times_S S'$. Let I' be an ideal of $\mathcal{O}_{X'}$ such that $I_{U'} = \mathcal{O}_{U'}$ and the norm map*

$$D_{I'}(X'/T) \to D_I(X/T)$$

is defined, and define Γ'_I and $\Phi_{I'}$ as before. Then if Proposition 6.25 is true for Γ'_I and $\Phi_{I'}$, it is true for Γ_I and Φ_I.

PROOF. For a closed point s' of S', let $X_{s'}$ be the closure of $U \times_S s'$ in X. Then $X_{s'}$ is an integral scheme finite over X_s. There exist ideals $I_{s'}$ of $\mathcal{O}_{X_s'}$ such that we have maps

$$D_{I_{s'}}(X_{s'}/T) \to D_{I'}(X'/T)$$
$$D_{I_{s'}}(X_{s'}/T) \to D_{I_s}(X_s/T),$$

where s is the closed point of S lying under s'. Consider the following commutative diagram

$$
\begin{array}{ccccc}
\bigoplus_{s \in S_0'} D_{I_{s'}}(X_{s'}/T)^0 & \longrightarrow & \Gamma_{I'} & \xrightarrow{\Phi_{I'}} & \pi_1^{ab}(U/k') \\
\downarrow & & \downarrow N & & \downarrow \phi \\
\bigoplus_{s \in S_0} D_{I_s}(X_s/T)^0 & \longrightarrow & \Gamma_I & \xrightarrow{\Phi_I} & \pi_1^{ab}(U/k)
\end{array}
$$

Here ϕ is the natural map. To prove the lemma, it will be enough to show
   (1) the norm map N is surjective;
   (2) the induced map $\mathrm{Ker}(N) \to \mathrm{Ker}(\phi)$ is surjective.

To prove (1), use the commutative diagram of exact sequences (Theorem 6.17):

$$
\begin{array}{ccccccc}
D_{I_{s'}}(X_{s'}/T) & \xrightarrow{N_{X_{s'}/X_s}} & D_{I_s}(X_s/T) & \longrightarrow & \mathrm{Gal}(\kappa(s')/\kappa(s)) & \longrightarrow & 0 \\
\downarrow & & \downarrow & & \downarrow & & \\
K_0(\kappa(s')) & \xrightarrow{N_{s'/s}} & K_0(\kappa(s)) & \longrightarrow & \mathrm{Gal}(\kappa(s')/\kappa(s)) & \longrightarrow & 0
\end{array}
$$

To prove (2), let k'' be a finite Galois extension containing k'. Then $\pi_1^{ab}(U/k)$ is equal to the group of coinvariants of $\pi_1^{ab}(U/k'')$ under the action of $\mathrm{Gal}(k''/k)$. Hence $\mathrm{Ker}(\phi)$ is generated by the image in $\pi_1^{ab}(U/k')$ of elements of the form $(g-1)(a)$ for $g \in G$ and $a \in \pi_1^{ab}(U/k'')$. Let S'' be the integral closure of S in k'' and define X'' and $X_{s''}$ for s'' a closed point of S'' in the same way as we did for $X', \ldots$ above. Similarly, take ideals I'', $I_{s''}$. Then by Čebotarev density, we can find a closed point s'' of S'' such that

(i) the map $S'' \to S$ is étale at s'', (ii) g is the Frobenius of s'', and (iii) $a \in \pi_1^{ab}(U/k'')$ lies in the image of the map

$$
D_{I_{s''}}(X_{s''}/T)^0 \to \Gamma_{I''} \xrightarrow{\Phi_{I''}} \pi_1^{ab}(U/k'').
$$

This proves assertion (2) and completes the proof of the lemma.

Now let l be a prime number different from char. K. We have shown that Γ_I is a torsion group and now we want to show that the map

$$
\Gamma_I(l) \to \pi_1^{ab}(U/k)(l)
$$

is an isomorphism.

Before stating the next lemma, we recall some terminology. Let $f \colon X \to S$ be a morphism of schemes and $\mathscr{F}$ an étale sheaf on X. Given a morphism $g \colon R \to S$, we have a diagram:

$$
\begin{array}{ccc}
X_R & \xrightarrow{f'} & R \\
g' \downarrow & & \downarrow g \\
X & \xrightarrow{f} & S
\end{array}
$$

We say the sheaf $R^i f_* \mathscr{F}$ commutes with any base change of S if for any such $g \colon R \to S$,

$$
g^* R^i f_* \mathscr{F} \cong R^i f'_*(g'^* \mathscr{F}).
$$

LEMMA 6.28. *There exists a nonempty subscheme S' of S such that for any m the étale sheaf $R^1 f_* \mathbf{Z}/l^m \mathbf{Z}$ is a locally constant constructible sheaf which commutes with any base change of S'.*

PROOF. This is an easy consequence of the finiteness theorem proved in [SGA $4\frac{1}{2}$, Th. Finitude, Théorème 1.9] and induction.

DEFINITION 6.29. A closed point s of S is said to be ordinary if there exists a neighborhood S' of s satisfying the conclusion of the lemma. An ordinary point is an l-Bloch point if the natural map

$$\pi_1^{ab}(U/k_s^h)(l) \to \pi_1^{ab}(U/k)(l)$$

is an isomorphism, where k_s^h is the henselization of S at s. We denote by B_S the set of all l-Bloch points of S (when l is fixed throughout the discussion).

Note that by Lemma 6.27 above, we may pass to a finite cover R of S at any time in the proof.

LEMMA 6.30. *Replacing S by a finite cover R if necessary, we may assume B_S is dense in S.*

PROOF. The fact that B_S is nonempty can be seen just as in the proof of Lemma 5.19 above, using the fact that $H^1(U_{\bar{k}}, \mathbf{Q}_l/\mathbf{Z}_l)^G$ is a finite group. We claim that B_S is dense if it is nonempty. To see this, let l^{m-1} be the order of $H^1(U_{\bar{k}}, \mathbf{Q}_l/\mathbf{Z}_l)^G$ and let H be the image of the Galois representation $\rho: G \to \mathrm{Aut}(M_{l^m})$. Then H is finite and the action of G on M is unramified at any ordinary point s. Let σ_s denote the Frobenius conjugacy class of s in H. Then the density follows from the Čebotarev density theorem and the following fact which the reader can verify: Suppose s, s' are two ordinary points of S and $\sigma_s = \sigma_{s'}^r$ for some positive integer r. If $s \in B_S$, $s' \in B_S$ (see [B2], p. 259).

Note that for any Bloch point s, the composite map

$$D_{I_s}(X_s/T)^0(l) \to \Gamma_I(l) \to \pi_1^{ab}(U/k)(l)$$

is an isomorphism, because it may be factored as

$$D_{I_s}(X_s/T)^0(l) \cong \pi_1^{ab}(U/s)(l) \cong \pi_1^{ab}(U/k_s^h)(l) \cong \pi_1^{ab}(U/k)(l).$$

Hence, it will be enough to show

PROPOSITION 6.31. *For any element $c = (c_i)$ of $\bigoplus_{s \in S_0} D_{I_s}(X_s/T)^0(l)$, its image in $\Gamma_I(l)$ lies in the image of $D_{I_{s_0}}(X_{s_0}/T)^0(l)$ for some Bloch point s_0.*

PROOF. We may assume that the support of $\mathcal{O}_{X_s}/I_s$ lies in $X_s - U_s$. Then by the induction hypothesis on X_s, we have an isomorphism of finite groups

$$D_{I_s}(X_s/T)^0(l) \cong \pi_1^{ab}(U/s)(l).$$

By Čebotarev density, there exist closed points x_i of X_{s_i} for each i such that the images of c_i and x_i in Γ_I are the same. Fix ordinary points t_j ($j = 1, \ldots, k$) different from the s_i such that the Frobenius conjugacy classes σ_{t_j} generate H. Let $y_j = i(s_j)$. We need the following lemma, which we will not prove.

LEMMA 6.32. *We may assume that there exists a one-dimensional integral subscheme C in X on which the x_j and y_j lie as regular points.*

Assuming this, we can finish the proof. By Lemma 6.13, there exists an ideal J of $\mathcal{O}_C$ such that the support of $\mathcal{O}_C/J$ does not contain any of the x_i or y_j and such that we have a natural map $\phi' : D_J(C/T) \to D_I(X/T)$; we denote by ϕ the composite of ϕ' with the natural map $D_I(X/T) \to \Gamma_I$. The image of ϕ contains the image of c in Γ_I and $\phi(y_j) = 0$. Let E be the function field of C. Then for any b in the image of ϕ and any $h \in H$, there exists a $\sigma \in \mathrm{Gal}(E^{\mathrm{sep}}/E)$ which maps to b under the composite

$$\mathrm{Gal}(E^{\mathrm{sep}}/E) \to \mathrm{Gal}(E^{\mathrm{ab}}/E) \cong \varprojlim_{J' \neq 0} D_{J'}(C/T) \to D_J(C/T) \xrightarrow{\phi} \Gamma_I$$

and to h under the map

$$\mathrm{Gal}(E^{\mathrm{sep}}/E) \to \mathrm{Gal}(k^{\mathrm{sep}}/k) \xrightarrow{\rho} H.$$

To complete the proof of the lemma, take b to be the image of c in Γ_I and for h any element in the Frobenius conjugacy class σ_{s_0} for some Bloch point s_0 and find σ as above. By Čebotarev density, there is a closed point s of S such that the image of σ in $D_J(C)$ is the same as that of x. Also, we have $\sigma_{s_0} = (\sigma_s)^r$, where $r = [\kappa(x) : \kappa(s)]$. Then s is a Bloch point and the image of c in Γ_I lies in the image of $C_{I_s}(X_s/T)$. This completes the proof of the proposition and of Step 2.

STEP 3. We prove the reciprocity law. The following lemma is easily verified:

LEMMA 6.33. *Let Y be an integral scheme which is finite over X such that the function field L is Galois over K and such that $V = U \times_X Y$ is finite étale over U. Then the reciprocity law for V implies the reciprocity law for U.*

Now replacing V with an open subset if necessary, we may assume that V is affine, smooth and geometrically connected over a nonempty open subscheme C of the spectrum of the ring of integers in a number field or a smooth connected curve over a finite field. Let k be the function field of C, and let k' be a finite extension of k over which V has a k'-rational point. Then there is a nonempty open subscheme S of C such that the integral closure of S in k' is finite étale over S and the map $V \times_C S' \to S'$ satisfies the conditions described at the beginning of Step 1. This proves the theorem.

Chapter 7. Complements and applications

In this chapter, we describe some "Hasse principles" for absolutely finitely generated fields. In some sense, these are really not applications of class field theory but applications of similar techniques which go into the proof of the reciprocity law in Chapter 6.

First, let us review the original Hasse principle from several different points of view. Let K be a global field. Then the Galois cohomology group $H^2(K, \mathbf{Q}/\mathbf{Z}(1))$ is the Brauer group of K. For each place v of K, there is a canonical map

$$\mathrm{inv}_v : \mathrm{Br}(K_v) \to \mathbf{Q}/\mathbf{Z}$$

which is an isomorphism if v is finite, an isomorphism onto the subgroup $\mathbf{Z}/2$ of $\mathbf{Q}/\mathbf{Z}$ if v is a real place and zero if v is complex. If $a \in \mathrm{Br}(K)$, its image in $\mathrm{Br}(K_v)$ is zero for almost all v. Let P be the set of all places of K. In the course of the proofs of the main theorems of class field theory for K, one derives an exact sequence

$$0 \to \mathrm{Br}(K) \xrightarrow{\mathrm{inv}} \bigoplus_{v \in P} \mathrm{Br}(K_v) \xrightarrow{\sum_v \mathrm{inv}_v} \mathbf{Q}/\mathbf{Z} \to 0.$$

That the map inv is injective is usually called the Brauer-Hasse-Noether theorem. Here is another interpretation of this sequence. Recall that for a nonarchimedean local field K_v with residue field $\mathbf{F}_v$, we have

$$\mathrm{Br}(K_v) \cong H^1(\mathbf{F}_v, \mathbf{Q}/\mathbf{Z});$$

this holds for real places if we take $K_v = \mathbf{F})v$. Then we can rewrite the exact sequence above as

$$0 \to H^2(K, \mathbf{Q}/\mathbf{Z}(1)) \to \bigoplus_{v \in P} H^1(\mathbf{F}_v, \mathbf{Q}/\mathbf{Z}) \to \mathbf{Q}/\mathbf{Z} \to 0.$$

We have an analogous sequence with $\mathbf{Z}/n$-coefficients. This also works for the p-primary part in characteristic $p > 0$ if we take flat cohomology instead of Galois cohomology. The advantage of this reformulation is that it is more suitable for generalization to higher dimensional fields.

Kato has made a very general conjecture which would imply that there are such sequences in any dimension [**K7**, Conjectures 0.3, 0.5]. We will summarize the work done in this direction.

Let X be an integral d-dimensional scheme of finite type over $\mathbf{Z}$. Then Kato has defined a complex which he calls C_n^0

$$C_n^0(X) : 0 \to H^{d+1}(K, \mathbf{Z}/n(d)) \xrightarrow{\partial} \bigoplus_{x \in X^1} H^d(\kappa(x), \mathbf{Z}/n(d-1)) \xrightarrow{\partial} \cdots$$

$$\xrightarrow{\partial} \bigoplus_{x \in X^{d-1}} H^2(\kappa(x), \mathbf{Z}/n(1)) \xrightarrow{\partial} \bigoplus_{x \in X^d} H^1(\kappa(x), \mathbf{Z}/n) \to \mathbf{Z}/n \to 0.$$

We regard the term involving points of codimension i as being of degree i. If l is a prime number, taking the direct limit over n a power of l gives the obvious analogue of Kato's conjecture with $\mathbf{Q}_l/\mathbf{Z}_l$-coefficients, which we call $C_{l^\infty}^0$ for short

$$C_{l^\infty}^0(X) : 0 \to H^{d+1}(K, \mathbf{Q}_l/\mathbf{Z}_l(d)) \xrightarrow{\partial} \bigoplus_{x \in X^1} H^d(\kappa(x), \mathbf{Q}_l/\mathbf{Z}_l(d-1)) \xrightarrow{\partial} \cdots$$

$$\xrightarrow{\partial} \bigoplus_{x \in X^{d-1}} H^2(\kappa(x), \mathbf{Q}_l/\mathbf{Z}_l(1)) \xrightarrow{\partial} \bigoplus_{x \in X^d} H^1(\kappa(x), \mathbf{Q}_l/\mathbf{Z}_l) \to \mathbf{Q}_l/\mathbf{Z}_l \to 0.$$

The definition of the maps ∂ is given in the last section of this chapter. We now state Kato's conjecture.

CONJECTURE 7.1. Assume X is a regular proper connected scheme of finite type over $\mathbf{Z}$ of pure dimension d. Then the complex $C_n^0(X)$ is exact.

We have the analogous conjecture with $\mathbf{Q}_l/\mathbf{Z}_l$-coefficients, and it is clear that exactness of C_n^0 implies exactness of $C_{l^\infty}^0$. It is usually easier to prove exactness of the latter sequence than the former. If we knew bijectivity of the Galois symbol for Milnor K-theory in degrees $\leq d + 1$, then we could prove the equivalence of Kato's conjecture with $\mathbf{Z}/n$ and $\mathbf{Q}_l/\mathbf{Z}_l$-coefficients.

The first major result in this direction was proved by Kato ([**K7**]):

THEOREM 7.2. *Let X be an integral proper scheme of dimension two over $\mathbf{Z}$ with function field K. Let n be a positive integer which is prime to* char. K. *If the total constant field of K has an embedding into $\mathbf{R}$, assume that n is odd. Then the complex C_n^0 is exact.*

Recently, there have been some other results in this direction for varieties over finite fields.

THEOREM 7.3 (Colliot-Thélène [**CT2**], Saito [**Sa11**]). *Let X be a smooth projective threefold over a finite field $\mathbf{F}$ and l a prime number different from* char. $\mathbf{F}$. *Then Kato's complex $C_{l^\infty}^0$ for X is exact.*

REMARK 7.4. Colliot-Thélène actually proves that for any smooth projective variety of dimension d over a finite field, the complex $C_{l^\infty}^0$ of Kato is exact in degrees $\geq d - 3$ and C_n^0 is exact in degrees $\geq d - 2$. Using logarithmic de Rham-Witt cohomology, Suwa has shown that the p-part of this is true as well, where $p = $ char. $\mathbf{F}$ [**Su**]. Saito has shown that Kato's conjecture with $\mathbf{Q}_l/\mathbf{Z}_l$-coefficients is true for certain threefolds over number fields.

Unfortunately, there is no Hasse principle for the Brauer group of the function field K of a scheme of finite type over $\mathbf{Z}$. Indeed, relations with the Tate conjecture for surfaces over finite fields and the conjecture on the finiteness of the Shafarevich-Tate group of an abelian variety over a number field indicate that the Brauer group of such a field K is a very tricky object.

Another type of Hasse principle first considered by Kato concerns the following situation. Let X be a smooth, proper, geometrically connected variety of dimension d over a number field k. Let K be the function field of X and for each place v of k, denote by K_v the function field of X_v. Let l be a prime number.

THEOREM 7.5 (Jannsen [J3]). *The natural map*

$$H^{d+2}(K, \mathbf{Q}_l/\mathbf{Z}_l(d+1)) \to \prod_v H^{d+2}(K_v, \mathbf{Q}_l/\mathbf{Z}_l(d+1))$$

is injective.

Taking $d = 0$, we have the Brauer-Hasse-Noether theorem for the Brauer group of a number field. The case $d = 1$ is due to Kato [K7] (who proves it with $\mathbf{Z}/n$-coefficients) and the case $d = 2$ was Jannsen's first result in this direction [J2]. Here are some applications:

THEOREM 7.6. *Let K be a function field in one variable over a number field k and D a central simple algebra over K of square-free index n. An element of K^* is a reduced norm from D if and only if it is a reduced norm from $D \otimes_K K_v$ for all places v of k.*

This result follows from Kato's version of Theorem 6.5 with $\mathbf{Z}/n$-coefficients and the following theorem of Merkur'ev-Suslin ([MS1], §12):

THEOREM 7.7. *Let D be a central simple algebra of square free index n over a field k of characteristic not dividing n, and denote by β its class in $\mathrm{Br}(k)_n = H^2(k, \mathbf{Z}/n(1))$. Then an element $\alpha \in k^*$ is a reduced norm for D iff the cup-product $\alpha \cup \beta$ is zero in $H^3(k, \mathbf{Z}/n(2))$.*

Using this result, Colliot-Thélène pointed out the following application to sums of squares in function fields:

THEOREM 7.8. *Let K be a function field in one variable over a number field. If K is not formally real, then -1 may be written as a sum of 4 squares in K. Any sum of squares in K may be written as a sum of 7 squares in K.*

For function fields in more than one variable, we have the following results, which are due to Colliot-Thélène and Jannsen [CTJ]:

THEOREM 7.9. *Let K be a function field in two variables over a number field.*

(i) *Any sum of squares in K is a sum of at most 8 squares.*

(ii) *Any sum of squares in the rational function field $K(t)$ is a sum of at most 16 squares.*

Recall the Galois symbol of degree n modulo 2 (see Chapter 1, §2):

$$K_n^M K/2 \to H^n(K, \mathbf{Z}/2).$$

THEOREM 7.10. *Let K be a function field in $d \geq 2$ variables over a number field k. Suppose the Galois symbol of degree $d + 1$ modulo 2 is an isomorphism for K and for all the K_v, for v a place of k. Then any sum of squares in K may be written as a sum of 2^{d+1} squares.*

We just briefly indicate the ingredients in the proofs of these results. First, the bijectivity of the Galois symbol of degree $d + 1$ modulo 2 for a field F implies that if $a_1, \ldots, a_{d+2} \in F^*$ and the cup product $a_1 \cup \cdots \cup a_{d+2}$ is trivial in $H^{d+2}(F, \mathbf{Z}/2)$, then the Pfister form

$$\langle\langle a_1, \ldots, a_{d+2} \rangle\rangle = \langle 1, -a_1 \rangle \otimes \cdots \otimes \langle 1, -a_{d+2} \rangle$$

in 2^{d+2} variables is totally hyperbolic over F. Applying this in the case where $F = K_v$ and using the Hasse principle, we see that if the form is totally hyperbolic over each K_v, then it is so over K. Now let $f \in K$ be a sum of squares and take the Pfister form Φ to be $\langle\langle f, -1, -1, \ldots, -1 \rangle\rangle$, where there are $d + 1$ copies of -1. Then for any field L containing K, f is a sum of 2^{d+1} squares in L iff this form is totally hyperbolic over L. A theorem of Pfister [Pf] says that for a real place v of k, f is a sum of 2^d squares and hence the form Φ above is totally hyperbolic over K_v. If v is a nonarchimedean place whose residue field is not of characteristic two (nondyadic), then -1 is a sum of two squares in k_v and one easily sees that Φ is totally hyperbolic if $d \geq 1$. If v is dyadic, then -1 is a sum of four squares, and the form Φ is totally hyperbolic over K_v if $d \geq 2$. By the Hasse principle and the assumption about the Galois symbol, Φ is totally hyperbolic over K and f is a sum of 2^{d+1} squares for $d \geq 2$. Now the Galois symbol modulo 2 is known to be an isomorphism in degrees ≤ 3, so for $d = 2$ we get part (i) of 7.9. Part (ii) follows from (i) and the following result:

THEOREM 7.11 (Pfister). *Let K be a formally real field, and let $L = K(t)$ be the rational function field in one variable over K. Then an element of L may be written as a sum of 2^{d+1} squares in L iff -1 is a sum of 2^d squares in any nonformally real extension of K.*

Definition of the maps in Kato's complex. To define the maps ∂, it is convenient to work in a more general context. Let A be an excellent discrete valuation ring with fraction field K and residue field k. Let $\widehat{K}$ be the completion of K with respect to the valuation and $\widehat{K}^{\mathrm{sh}}$ the strict henselization of $\widehat{K}$. Then there is a Hochschild-Serre spectral sequence

$$H^s(\widehat{K}^{\mathrm{sh}}/\widehat{K}, H^t(\widehat{K}^{\mathrm{sh}}, \mathbf{Z}/n(i))) \Rightarrow H^{s+t}(\widehat{K}, \mathbf{Z}/n(i))$$

and $\mathrm{Gal}(\widehat{K}^{\mathrm{sh}}/\widehat{K}) \cong \mathrm{Gal}(k^{\mathrm{sep}}/k)$.

We consider separately the cases where n is invertible in k and n is a power of $p = \mathrm{char.}\ k$.

(i) Suppose n is invertible in k. From the Hochschild-Serre spectral sequence and the fact that $\widehat{K}^{\mathrm{sh}}$ has cohomological dimension one for n-torsion modules, we get a map

$$H^j(\widehat{K}, \mathbf{Z}/n(i)) \to H^{j-1}(\widehat{K}^{\mathrm{sh}}/\widehat{K}, H^1(\widehat{K}^{\mathrm{sh}}, \mathbf{Z}/n(i))).$$

If n is invertible in the residue field k, then $\widehat{K}^{\text{sh}}$ contains the nth roots of unity, and then the valuation of $\widehat{K}^{\text{sh}}$ together with Hilbert's theorem 90 gives a natural map

$$H^1(\widehat{K}^{\text{sh}}, \mathbf{Z}/(n(i)) \to \mathbf{Z}/n(i-1);$$

hence, a map

$$H^j(\widehat{K}, \mathbf{Z}/n(i)) \to H^{j-1}(k, \mathbf{Z}/n(i-1)).$$

Now we define

$$\partial: H^j(K, \mathbf{Z}/n(i)) \to H^{j-1}(k, \mathbf{Z}/n(i-1))$$

to be the composite of this map with the natural map $H^j(K, \mathbf{Z}/n(i)) \to H^j(\widehat{K}, \mathbf{Z}/n(i))$. Note that this map exists for all i and j.

(ii) Suppose $n = p^r$, where $p = \text{char. } k$. We assume $[k : k^p] \leq p^{j-1}$. Then $H^j(\widehat{K}^{\text{sh}}, \mathbf{Z}/p^r(j-1)) = 0$ ([**K1**, §3.2, Lemma 3] and [**K3**, §0]) and the p-cohomological dimension of k is at most one [**S3**, Chapter II, §2]. Then the Hochschild-Serre spectral sequence mentioned above gives an isomorphism

$$(*) \quad H^j(\widehat{K}, \mathbf{Z}/p^r(j-1)) \cong H^1(k, H^{j-1}(\widehat{K}^{\text{sh}}, \mathbf{Z}/p^r(j-1))).$$

By the theorem of Bloch-Gabber-Kato (see Theorem 1.6, (iv), (v)), we have isomorphisms

$$K_{j-1}^M(\widehat{K}^{\text{sh}})/p^r \cong H^{j-1}(\widehat{K}^{\text{sh}}, \mathbf{Z}/p^r(j-1))$$

and

$$K_{j-2}^M(k^{\text{sep}})/p^r \cong H^{j-2}(k^{\text{sep}}, \mathbf{Z}/p^r(j-2)).$$

These and the tame symbol $K_{j-1}^M(\widehat{K}^{\text{sh}}) \to K_{j-2}^M(k^{\text{sep}})$ then give a map

$$H^{j-1}(\widehat{K}^{\text{sh}}, \mathbf{Z}/p^r(j-1)) \to H^{j-2}(k^{\text{sep}}, \mathbf{Z}/p^r(j-2)).$$

Now $H^1(k, H^{j-2}(k^{\text{sep}}, \mathbf{Z}/p^r(j-2))) \cong H^{j-1}(k, \mathbf{Z}/p^r(j-2))$, and putting everything together and composing with the natural map

$$H^j(K, \mathbf{Z}/p^r(j-1)) \to H^j(\widehat{K}, \mathbf{Z}/p^r(j-1)),$$

we finally get

$$\partial: H^j(K, \mathbf{Z}/p^r(j-1)) \to H^{j-1}(k, \mathbf{Z}/p^r(j-2)).$$

Now we can define the boundary maps in Kato's complex. Let x be a point of X of codimension $j-1$. Let Z be its closure and $\widetilde{Z}$ the normalization of Z. If y is a codimension j point of X which lies on Z, define

$$\partial_{xy} = \sum_{v \mapsto y} \text{Cor}_{\kappa(v)/\kappa(y)} \partial_v,$$

where the sum runs over all v in $\widetilde{Z}$ which map to y under the natural map $\widetilde{Z} \to Z$. If y is not in the closure of x, define ∂_{xy} to be zero. Then $\partial_x = \sum_{y \in Z^1} \partial_{xy}$ and ∂ is the direct sum over all codimension $(j-1)$-points x of ∂_x.

REFERENCES

[AK]   A. Altman and S. Kleiman, *Bertini theorems for hypersurface sections containing a subscheme*, Comm. Algebra **7** (1979), 775–790.

[A1]   E. Artin, *Über eine neue Art von L-Reihen*, Hamb. Abh. **3** (1923) 89–108; also in Collected Papers (S. Lang and J. Tate, eds.), Springer, New York, 1982.

[A2]   _____, *Beweis des allgemeinen Reziprozitätsgesetzes*, Hamb. Abh. **5** (1927) 353–363; also in Collected Papers (S. Lang and J. Tate, eds.), Springer, New York, 1982.

[AT]   E. Artin and J. Tate, *Class field theory*, Benjamin, New York, 1967.

[BT]   H. Bass and J. Tate, *The Milnor ring of a global field*, Algebraic K-theory II-"classical" K-theories and connections with arithmetic (H. Bass, ed.), Lecture Notes in Mathematics, vol. 342, Springer-Verlag, Berlin and New York, 1973.

[B1]   S. Bloch, *Algebraic K-theory and crystalline cohomology*, Inst. Hautes Études Sci. Publ. Math. **47** (1977), 181–278.

[B2]   _____, *Algebraic K-theory and class field theory for arithmetic surfaces*, Ann. of Math. **114** (1981), 229–265.

[BK]   S. Bloch and K. Kato, *p-adic étale cohomology*, Inst. Hautes Études Sci. Publ. Math. **63** (1986), 107–152.

[BO]   S. Bloch and A. Ogus, *Gersten's conjecture and the homology of schemes*, Ann. Sci. École Norm. Sup. **7** (1974), 181–202.

[Br]   J.-L. Brylinski, *Théorie du corps de classes de Kato et revêtements abéliens de surfaces*, Ann. Inst. Fourier **33** (1983), 23–38.

[CE]   H. Cartan and S. Eilenberg, *Homological algebra*, Princeton University Press, Princeton, NJ, 1956.

[CF]   J. Cassels and A. Fröhlich, *Algebraic number theory*, Academic Press, New York, 1986.

[CW]   S. Chase and W. Waterhouse, *Moore's theorem on uniqueness of reciprocity laws*, Invent. Math. **16** (1972), 267–279.

[C1]   C. Chevalley, *Sur la théorie du corps de classes dans les corps finis et les corps locaux*, Fac. Sci. Imp. Univ. Tokyo **2** (1933), 365–476.

[C2]   _____, *La théorie du corps de classes*, Ann. of Math. **41** (1940), 394–418.

[CT1]   J.-L. Colliot-Thélène, *Appendix to a paper of K. Kato*, J. reine angew. Math. **366** (1986), 181–183.

[CT2]   _____, *On the reciprocity sequence in the higher class field theory of function fields*, Algebraic K-Theory and Algebraic Topology (P. G. Goerss and J. F. Jardine, eds.), NATO Adv. Sci. Inst. Ser., Kluwer, Dordrecht, 1993.

[CTJ]   J.-L. Colliot-Thélène and U. Jannsen, *Sommes de carrés dans les corps de fonctions*, C. R. Acad. Sci. Paris **312** (1991), 759–762.

[CTR1]   J.-L. Colliot-Thélène and W. Raskind, $\mathcal{H}_2$-*cohomology and the second Chow group*, Math. Ann. **270** (1985), 165–199.

[CTR2]   _____, *On the reciprocity law for surfaces over finite fields*, J. Fac. Sci. Univ. Tokyo **33** (1986), 283–294.

[CTSS]   J.-L. Colliot-Thélène, J.-J. Sansuc, and C. Soulé, *Torsion dans le groupe de Chow de codimension deux*, Duke Math. J. **50** (1983), 763–801.

[Co]   K. Coombes, *Local class field theory for curves*, Applications of Algebraic K-Theory to Algebraic Geometry and Number Theory (S. Bloch, R. K. Dennis, E. M. Friedlander, and M. Stein, eds.), Contemp. Math., vol. 55, pp. 117–134, Amer. Math. Soc., Providence, RI, 1986.

[Cor]   D. Coray, *On an argument of Spencer Bloch*, 1980, typescript.

[D]   P. Deligne, *La conjecture de Weil*. II, Inst. Hautes Études Sci. Publ. Math. **52** (1980), 137–252.

[DM]   P. Deligne and D. Mumford, *The irreducibility of the moduli space of curves of given genus*, Inst. Hautes Études Sci. Publ. Math. **36** (1969), 75–109.

[De1]   C. Deninger, *On Artin-Verdier duality for function fields*, Math. Z. **188** (1984), 91–100.

[De2]   _____, *An extension of Artin-Verdier duality to nontorsion sheaves*, J. reine angew. Math. **366** (1986), 18–31.

[De3]   ———, *Duality in the étale cohomology of one-dimensional proper schemes and generalizations*, Math. Ann. **277** (1987), 529–541.

[DeW]   C. Deninger and K. Wingberg, *Artin-Verdier duality for n-dimensional local fields involving higher algebraic K-sheaves*, J. Pure Appl. Algebra **43** (1986), 243–255.

[DS]   R. K. Dennis and M. Stein, K_2 *of split radicals and semilocal rings revisited*, Algebraic K-Theory. II: "Classical" K-Theories and Connections with Arithmetic (H. Bass, ed.), Lecture Notes in Math., vol. 342, Springer-Verlag, Berlin and New York, 1973.

[FC]   G. Faltings and C.-L. Chai, *Degeneration of abelian varieties*, Springer-Verlag, Berlin and New York, 1990.

[FSS]   B. Fein, D. Saltman, and M. Schacher, *Heights of cyclic field extensions*, Bull. Soc. Math. Belg. **40** (1988), 213–223.

[Fe1]   I. Fesenko, *On class field theory of multidimensional local fields of positive characteristic*, Algebraic K-Theory, Adv. in Soviet Math., vol. 4, Amer. Math. Soc., Providence, RI, 1991, pp. 103–127.

[Fe2]   ———, *A multidimensional local theory of class fields*, Dokl. Akad. Nauk. SSSR **318** (1991), 47–50; English transl. in Soviet Acad. Sci. Dokl. Math. **43** (1991).

[Fe3]   ———, *A multidimensional local class field theory. II*, Algebra i Analiz **3** (1991), 168–189; English transl. in St. Petersburg Math. J. **3** (1992).

[Fe4]   ———, *Class field theory of multidimensional local fields of characteristic zero with residue field of positive characteristic*, Algebra i Analiz **3** (1991), 165–196; English transl. in St. Petersburg Math. J. **3** (1992).

[Fe5]   ———, *Local fields, local class field theory, higher local class field theory via algebraic K-theory*, Algebra i Analiz **4** (1992), 1–41; English transl. in St. Petersburg Math. J. **4** (1993), (also to appear in Handbook of Algebra (M. Hazewinkel, ed.)), vol. 1, 1993.

[Fe6]   ———, *Local class field theory: perfect residue field case*, Izv. Russian Akad. Nauk. Ser. Mat **57** (1993), 72–91; English transl. in Russian Acad. Sci. Izv. Math. (1993).

[Fe7]   ———, *Abelian local p-class field theory*, 1993, preprint.

[Fe/V]   I. Fesenko and S. Vostokov, *Local fields and their extensions*, Transl. Math. Monographs, vol. 121, Amer. Math. Soc., Providence, RI, 1993.

[Fe/V/Z]   I. Fesenko, S. Vostokov, and I. Zhukov, *On the theory of multidimensional local fields: methods and constructions*, Algebra i Analiz **2** (1990), 91–118; English transl. in Leningrad Math. J. **2** (1991), 775–800.

[FP]   T. Fimmel and A. Parshin, *Adeles*, to appear.

[Fo]   J.-M. Fontaine, *Groupes finis commutatifs sur les vecteurs de Witt*, C. R. Acad. Sci. Paris **280** (1975), 1423–1425.

[FL]   J.-M. Fontaine et G. Laffaille, *Constructions de répresentations p-adiques*, Ann. Sci. École Norm. Sup. **15** (1982), 547–608.

[Fu1]   W. Fulton, *Rational equivalence on singular varieties*, Inst. Hautes Études Sci. Publ. Math. **45** (1975), 147–167.

[Fu2]   ———, *Intersection theory*, Springer-Verlag, Berlin and New York, 1983.

[Furt]   P. Furtwängler, *Allgemeiner Existenzbeweis für den Klassenkörper eines beliebigen algebraischen Zahlkörpers*, Math. Ann. **63** (1907), 1–37.

[Gra1]   D. Grayson, *The K-theory of hereditary categories*, J. Pure Appl. Algebra **11** (1977), 67–74.

[Gra2]   ———, *Universal exactness in algebraic K-theory*, J. Pure Appl. Algebra **36** (1985), 139–141.

[Gro]   M. Gros, *Sur la partie p-primaire du groupe de Chow de codimension deux*, Comm. Algebra **13** (1985), 2407–2420.

[Ha1]   H. Hasse, *Die Normenresttheorie relativ-Abelscher Zahlkörper als Klassenkörpertheorie in Kleinen*, J. Reine Angew. Math. **162** (1930), 145–154.

[Ha2]   ———, *Theorie der relativ-Zyklischen algebraischen Funtionenkörper inbesondere bei endlichen Konstantenkörper*, J. Reine Angew. Math. **172** (1934), 37–54.

[Ha3]   ———, *Bericht über neure Untersuchungen und Probleme aus der Theorie der algebraischen Zahlkörper*, Jber. Deutsch. Math. Verein I **35** (1926), 1–55; Ia **36** (1927), 231–311; II, 1930 (Physica Verlag, 1965).

[Ha4]   ———, *History of class field theory*, Algebraic Number Theory (J.W.S. Cassels and A. Fröhlich, eds.), Academic Press, New York, 1967, pp. 266–279.

[He]   E. Hecke, *Über eine neue Art von Zetafunktionen und ihre Beziehungen zur Verteilung der Primzahlen*, Math. Z. **1** (1918), 357–376; Math. Z. **4** (1920), 11–21; reprinted in Mathematische Werke, Vandenhoeck and Ruprecht, Göttingen, 1959.

[Hi]   D. Hilbert, *Über die Theorie der relativ-Abelschen Zahlkörper*, Acta Math. **26** (1902), 99–132; reprinted in Gesammelte Abhandlungen, Band I: Zahlentheorie, Springer-Verlag, Berlin and New York, 1970; reprinted by Chelsea Publishing Company, New York..

[HN]   G. Hochschild and T. Nakayama, *Cohomology of class field theory*, Ann. of Math. **55** (1952), 348–366.

[I]   L. Illusie, *Complexe de de Rham-Witt et cohomologie cristalline*, Ann. Sci. École Norm. Sup. **12** (1979), 501–661.

[Iw1] K. Iwasawa, *Local class field theory*, Oxford Univ. Press, London and New York, 1986.

[Iw2] ———, *On papers of Takagi in number theory*, in Teiji Takagi: Collected Papers (S. Iyanaga, ed.), 342–351, Springer-Verlag, Berlin and New York, 1990.

[J1]   U. Jannsen, *On the l-adic cohomology of varieties over number fields and its Galois cohomology*, Galois Groups over Q (Y. Ihara, K. Ribet, and J.-P. Serre, eds.), Math. Sci. Res. Inst. Publ., vol. 16, Springer-Verlag, Berlin and New York, 1989, 315–360.

[J2]   ———, *Principe de Hasse cohomologique*, in Séminaire de Théorie des Nombres, Paris, 1989–1990 (S. David, ed.), Birkhäuser, Boston, 1992, 121–140.

[J3]   ———, *A Hasse principle for function fields over number fields*, in preparation.

[J4]   ———, *Rigidity theorems for K-theory and other functors*, in preparation.

[Ka]   M. Kapranov, *Analogies between the Langlands correspondence and topological quantum field theory*, 1993, preprint.

[K1]   K. Kato, *A generalization of local class field theory by using K-groups*. I, II, III, J. Fac. Sci. Univ. Tokyo **26** (1979), 303–376; **27** (1980), 603–683; **29** (1982), 31–43.

[K2]   ———, *The existence theorem for higher local class field theory*, Inst. Hautes Études Sci., 1980, preprint.

[K3]   ———, *Galois cohomology of complete discrete valued fields*, Algebraic K-Theory, Oberwolfach 1980, Lecture Notes in Math., vol. 967, Springer-Verlag, Berlin and New York, 1982, 215–238.

[K4]   ———, *Class field theory and algebraic K-theory*, Algebraic Geometry: Tokyo/Kyoto, 1982, Lecture Notes in Math., vol. 1016, Springer-Verlag, Berlin and New York, 1983, 109–126.

[K5]   ———, *Milnor K-theory and the Chow group of zero-cycles*, Applications of Algebraic K-Theory to Algebraic Geometry and Number Theory (S. Bloch, R. K. Dennis, E. Friedlander, and M. Stein, eds.), Contemp. Math., vol. 55, Amer. Math. Soc., Providence, RI, 1986, 241–253.

[K6]   ———, *Residue homomorphisms in Milnor's K-theory*, Galois Groups and Their Representations, Adv. Stud. Pure Math., vol. 2, Kinokuniya-North Holland, Amsterdam, 1983, 153–172.

[K7]   ———, *A Hasse principle for two-dimensional global fields*, J. reine angew. Math. **366** (1986), 142–183.

[K8]   ———, *Generalized class field theory*, Proceedings of the International Congress of Mathematicians, Kyoto, 1990, Amer. Math. Soc., Providence, RI, 1991, pp. 419–428.

[K9]   ———, *A generalization of class field theory*, Sûgaku **40** (1988), 289–311, (to appear in AMS translation of Sûgaku).

[KS1] K. Kato and S. Saito, *Two-dimensional class field theory*, Galois Groups and their Representations, Adv. Stud. Pure Math., vol. 2 (Y. Ihara, ed.), Kinokuniya-North Holland, Amsterdam, 1983, 103–152.

[KS2] ———, *Unramified class field theory of arithmetical surfaces*, Ann. of Math. **118** (1983), 241–275.

[KS3] ———, *Global class field theory of arithmetic schemes*, Applications of Algebraic K-Theory to Algebraic Geometry and Number Theory (S. Bloch, R. K. Dennis, E. Friedlander, and M. Stein, ed.), Contemp. Math., vol. 55, Amer. Math. Soc., Providence, RI, 1986, 255–331.

[KL]   N. Katz and S. Lang, *Finiteness theorems in geometric class field theory*, Enseign. Math. **27** (1981), 285–319.

[Ke]   I. Kersten, *Brauergruppen von Körpern*, Aspekte der Math., Vieweg, Braunschweig and Wiesbaden, 1990.

[Ko1]   Y. Koya, *A generalization of class formation by using hypercohomology*, Invent. Math. **101** (1990), 705–715.

[Ko2]   _____, *Hasse's norm theorem for K_2*, Proc. Japan Acad. Ser. A **69** (1993), 13–15.

[Ko3]   _____, *Hasse's norm theorem for K_2*, K-Theory **7** (1993), 41–54.

[Ko4]   _____, *Class field theory without Theorem* 90, 1994, preprint.

[Kr]   L. Kronecker, *Über die algebraisch auflösbaren Gleichungen*. I, Sbr. Preuss Akad. Wiss. (1853), 365–374; also in Mathematische Werke, vol. 4, pp. 1–13, Chelsea Publishing Company, 1968.

[L1]   S. Lang, *Unramified class field theory over function fields in several variables*, Ann. of Math. **64** (1956), 285–325.

[L2]   _____, *Sur les séries L d'une variété algébrique*, Bull. Soc. Math. France **84** (1956), 385–401.

[L3]   _____, *Algebraic groups over finite fields*, Amer. J. Math. **78** (1956), 555–563.

[LS]   S. Lang and J.-P. Serre, *Sur les revêtements non-ramifiés des variétés algébriques*, Amer. J. Math. **79** (1957), 319–330.

[LL]   R. Langlands, *On the functional equations of Artin L-series*, 1967, mimeographed notes.

[La]   G. Laumon, *Correspondance de Langlands géométrique pour les corps de fonctions*, Duke Math. J. **54** (1987), 309–359.

[Li1]   S. Lichtenbaum, *Duality theorems for curves over p-adic fields*, Invent. Math. **7** (1969), 120–136.

[Li2]   _____, *The construction of weight-two arithmetic cohomology*, Invent. Math. **88** (1987), 183–215.

[Li3]   _____, *New results on weight-two arithmetic cohomology*, Grothendieck Festschrift (P. Cartier, ed.), Birkhäuser, Boston, 1990, 35–55.

[Ma]   B. Mazur, *Notes on étale cohomology of number fields*, Ann. Sci. École Norm. Sup. **6** (1973), 521–552.

[MS1]   A. S. Merkur'ev and A. A. Suslin, *K-cohomology of Severi-Brauer varieties and norm residue homomorphism*, Izv. Akad. Nauk. SSSR Ser. Mat. **46** (1982), 1011–1046; English transl. in Mathematics USSR Izv. **21** (1983).

[MS2]   _____, *On the norm residue homomorphism of degree* 3, Izv. Akad. Nauk SSSR **54** (1990), 339–356; English transl. in Math. USSR Izv. **36** (1991), 349–367.

[M1]   J. Milne, *Duality in the flat cohomology of a surface*, Ann. Sci. École Norm. Sup. **9** (1976), 171–202.

[M2]   _____, *Étale cohomology*, Princeton Math. Ser., vol. 33, Princeton Univ. Press, Princeton, NJ, 1980.

[M3]   _____, *Zero-cycles on algebraic varieties in nonzero characteristic: Rojtman's theorem*, Compositio Math. **47** (1982), 271–287.

[M4]   _____, *Arithmetic duality theorems*, Academic Press, New York, 1986.

[M5]   _____, *Values of zeta functions of varieties over finite fields*, Amer. J. Math. **108** (1986), 297–360.

[Mi]   J. Milnor, *Algebraic K-theory and quadratic forms*, Invent. Math. **9** (1970), 318–344.

[Mo]   C. Moore, *Group extensions of p-adic and adelic linear groups*, Inst. Hautes Études Sci. Publ. Math. **35** (1968), 5–70.

[Mu]   D. Mumford, *Abelian varieties*, Oxford University Press, London and New York, 1970.

[Na1]   M. Nagata, *Remarks on a paper of Zariski*, Proc. Nat. Acad. Sci. USA **44** (1958), 796–799.

[Na2]   _____, *Purity of branch locus in regular local rings*, Illinois J. Math. **3** (1959), 328–333.

[Na3]   _____, *Local rings*, Interscience Publ., vol. 13, Wiley, New York, 1962.

[Na4]   _____, *Imbedding of an abstract variety in a complete variety*, J. Math. Kyoto Univ. **2** (1962), 1–10.

[Nak1] T. Nakayama, *Idèle class factor sets and class field theory*, Ann. of Math. **55** (1952), 73–84.

[Nak2] _____, *Cohomology of class field theory and tensor product modules*, Ann. of Math. **65** (1957), 255–267.

[N1] J. Neukirch, *Class field theory*, Grundlehren Math. Wiss., vol. 280, Springer-Verlag, Berlin and New York, 1986.

[N2] _____, *Algebraische Zahlentheorie*, Springer-Verlag, Berlin and New York, 1992.

[N3] _____, *Microprimes*, Math. Annalen **298** (1994), 629–666.

[Ni] Y. Nisnevich, *The completely decomposed topology on schemes and associated descent spectral sequences in algebraic K-theory*, Algebraic K-theory: Connections with Geometry and Topology (J. F. Jardine and V. Snaith, eds.), NATO Adv. Sci. Inst. Kluwer, Dordrecht, 1989, 241–342.

[O] F. Oort, *Finite commutative group schemes*, Lecture Notes in Math., vol. 15, Springer-Verlag, Berlin and New York, 1966.

[Pan] I. Panin, *Fields whose K_2 is zero; torsion in $H^1(X; \mathscr{K}_2)$ and $CH^2(X)$*, Zap. Nauchn. Sem. Leningrad. Otdel. Mat. Inst. Steklov. (LOMI) **116**, 108–118.

[P1] A. Parshin, *Class fields and algebraic K-theory*, Uspekhi. Mat. Nauk **30** (1975), 253–254.

[P2] _____, *On the arithmetic of two-dimensional schemes. I: repartitions and residues*, Izv. Akad. Nauk SSSR **40** (1976), 736–773; English transl. in Math. USSR Izv. **10** (1976), 695–747.

[P3] _____, *Abelian coverings of arithmetic schemes*, Dokl. Akad. Nauk. SSSR **243** (1978), no. 4, 855–858; English transl. in Soviet Math. Dokl. **19** (1978), 1438–1442.

[P4] _____, *Local class field theory*, Trudy Mat. Inst. Akad. Nauk SSSR **165** (1985), 143–170; English transl. in Proc. Steklov Inst. Math. **183** (1985).

[P5] _____, *Galois cohomology and the Brauer group of local fields*, Trudy Mat. Inst. Akad. Steklova **183** (1990), 159–169; English transl. in Proc. Steklov. Inst. Math. (1991).

[Pf] A. Pfister, *Zur Darstellung definiter Funktionen als Summe von Quadraten*, Invent. Math. **4** (1967), 229–237.

[Po] G. Poitou, *Cohomologie des modules finis*, Dunod, Paris, 1967.

[Q] D. Quillen, *Higher algebraic K-theory*. I, Algebraic K-Theory. I: Higher K-Theories (H. Bass, ed.), Lecture Notes in Math., vol. 341, Springer-Verlag, Berlin and New York, 1973.

[Ro] M. Rosenlicht, *Generalized Jacobian varieties*, Ann. of Math. **59** (1954), 505–530.

[Sa1] S. Saito, *Functional equation of L-functions of varieties over finite fields*, J. Fac. Sci. Univ. Tokyo **31** (1984), 287–296.

[Sa2] _____, *Class field theory for curves over local fields*, J. Number Theory **21** (1985), 44–80.

[Sa3] _____, *Unramified class field theory of arithmetical schemes*, Ann. of Math. **121** (1985), 251–281.

[Sa4] _____, *Arithmetic on two-dimensional local rings*, Invent. Math. **85** (1986), 379–414.

[Sa5] _____, *General fixed point formula for algebraic surfaces and the theory of Swan representations for two-dimensional local rings*, Amer. J. Math. **109** (1987), 1009–1042.

[Sa6] _____, *A Global duality theorem for varieties over number fields*, Algebraic K-Theory: Connections with Geometry and Topology (J. F. Jardine and V. P. Snaith, eds.), Kluwer, Dordrecht, 1989.

[Sa7] _____, *Class field theory for two-dimensional local rings*, Galois Groups and Their Representations, Kyoto 1985/Tokyo 1986, Adv. Stud. Pure Math., vol. 12, Kinokuniya-North Holland, Amsterdam, 1987, 343–373.

[Sa8] _____, *Some observations on motivic cohomology of arithmetic schemes*, Invent. Math. **98** (1989), 371–404.

[Sa9] _____, *Arithmetic theory of arithmetic surfaces*, Ann. of Math. **129** (1989), 547–589.

[Sa10] _____, *Torsion zero-cycles and étale homology of singular schemes*, Duke Math. J. **64** (1991), 71–83.

[Sa11] _____, *Cohomological Hasse principle for a threefold over a finite field*, Algebraic K-Theory and Algebraic Topology (P. G. Goerss and J. F. Jardine, eds.), NATO Adv. Sci. Inst., Kluwer, Dordrecht, 1993.

[Sa12]   _____, *Class field theory for surfaces over local fields*, in preparation.

[Sal1]   P. Salberger, *Torsion cycles of codimension two and l-adic realizations of motivic co-homology*, Séminaire de Théorie des Nombres 1991/1992 (S. David, ed.), Birkhäuser, Boston, 1993.

[Sal2]   _____, *Class field theory for surfaces over local fields*, in preparation.

[Schmid] H. Schmid, *Über das Reziprozitätsgesetz in relativ-zyklischen algebraischen Functionenkörpern mit endlichen Konstantkörper*, Math. Z. **40** (1936), 94–109.

[Schmidt1] F. Schmidt, *Zur Klassenkörpertheorie im Kleinen*, J. Reine Angew. Math. **162** (1930), 155–168.

[Schmidt2] F. K. Schmidt, *Die Theorie der Klassenkörper über einem Körper algebraischer Funktionen in einer Unbestimmten und mit endlichem Koeffizientenbereich*, S.-B. Phy.-Med. Soz. Erlangen **62** (1931), 267–284.

[S1]   J.-P. Serre, *Groupes algébriques et corps de classes*, Hermann, Paris, 1959; English transl., *Algebraic groups and class field theory*, Graduate Texts in Math., vol. 117, Springer-Verlag, Berlin and New York, 1988.

[S2]   _____, *Zeta and L-functions*, Arithmetical Algebraic Geometry (P. Schilling, ed.), Harper and Row, New York, 1965, pp. 82–92, reprinted in Oeuvres, Volume II, Springer-Verlag, Berlin and New York, 1986.

[S3]   _____, *Cohomologie Galoisienne*, Lecture Notes in Math., vol. 5, Springer-Verlag, Berlin and New York, 1965.

[S4]   _____, *Corps locaux*, 2ème ed., Hermann, Paris, 1968; English transl. *Local fields*, Graduate Texts in Math., vol. 67, Springer-Verlag, Berlin and New York, 1979.

[Sh]   C. Sherman, *Some theorems on coherent sheaves*, Comm. Algebra **7** (1979), 1489–1508.

[So]   M. Somekawa, *On Milnor K-groups attached to semi-abelian varieties*, K-Theory **4** (1990), 105–119.

[Sp]   M. Spiess, *Artin-Verdier duality for arithmetic surfaces*, Ph.D. thesis, Universität Regensburg, 1994.

[Sus]   A. Suslin, *Torsion in K_2 fields*, K-Theory **1** (1987), 5–29.

[Su]   N. Suwa, *A note on Gersten's conjecture for logarithmic Hodge-Witt sheaves*, 1993, preprint.

[Ta1]   T. Takagi, *Über eine Theorie des relativ Abel'schen Zahlkörpers*, J. College of Sci. Univ. Tokyo **41** (1920), 1–133; also in Collected Papers (S. Iyanaga, ed), Springer-Verlag, Berlin and New York, 1990.

[Ta2]   _____, *Über das Reciprocitätsgesetz in einem beliebigen algebraischen Zahlkörper*, J. College of Sci. Imp. Univ. Tokyo **44** (1922), 1–50; also in Collected Papers (S. Iyanaga, ed.), Springer-Verlag, Berlin and New York, 1990.

[T1]   J. Tate, *Fourier analysis in number fields and Hecke's zeta-functions*, Ph.D. Thesis, Princeton Univ., Princeton, NJ, 1950; reprinted in *Algebraic number theory* (J. W. S. Cassels and A. Fröhlich, eds.), Academic Press, New York, 1967.

[T2]   _____, *The higher dimensional cohomology groups of class field theory*, Ann. of Math. **56** (1952), 294–297.

[T3]   _____, *WC-groups over p-adic fields*, Sém. Bourbaki, 10ème année, 1957/58.

[T4]   _____, *Duality theorems in the Galois cohomology of number fields*, Proceedings of the International Congress of Mathematicians, Stockholm 1962, Institut Mittag-Leffler.

[T5]   _____, *p-divisible groups*, Proceedings of a Conference on Local Fields (T. Springer, ed.), Springer-Verlag, Berlin and New York, 1966.

[T6]   _____, *Global class field theory*, Algebraic number theory, Academic Press, New York, 1986, 162–203.

[T7]   _____, *Relations between K_2 and Galois cohomology*, Invent. Math. **36** (1976), 257–274.

[Th1]   R. Thomason, *Algebraic K-theory and étale cohomology*, Ann. Sci. École. Norm. Sup. **18** (1985), 437–552.

[Th2]   _____, *Higher algebraic K-theory of schemes and of derived categories*, Grothendieck Festschrift. III, (P. Cartier, ed.), Birkhäuser, Boston, 1990, 247–435.

[V]   S. Vostokov, *Explicit construction of class field theory for a multidimensional local field*, Izv. Akad. Nauk. SSSR **49** (1985), 283–308; English transl. in Math. USSR Izv. **26** (1986), 263–287.

[Wa]   F. Waldhausen, *Algebraic K-theory of spaces*, Algebraic and Geometric Topology, Lecture Notes in Math. 1126, Springer-Verlag, Berlin and New York, 1985, 318–419.

[We1] H. Weber, *Theorie der Abel'schen Zahlkörper*. I, Acta. Math. Stockholm **8** (1886), 193–263; II, Acta. Math. Stockholm **9** (1887), 105–130.

[We2] H. Weber, *Über Zahlgruppen in algebraischen Körpern*. I, Math. Ann. **48** (1897), 433–473; II, Math. Ann. **49** (1897), 83–100; III, Math. Ann. **50** (1898), 1–26.

[We3] H. Weber, *Lehrbuch der Algebra, III: Elliptische Funktionen und algebraische Zahlen*, Vieweg, Braunschweig, 1908.

[W1]   A. Weil, *Sur les courbes algébriques et les variétés qui s'en déduisent*, Hermann, Paris, 1948.

[W2]   ——, *Sur la théorie du corps de classes*, J. Math. Soc. Japan **3** (1951), 1–35; reprinted in Collected Works, vol. I, Springer-Verlag, Berlin and New York, 1979.

[Wi]   E. Witt, *Zyklische Körper und Algebren der Charakteristik p vom Grade p^n*, J. reine angew. Math. **176** (1936), 126–140.

[Z]   O. Zariski, *On the purity of the branch locus of algebraic functions*, Proc. Nat. Acad. Sci. USA **44** (1958), 791–796.

[SGA 1] A. Grothendieck, *Revêtements Étales et Groupe Fondamental*, Lecture Notes in Math., vol. 224, Springer-Verlag, Berlin and New York, 1971.

[SGA 2] ——, *Cohomologie locale des faisceaux cohérents et théorèmes de Lefschetz locaux et globaux*, Masson et Cie, North-Holland, Amsterdam, 1968.

[SGA 4] (par P. Deligne avec la collaboration de J.-F. Boutot, A. Grothendieck, L. Illusie et J.-L. Verdier), *Théorie des Topos et Cohomologie Étale des Schémas*, Lecture Notes in Math., vols. 269, 270, and 305, Springer-Verlag, Berlin and New York, 1972.

[SGA $4\frac{1}{2}$] (dirigé par M. Artin, A. Grothendieck et J.-L. Verdier, avec la collaboration de P. Deligne et B. Saint-Donat), *Cohomologie étale*, Lecture Notes in Math., vol. 569, Springer-Verlag, Berlin and New York, 1977.

[SGA 7] (avec la collaboration de M. Raynaud et D. S. Rim), *Groupes de monodromie en géométrie algébrique (première partie)*, Lecture Notes in Math., vol. 288, Springer-Verlag, Berlin and New York, 1972.

DEPARTMENT OF MATHEMATICS, UNIVERSITY OF SOUTHERN CALIFORNIA, LOS ANGELES, CALIFORNIA 90089-1113

E-mail address: raskind@mtha.usc.edu

Proceedings of Symposia in Pure Mathematics
Volume **58.1** (1995)

Brauer Groups of Invariant Fields, Geometrically Negligible Classes, an Equivariant Chow Group, and Unramified H^3

DAVID J. SALTMAN

Introduction

As the title suggests, this paper will touch on a variety of topics connected with Galois cohomology and the Brauer group. The basic situation is the following. Let G be a finite group and F an algebraically closed field of characteristic 0. Let V be a faithful representation of G over F. We form the field of rational functions $F(V)$ and note that G acts naturally on this field. If $K = F(V)$ and $L = F(V)^G$ is the invariant field, we consider a number of questions about these objects. First of all, what can we say about the Brauer group $\mathrm{Br}(F(V)^G)$? More specifically, can we describe the image of $\mathrm{Br}(F(V)^G)$ in $\mathrm{Br}(F(V))$? Secondly, let G_L be the absolute Galois group of L and similarly for G_K. Denote by $\mu \subset F^*$ the group of roots of one. We have an exact sequence

(1) $$1 \to G_K \to G_L \to G \to 1$$

and so we can ask for a description of the kernel $H^3(G, \mu) \to H^3(G_L, \mu)$. Note that $H^3(G_L, \mu)$ is the natural étale cohomology group of L with coefficients in μ. Denote this kernel by $H^3(G, \mu)_n$. This is the group of geometrically negligible classes, a concept inspired by the negligible classes of Serre (which are "negligible" over all fields and not just those containing F). Finally, Colliot-Thélène and Ojanguren introduced an unramified cohomology group H^3 of a field and used that group to show certain fields were nonrational. We are interested in the unramified H^3 of invariant fields $F(V)^G$ and prove some results about them.

1991 *Mathematics Subject Classification.* Primary 12G05, 14L30, 14C99; Secondary 12E15, 13A20, 14C10, 14D25, 14M20.

The author is grateful for support under National Science Foundation grant DMS-8901778.

This paper is in final form and no version of it will be submitted for publication elsewhere.

189

In more detail, section one concerns the Brauer group of $F(V)$ ignoring the group action. We show:

THEOREM 1.4. *There is an exact sequence*

(4) $$0 \to \mathrm{Br}(K) \to \oplus_P \chi(P) \to \oplus_C \mathbb{Q}/\mathbb{Z} \to 0$$

where $\mathrm{Br}(K) \to \oplus_p \chi(P)$ *is the map* r *above, and* $\oplus_P \chi(P) \to \oplus_C \mathbb{Q}/\mathbb{Z}$ *is the sum of all the maps* ram_C.

If we now consider the G action on $F(V)$ it is trivial to see that the above is an exact sequence of G modules. The goal of sections two and three is some technical results giving an alternate description of the ramification map on Galois cohomology and alternate descriptions of some of the maps associated with the Hochschild-Serre spectral sequence of the exact sequence (1). With this in hand, we begin to try to describe the image of $\mathrm{Br}(F(V)^G)$ in $\mathrm{Br}(F(V))$. Of course this image lies in in the invariant group $\mathrm{Br}(F(V))^G$. Moreover, any α in this image must have all its ramification $r_p(\alpha) \in \chi(P)$ in the image of the corresponding character groups of primes in $F(V)^G$. Let $\mathrm{Br}(F(V))_0^G \subset \mathrm{Br}(F(V))$ be the subgroup of all such α. It is reasonable, then, to consider the quotient $\mathrm{Br}(F(V))_0^G / \mathrm{Res}(\mathrm{Br}(F(V)^G))$.

Before saying more, we turn to geometrically negligible cocycles. Let $H^3(G, \mu)_n$ be the group of geometrically negligible cocycles. In this group is a subgroup of more easily described elements as follows. We call a G module M H^1 trivial if $H^1(H, M) = 0$ for all subgroups $H \subset G$. For example, Hilbert's Theorem 90 shows that the multiplicative group $F(V)^*$ is H^1 trivial. On the other hand, $\mu \subset F(V)^*$ is not H^1 trivial. There is a "generic" way to enlarge μ and make it H^1 trivial that looks like the following.

LEMMA 4.6.

(b) *There is an extension* $0 \to \mu \to P^* \to P \to 0$ *such that* P *is a permutation lattice,* $H^1(H, \mu) \to H^1(H, P^*)$ *is the zero map for all* $H \subset G$, *and for all* $\phi : \mu \to N$, ϕ *extends to* P^* *if and only if* $H^1(H, \mu) \to H^1(H, N)$ *is the zero map for all* $H \subset G$. *Furthermore* $H^1(H, P^*) = 0$ *for all* $H \subset G$.

Because of Hilbert's Theorem 90, it is easy to see that if $H^3(G, \mu)_p$ is the kernel of $H^3(G, \mu) \to H^3(G, P^*)$, then $H^3(G, \mu)_p \subset H^3(G, \mu)_n$. We can think of $H^3(G, \mu)_p$ as those elements geometrically negligible because of Hilbert's Theorem 90. Again, it is therefore natural to consider the quotient $H^3(G, \mu)_n / H^3(G, \mu)_p$. The previous two problems are really the same because:

COROLLARY 4.8. $(d_3)_{0,2}$ *induces an isomorphism*

$$\mathrm{Br}(F(V))_0^G / \mathrm{Res}(\mathrm{Br}(F(V)^G)) \cong H^3(G, \mu)_n / H^3(G, \mu)_p.$$

We would like to further study either or both of the groups in 4.8. It turns out there is a geometrical description that might be useful. But first, we change our focus a bit. Among the elements of $H^3(G, \mu)_n/H^3(G, \mu)_p$ are some that are corestrictions of elements of $H^3(H, \mu)_n$ for a proper subgroup $H \subset G$. Under the theory that these elements are simpler, we set all these to 0 as follows.

DEFINITION. Let G be a finite group and $M \subset H^3(G, \mu)_n$ the subgroup generated by all images $\mathrm{Cor}^G_H(H^3(H, \mu)_n)$ for all proper subgroups $H \subset G$. Set $N^3(G) = H^3(G, \mu)_n/(M + H^3(G, \mu)_p)$.

The group $N^3(G)$ has a geometric description as a sort of equivariant Chow group as follows. First:

LEMMA 4.12. *If G is not a p-group for some p, $N^3(G) = 0$. If G is a p-group, $pN^3(G) = 0$.*

Thus we only consider p groups G. Let F be the free group with basis $\{\gamma(C)|C \in \mathscr{C}^G\}$ where $\mathscr{C}^G$ is the set of all irreducible closed G invariant codimension 2 subvarieties of V. Suppose $P \subset V$ is irreducible, closed, codimension 1, and G invariant. Assume $f \in F(P)$ is such that f^n is G invariant for some n. Consider $(f) = \sum_{C \in \mathscr{C}^G} \mathrm{ord}_C(f)\gamma(C) \in F$ and let $R \subset F$ be generated by all such (f). Note that only the irreducible G invariant zeros or poles appear in (f). We have:

THEOREM 4.13 (RESTATED). $N^3(G) \cong F/R$.

We cannot compute $N^3(G)$ very often but at least we know it can be nonzero because:

THEOREM 4.14. *Suppose G is a nonabelian 2 group with a cyclic subgroup of index 2. Then $N^3(G) = \mathbb{Z}/2\mathbb{Z}$.*

REMARK. One can show that $H^3(A, \mu)_n/H^3(A, \mu)_p = 0$ for all abelian groups A. Thus if G is, say, dihedral of order 8, then

$$\mathbb{Z}/2\mathbb{Z} = \mathrm{Br}(F(V))^G_0/\mathrm{Res}(\mathrm{Br}(F(V)^G)) = H^3(G, \mu)_n/H^3(G, \mu)_p.$$

However, if G is a 3 group with a cyclic normal subgroup of index 3 then $N^3(G) = 0$.

Let us now return to unramified H^3. Let $H^3_u(F(V)^G)$ be the unramified H^3 of the invariant field $L = F(V)^G$. There is an inflation map $H^3(G, \mu) \to H^3(G_L, \mu)$. We show:

THEOREM 5.3. $H^3_u(L)$ *is in the image of the inflation map $H^3(G, \mu) \to H^3(G_L, \mu)$.*

Of course, we would like to be able to compute some $H^3_u(F(V)^G)$. A case of particular interest is the center $Z(F, n, r)$ of the generic division algebra $UD(F, n, r)$. $Z(F, n, r)$ can also be described as $F(V)^G$ for G the (nonfinite) projective linear group PGL_n and V a certain representation. Procesi showed $Z(F, n, r)$ also had the form of a multiplicative invariant field $F(Y)^{S_n}$ for a certain lattice Y over the symmetric group S_n. Such multiplicative invariant fields can often be seen to be equivalent to $F(V')^H$ for finite H (and different V') and so 5.3. applies. With this hint and a bunch of cohomology computations we show:

THEOREM 6.7 (RESTATED). *Let n be odd. Then the unramified cohomology group* $H_u^3(Z(F, n, r)) = 0$.

Before we start the paper proper, let us define some notation etc. If G is a finite group, then $|G|$ will denote the order of G. A lattice over a finite G is a finite generated $\mathbb{Z}[G]$ module that is torsion free over $\mathbb{Z}$. A permutation lattice is a lattice with a $\mathbb{Z}$ basis that is permuted by G. A permutation G module is a $\mathbb{Z}[G]$ module that is $\mathbb{Z}$ free and has a basis permuted by G, but may not be finitely generated. If $H \subset G$ is a subgroup and M is an H module we define the induced module $\mathrm{Ind}_H^G(M)$ to be $\mathbb{Z}[G] \otimes_H M$. In a similar manner, if L is a field and V is a representation of H over L, then the induced representation $\mathrm{Ind}_H^G(V) = L[G] \otimes_{L[H]} V$.

For a G module M, $H^n(G, M)$ will denote the usual cohomology groups while $\check{H}^n(G, M)$ will denote the Tate cohomology groups. We will make frequent use of Shapiro's lemma which says that $H^n(G, \mathrm{Ind}_H^G(M)) \cong H^n(H, M)$ (e.g. [**B**, p. 73]). We will also make frequent use of the corestriction map as follows. Let $H \subset G$ be a subgroup of finite index and M a G module. The corestriction map $\mathrm{Cor}_H^G : H^n(H, M) \to H^n(G, M)$ is defined in, for example, [**B**, p. 80] and we will use its properties developed there.

Next, let K be a field. In this paper, all fields will have characteristic 0. Denote by K_a the algebraic closure of K and by G_K the absolute Galois group. That is, G_K is the Galois group of K_a/K. For virtually all of this paper, we will be working over a ground field F that is algebraically closed and of characteristic 0. We denote by $\mu \subset F^*$ the group of roots of one. We fix once and for all an isomorphism $\mu \cong \mathbb{Q}/\mathbb{Z}$. That is, for all positive integers n we fix a primitive n^{th} root of unity $\rho(n)$ such that $\rho(nm)^m = \rho(n)$ for all n, m.

Of considerable importance to us will be the notion of the Brauer group $\mathrm{Br}(L)$ of a field L. Of course, $\mathrm{Br}(L)$ consists of Brauer classes $[A]$ of central simple algebras A/L. If K/L is a finite G Galois extension then $\Delta(K/L, G, c)$ will denote the crossed product with respect to the 2 cohomology element $c \in H^2(G, K^*)$. Of particular interest is the case $G = C_n$ is a cyclic group of order n. Then any G crossed product is a cyclic algebra $\Delta(K/L, \sigma, a)$ where $\sigma \in G$ is a generator and $a \in L^*$. Finally, assume L contains $\rho(n)$. Then $K = L(\beta)$ where $\sigma(\beta) = \rho(n)\beta$ and $b = \beta^n \in L^*$. Then the cyclic algebra above can also be written as the symbol algebra $(b, a)_{L, n}$ or $(b, a)_n$ where L is clear.

If L contains all roots of one, then G_L acts trivially on μ and $H^1(G_L, \mu)$ is the group of continuous homomorphisms $\chi(L) = \text{Hom}_c(G_L, \mu)$ we call the character group of G_L. Any character $\chi \in \chi(L)$ has a kernel, N, of finite index and image a finite cyclic group. If $K = (L_a)^N$ is the fixed field of N, we say χ defines K and we note that K/L is a cyclic extension of degree the order of χ. Also associated to χ is the generator $\sigma \in \text{Gal}(K/L)$ defined by $\chi(1/n + \mathbb{Z}) = \sigma$ where χ has order n. Thus given a character χ it makes sense to define the cyclic algebra $\Delta(\chi, a) = \Delta(K/L, \sigma, a)$. If L contains all roots of unity, and χ has order n, then $K = L(\beta)$ where $\sigma(\beta) = \rho(n)\beta$ for σ as above. Of course, $b = \beta^n$ is a member of L^* and b and n completely determine χ. We say $b^{1/n}$ defines χ. Note that this definition even makes sense if $b = c^m$ for m dividing n. In that case, if $q = n/m$, $b^{1/n}$ and $c^{1/q}$ define the same character χ which has order q if c is not an eth power for any e dividing q.

The crossed product construction induces, for a field K, an isomorphism $\text{Br}(K) \cong H^2(G_K, K_a^*) \cong H^2(G_K, \mu)$ (if $\mu \subset K^*$). Thus if K/L is a finite separable extension of fields, there is a corestriction map $\text{Cor}_K^L : \text{Br}(K) \to \text{Br}(L)$. Since $\chi(L)$ is also a cohomology group, there is a corestriction map $\text{Cor}_K^L : \chi(K) \to \chi(L)$. The corestriction map on Brauer groups is an important but mysterious map. On characters it is much clearer as the following well-known result shows.

LEMMA 0.1. *Suppose K/L is a finite separable field extension and $\chi \in \chi(K)$ is defined by $b^{1/n}$. Then the corestriction $\text{Cor}_K^L(\chi)$ is defined by $N_{K/L}(b)^{1/n}$ where $N_{K/L} : K^* \to L^*$ is the norm map.*

PROOF. χ can be viewed as a member of $H^1(G_K, \mu_n)$ where $\mu_n \subset \mu$ is the subgroup of order n. The Kummer exact sequence

$$0 \to \mu_n \to K_a^* \xrightarrow{n} K^* \to 0$$

defines a boundary map $\delta : K^* = H^0(G_K, K_a^*) \to H^1(G_K, \mu)$ from the long exact cohomology sequence. The relationship of χ and b is then just that $\delta(b) = \chi$. 0.1 now follows from the description of the corestriction in, for example, [AW, p. 104]. $\square$

We will also have need of some simple notation from commutative rings and algebraic geometry. If R is a commutative domain, $q(R)$ will be the field of fractions of R and $\widetilde{R}$ the integral closure of R in $q(R)$. As we see in the very beginning of section 1, we will require a bit about the Brauer group of a commutative ring (e.g. [OS]) and in particular about the Brauer group of a discrete valuation domain (e.g. [AB]). We will introduce more geometry as the paper proceeds, but let us mention that $\mathbf{A}^n$ will represent affine space and $\mathbb{P}^n$ projective space. We leave to the standard sources (e.g. [**Ha**]) the definition of varieties, etc.

1. The Brauer group of a rational field

The Brauer group of a discrete valuation R ring and its field of fractions $q(R)$ can be related as follows. Let K be a field of characteristic 0 and let $R \subset K$ be a discrete valuation ring with $q(R) = K$. Assume the residue field $\bar{R}$ of R also has characteristic 0. Then there is an exact sequence (e.g. [**Se**, p. 186] or [**AB**]):

$$(1) \qquad 0 \to \mathrm{Br}(R) \to \mathrm{Br}(K) \to \chi(\bar{R}) \to 0$$

where $\chi(\bar{R}) = \mathrm{Hom}_c(G_R, \mathbb{Q}/\mathbb{Z})$ is defined as in the introduction. Note that $\mathrm{Br}(R) \to \mathrm{Br}(K)$ is the map induced by the inclusion $R \subset K$. The map $\mathrm{Br}(K) \to \chi(\bar{R})$ is called the ramification map. The elements of $\mathrm{Br}(K)$ in the image of $\mathrm{Br}(R)$ are said to be unramified at R.

The goal of this section is to show how this ramification map can be used to describe the Brauer group of a rational field extension. To begin with, suppose $K = L(x)$ is the rational field in one variable over L. The well-known Faddeev-Auslander-Brumer exact sequence says (e.g. [**FS**, p. 51])

$$(2) \qquad 0 \to \mathrm{Br}(L) \to \mathrm{Br}(K) \to \oplus_P \chi(P) \to 0$$

is exact where the notation and maps are as follows. The map $\mathrm{Br}(L) \to \mathrm{Br}(K)$ is the map induced by $L \subset K$. The direct sum is over all nonzero primes $P \subset L[x]$. If P is such a prime, set R_P to be the localization of $L[x]$ at P so R_P is a discrete valuation ring. We set $\chi(P) = \chi(\bar{R}_P)$. Finally, the map $\mathrm{Br}(K) \to \oplus_P \chi(P)$ is the sum of the ramification maps $r_P : \mathrm{Br}(K) \to \chi(P)$. If $\alpha \in \mathrm{Br}(K)$ maps to 0 in $\chi(P)$ we say α is unramified at P.

From now on, we will assume F is an algebraically closed field of characteristic 0. Obviously, using equation (2) we could describe the Brauer group of a rational field $K = F(x_1, \ldots, x_n)$ by induction. Unfortunately, such a description would not be very natural, depending as it would on an ordered choice of transcendence base $x_1, \ldots, x_n$. In the next section we will view $K = F(x_1, \ldots, x_n)$ as $F(V)$, the function field of an affine F space $V \cong \mathbb{A}^n$. We will require a description of the Brauer group of K that behaves well with respect to the action of the general linear group $GL(V)$ on $F(V)$. To achieve this let us take a closer look at the exact sequence (2).

To begin we summarize the discussion of [**FSS**, Proposition 2.1], which gives a precise description of elements $\alpha \in \mathrm{Br}(L(x))$ which have prescribed ramification at P and are unramified elsewhere. To this end fix a prime P and set $L(P) = L[x]/P$. Let G_P be the absolute Galois group of $L(P)$. If $\chi : G_P \to \mathbb{Q}/\mathbb{Z}$ is in $\chi(P)$, let $N \subset G_P$ be the kernel of χ and $M/L(P)$ the cyclic extension of $L(P)$ defined by N or χ. Suppose $\sigma \in G_P/N$ satisfies $\chi(\sigma) = 1/n + \mathbb{Z}$, and consider the natural embedding of fields $L \subset L(P) \subset M$ which induces $K = L(x) \subset L(P)(x) \subset M(x)$. We form the cyclic algebra $\Delta(M(x)/L(P)(x), \sigma, x)$ and use the corestriction map (e.g. [**B**, p. 80]) to define

$$\alpha(P, \chi) = \text{Cor}_{L(P)(x)/K}([\Delta(M(x)/L(P)(x), \sigma, x)])$$

which is an element of the Brauer group $\text{Br}(K)$. It was shown in [FSS, Proposition 2.1] that:

PROPOSITION 1.1. $\alpha(P, \chi)$ *ramifies only at* P *and* $r_P(\alpha(P, \chi)) = \chi$.

We will not repeat the proof of the above proposition but we must quote a key result in the proof comparing corestriction and ramification.

LEMMA 1.2. *Suppose* N/L *is a finite extension of fields of characteristic* 0, *and* $R \subset L$ *is a discrete valuation ring with* $q(R) = L$. *Let* $S_i \subset$, $i = 1, \ldots, n$, *be the full set of extensions of* R *in* N. *Then the following diagram commutes*:

$$
\begin{array}{ccc}
\text{Br}(N) & \longrightarrow & \oplus \chi(\bar{S}_i) \\
\text{Cor} \downarrow & & \downarrow \sum \text{Cor}_{S_i/R} \\
\text{Br}(L) & \longrightarrow & \chi(\bar{R}).
\end{array}
$$

Note here that $\chi(\bar{S}_i) = H^1(G_{\bar{S}_i}, \mathbb{Q}/\mathbb{Z})$ *so that the corestrictions in the right vertical arrow are defined.*

It is immediate from Proposition 1.1 that the map $\text{Br}(K) \to \oplus\chi(P)$ is a surjection. The main goal of this section is to prove a generalization of this to the higher-dimensional case $K = F(x_1, \ldots, x_n)$. To this end, let $X = \mathbb{A}^n$, and let $C \subset P \subset X$ be such that C, P are closed irreducible subvarieties of codimensions 2 and 1. Let $F(P)$ be the function field of P, G_P the absolute Galois group of $F(P)$, and $\chi(P) = \text{Hom}_c(G_P, \mathbb{Q}/\mathbb{Z})$. We next define the ramification of $\chi \in \chi(P)$ at C.

Let $R = (\mathscr{O}_P)_C$ be the localization of the structure sheaf $\mathscr{O}_P$ of P at C. Let $\pi : \tilde{P} \to P$ be the integral closure of P in $F(P)$, assume $C_1, \ldots, C_n$ are the preimages of C, and R_i the localization of $\mathscr{O}_{\tilde{P}}$ at C_i. Thus the R_i are discrete valuation rings with $q(R_i) = F(P)$. Let $v_i : F^* \to \mathbb{Z}$ be the associated valuation. Finally, let f_i be the degree of the extension of residue fields $\bar{R}_i/\bar{R}$ (or function fields $F(C_i)/F(C)$). As is standard, for $a \in F(P)^*$ define $\text{ord}_C(a) = \sum v_i(a)f_i$. We note (e.g. [Fu, p. 8]) that $\text{ord}_C(a)$ is the length of R/aR and that the map $\text{ord}_C : F(P)^* \to \mathbb{Z}$ is a homomorphism. Geometrically, if $a \in F(P)$ is the image of $b \in \mathscr{O}_X$, then $\text{ord}_C(a)$ is the intersection multiplicity of P and the divisor defined by "$b = 0$" at C (e.g. [Fu, p. 8]).

Suppose $\chi \in \chi(P)$ has order n and is defined by $a^{1/n}$. Set $r = \text{ord}_C(a)$ and define the ramification of χ at C by $\text{ram}_C(\chi) = r/n \in \mathbb{Q}/\mathbb{Z}$. If C is not a subset of P we set $\text{ram}_C(\chi) = 0$.

If $P \subset \mathbb{A}^n$ is as above, the localization $(\mathscr{O}_X)_P$ is a discrete valuation ring and so defines a ramification map $r_P : \text{Br}(K) \to \chi(P)$. Adding these maps up, we have:

$$r : \text{Br}(K) \longrightarrow \oplus_P \chi(P).$$

Our goal is to describe the image of r. We will find that the difference between $L(x)$ and $F(x_1, \ldots, x_n)$ shows up because the existence of codimension 2 subvarieties force the following (well-known) restriction on the image of r.

PROPOSITION 1.3. *Suppose* $K = F(x_1, \ldots, x_n)$ *and* $\alpha \in \mathrm{Br}(K)$ *ramifies at* $P_1, P_2, \ldots, P_r \subset \mathbf{A}^n$ *with* $r_{P_i}(\alpha) = \chi_i \in \chi(P_i)$. *If* $C \subset \mathbf{A}^n$ *has codimension 2, then* $\sum \mathrm{ram}_C(\chi_i) = 0$.

PROOF. This result is very well known, so we just sketch a proof. By Merkuriev-Suslin, it suffices to consider the case α is the class of the symbol algebra $(a, b)_{K,n}$ where $a, b \in F[\mathbf{A}^n] = F[x_1, \ldots, x_n]$ are irreducible. Let $P, Q \subset \mathbf{A}^n$ be defined by $a = 0$ and $b = 0$ respectively. Clearly P and Q are the only primes where α ramifies. Denote by $\bar{a}$ the image of a in $F(Q)$ and by $\bar{b}$ the image of b in $F(P)$. By, e.g., 1.1, $r_P(\alpha) \in \chi(P)$ is defined by $(\bar{b}^{-1})^{1/n}$ and $r_Q(\alpha) \in \chi(Q)$ is defined by $\bar{a}^{1/n}$. We have $\mathrm{ram}_C(r_P(\alpha)) = -\mathrm{ord}_C(\bar{b})/n$ and $\mathrm{ram}_C(r_Q(\alpha)) = \mathrm{ord}_C(\bar{a})/n$. Finally, $\mathrm{ord}_C(\bar{a}) = \mathrm{ord}_C(\bar{b})$ by the symmetry of the intersection multiplicity of P and Q at C (e.g., [**Fu**, p. 35]). $\square$

Our ultimate goal in this section is to prove:

THEOREM 1.4. *Let* F *be an algebraically closed field of characteristic* 0, *and let* $K = F(x_1, \ldots, x_n)$ *There is an exact sequence*

$$(3) \qquad 0 \to \mathrm{Br}(K) \to \oplus_P \chi(P) \to \oplus_C \mathbb{Q}/\mathbb{Z} \to 0$$

where $\mathrm{Br}(K) \to \oplus_P \chi(P)$ *is the map* r *above, and* $\oplus_P \chi(P) \to \oplus_C \mathbb{Q}/\mathbb{Z}$ *is the sum of all the maps* ram_C.

REMARK. Colliot-Thélène informs me that this result follows from the Bloch-Ogus spectral sequence and a "homotopy invariance" argument. John Tate has pointed out that one can quite easily prove this theorem from the $n = 1$ case and induction. Though longer, the argument we present here has the advantage of explicitly exhibiting the required Brauer classes in a more efficient way. The $n = 2$ case of this theorem is a special case of the result in [**AM**, p. 84] or more explicitly [**F**, p. 8].

Though the full proof of the above theorem will take a bit of work, let us begin with the easy parts. Since F is algebraically closed, $\mathrm{Br}(F) = 0$. By [**AG**, p. 391], $\mathrm{Br}(F[x_1, \ldots, x_n]) = \mathrm{Br}(F) = 0$. By [**H**, p. 90], $\mathrm{Br}(F[x_1, \ldots, x_n]) = \cap_P \mathrm{Br}(F[x_1, \ldots, x_n]_P)$, the intersection being over all height one primes. By (1), $\mathrm{Br}(K) \to \oplus_P \chi(P)$ is an injection. Of course Proposition 1.3 above shows that (3) is a complex.

To show exactness at the remaining two places, we consider projections defined on $\mathbf{A}^n$. In detail, suppose $L \subset \mathbf{A}^n$ is a one-dimensional linear subspace. Then L defines a projection $\pi_L : \mathbf{A}^n \to \mathbf{A}^n/L = \mathbf{A}^{n-1}$. If we embed $\mathbf{A}^n$ into projective space $\mathbb{P}^n$ in the standard way, the hyperplane at infinity $\mathbb{H} = \mathbb{P}^n - \mathbf{A}^n$ can be identified with the one-dimensional subspaces of A^n. If $D \subset A^n$ is a closed subvariety, let $\bar{D} \subset \mathbb{P}^n$ be its Zariski closure. The following result appears in Shafarevich's book [**Sh**, p. 50] in a slightly different form.

LEMMA 1.5. *Suppose* $L \in \mathbb{H}$ *is a one-dimensional subspace of* A^n. *If* $L \notin \bar{D}$, *then the restriction* $\pi_L : D \to \pi_L(D) \subset A^{n-1}$ *is a finite morphism.*

REMARK. The above is actually the restriction of the result stated in [**Sh**] to an affine piece.

If $D \subset A^n$ is as above and has codimension one, then $\pi_L(D) = A^{n-1}$ because $\pi_L(D)$ and A^{n-1} have the same dimension and π_L is a closed morphism on D.

Algebraically, a projection $\pi_L : A^n \to A^{n-1}$ corresponds to writing $F[A^n] = F[A^{n-1}][y]$ where y is an indeterminant. We can therefore mimic the above corestriction construction, but we will have to be more careful about ramification. Let $C \subset P \subset A^n$ be such that C, P are closed irreducible of codimension 2 and 1 respectively. Choose $L \notin \bar{P}$ so that $\pi_L : P \to A^{n-1}$ is finite. Set $C' = \pi_L(C)$ and $E = \pi_L^{-1}(C') \subset A^n$. Since π_L is finite on P, C' has dimension $n-2$ and so E has dimension $n-1$. Algebraically, $F[E] = F[C'][y]$ and so E is irreducible. We call E a vertical prime. Since $\pi_L(P) = A^{n-1}$, P is a nonvertical prime. Algebraically, if we associate to P the ideal $I(P) \subset F[A^n]$, then $I(P) \cap F[A^{n-1}] = (0)$. The projection $\pi_L : P \to A^{n-1}$ corresponds to an injection $\pi_L^* : F[A^{n-1}] \to F[P]$ which we take to be an inclusion. This, of course, yields an inclusion $F(A^{n-1}) \subset F(P)$ of function fields. P induces an ideal we call $I(P) \subset F(A^{n-1})[y]$ which is generated by an element $f_0 + f_1 y + \cdots + f_n y^n$.

We next observe that the f_i can be assumed to be integral over $F[A^{n-1}]$ and hence elements of $F[A^{n-1}]$. This is an argument we will use a few times more. Of course, we may assume $f_n = 1$. Since $\pi_L : P \to A^{n-1}$ is finite, the image, c, of y in $F[P]$ is integral over $F[A^{n-1}]$. This also applies to all conjugates of c and so all the f_i are integral over $F[A^{n-1}]$. Since $F[A^{n-1}]$ is integrally closed, $f_i \in F[A^{n-1}]$.

With $E = \pi_L^{-1}(C')$ as above, we can make a few observations about nonvertical primes on E. Here, of course, we mean primes $I(Q) \subset F[E]$ such that $I(Q) \cap F[C'] = (0)$. Clearly, the nonvertical primes are in one-to-one correspondence with the primes of $F(C')[y]$, or equivalently, with the monic irreducible elements in $F(C')[y]$. Let $\widetilde{F[C']}$ be the integral closure of $F[C']$ in its field of fractions $F(C')$. Clearly $\widetilde{F[C']}[y]$ is the integral closure of $F[C'][y]$. Our description of nonvertical primes also applies to $\operatorname{Spec}(\widetilde{F[C']}[y])$ and so every nonvertical prime $I(Q) \subset F[C'][y]$ extends to a unique nonvertical prime $I(\widetilde{Q}) \subset \widetilde{F[C']}[y]$ and Q and $\widetilde{Q}$ have equal residue fields.

To refine our discussion further, suppose $f \in F[\widetilde{C'}][y] = \widetilde{F[E]}$ is monic irreducible in $F(C')[y]$. If $I(\widetilde{P}) \subset \widetilde{F[E]}$ is a vertical prime, then $I(\widetilde{P}) = I(\widetilde{Q})[y]$ for $I(\widetilde{Q}) \subset \widetilde{F[C']}$ prime. Thus $f \notin I(\widetilde{P})$. It follows that f is contained in a unique prime $I(P_f) \subset F[E]$, P_f is nonvertical, $I(P_f)$ extends to a unique prime $I(\widetilde{P}_f) \subset \widetilde{F[E]}$, and P_f and $\widetilde{P}_f$ have equal residue fields. Since $fF(C')[y] = I(\widetilde{P}_f)F(C')[y]$, f is a local generator of $I(\widetilde{P}_f)$. Thus $\mathrm{ord}_{P_f}(f) = 1$. We say f defines P_f.

Next we apply this to $C \subset E = \pi_L^{-1}(C')$. Clearly, $F[C]$ is generated over $F[C']$ by c, the image of y. Since $F[C]$ is integral over $F[C']$, c is the root of a monic polynomial $f' \in F[C'][y]$. Let f be the irreducible monic element of $F(C')[y]$ with c as a root. Since all conjugates of c are integral over $F[C']$, $f \in \widetilde{F[C']}[y]$. Clearly f defines the prime ideal $I(P_f)$ that defines C as a subvariety of E. We can now show:

PROPOSITION 1.6. *The map*

$$\oplus_P \chi(P) \longrightarrow \oplus_C \mathbb{Q}/\mathbb{Z}$$

in 1.3 *is a surjection.*

PROOF. Given $C \subset \mathbb{A}^n$ of codimension 2, choose a projection $\pi_L : \mathbb{A}^n \to \mathbb{A}^{n-1}$ such that π_L is finite on C. Set $C' = \pi_L(C)$ and $E = \pi_L^{-1}(C')$. Let $f \in \widetilde{F[E]} = F[C'][y]$ be the monic irreducible defining C as above. If $\chi \in \chi(E)$ is defined by $f^{1/n}$, then χ has image $1/n$ in the copy of $\mathbb{Q}/\mathbb{Z}$ corresponding to C and 0 elsewhere. □

REMARK. The above argument is very well known and is the proof that $\mathbb{A}^n$ has trivial Chow ring found, for example, in [**Fu**, p. 23].

More generally, let $E = \pi_L^{-1}(C')$ be as above, and $f \in \widetilde{F[E]} = \widetilde{F[C']}[y]$ a monic polynomial. If $\chi \in \chi(E)$ is defined by $f^{1/n}$, we say χ is monically defined. Note that the monically defined χ form a subgroup of $\chi(E)$. For future use we observe:

LEMMA 1.7. *Suppose $\chi \in \chi(E)$ is monically defined and unramified at all nonvertical primes of E. Then $\chi = 1$.*

PROOF. Suppose χ is defined by $f^{1/n}$, with $f \in \widetilde{F[E]} = \widetilde{F[C']}[y]$ monic. Write $f = f_1^{n_1} f_2^{n_2} \cdots f_r^{n_r}$ where all the f_i are distinct, monic, and irreducible. As argued above, all $f_i \in \widetilde{F[E]}$. Now if f_i defines the nonvertical prime P_i, $\mathrm{ord}_{P_i}(f_i) = 1$. Thus $\mathrm{ram}_{P_i}(\chi) = n_i/n$, and so all n_i are divisible by n. □

We are ready to complete the proof of Theorem 1.4 via the construction of specific Brauer group elements. Let $P \subset \mathbb{A}^n$ be irreducible of codimension 1, and let $\chi \in \chi(P)$ be a character. Choose a projection $\pi_L : \mathbb{A}^n \to \mathbb{A}^{n-1}$ such that π_L is finite on P. Then $F(\mathbb{A}^{n-1}) \subset F(P)$ and so $F(\mathbb{A}^n) = F(\mathbb{A}^{n-1})(y) \subset F(P)(y)$. Let $c \in F[P]$ be the image of $y \in F[\mathbb{A}^n]$ under the surjection $F[A^n] \to F[P]$ induced by the closed immersion $P \subset A^n$. Since π_L is finite on P, c is a root of a monic irreducible $f(y) = f_0 + f_1 y + \cdots + f_{n-1} y^{n-1} + y^n \in F[A^{n-1}][y]$. In $F(P)$, $f(y)$ has root c so $f(y) = (y-c)f'(y)$ where $f'(y) \in F(P)[y]$. Once again, since all the roots of $f(y)$ are integral over $F[P]$ we have $f'(y) \in \widetilde{F[P]}[y]$. The character $\chi \in \chi(F(P))$ induces via the inclusion $F(P) \subset F(P)(y)$ a character we write as $\chi(y) \in \chi(F(P)(y))$. Form

$$\beta(\chi) = \mathrm{Cor}_{F(P)(y)/F(A^n)}(\Delta(\chi(y), y-c)) \in \mathrm{Br}(F(A^n)).$$

Our main argument rests on describing the ramification of β.

THEOREM 1.8. *Suppose P, χ, π_L, and $\beta(\chi)$ are as above. Let χ ramify along $C_i \subset P$ for $i = 1, \ldots, r$. Set $C_i' = \pi_L(C_i)$ and $E_i = \pi_L^{-1}(C_i')$. Then $\beta(\chi)$ only ramifies along P and the E_i. Furthermore, $r_P(\beta(\chi)) = \chi$. If $\chi_i = r_{E_i}(\beta(\chi))$, then χ_i only ramifies at $C_i \subset E_i$ and is defined monically.*

PROOF. Before we begin the proof proper, we make some observations about our setup. By definition, the residue field $F(E_i)$ can be identified with $F(C_i')(y)$. Let $\bar{c}_i$ denote the image of c in $F[C_i]$. Since c generates $F[P]$ over $F[A^{n-1}]$, $\bar{c}_i$ generates $F[C_i]$ over $F[C_i']$. Since c is integral over $F[A^{n-1}]$, $\bar{c}_i$ is integral over $F[C_i']$. Thus if $g_i \in F(C_i')[y]$ is the monic irreducible with $\bar{c}_i$ as a root, $g_i \in \widetilde{F[C_i']}[y]$ and in the terminology introduced above g_i defines C_i as a nonvertical prime in E_i.

Next we find the primes in $\widetilde{F[P]}[y]$ where $\Delta(\chi(y), y-c)$ ramifies. $y-c$ is a unit with respect to all primes except $I(P_c)$, the prime defined by $y-c$. The residue field of P_c is $F(P)$ and the ramification of $\Delta(\chi(y), y-c)$ is clearly χ (e.g. 1.1). Let $C_{i,1}, \ldots, C_{i,s}$ be all the extensions of C_i to $\tilde{P} = \mathrm{Spec}(\widetilde{F[P]})$, and let $F_{i,j} = F(C_{i,j})$ be the residue fields. Denote by $E_{i,j} \subset \mathrm{Spec}(\widetilde{F[P]}[y])$ the inverse image of $C_{i,j}$, so $F(E_{i,j}) = F(C_{i,j})(y) = F_{i,j}(y)$. $y-c$ is a unit with respect to $E_{i,j}$ (being monic). Since $C_{i,j}$ is an extensions of C_i, there is a natural inclusion $F[C_i] \subset F[C_{i,j}]$ and we can view $\bar{c}_i$ as an element of $F[C_{i,j}]$. If $\chi(y)$ is defined by $f^{1/n}$, $\Delta(\chi(y), y-c)$ can be rewritten as the symbol algebra $(f, y-c)_n = (y-c, f)_m^\circ$. Thus the ramification of $\Delta(\chi(y), y-c)$ at $E_{i,j}$ is defined by a power of $y - \bar{c}_i$.

We compute the ramification of the corestriction $\beta(\chi)$ by using Lemma 1.2. If $P' \subset A^n$ is nonvertical and not the restriction of P_c, $\Delta(\chi(y), y-c)$ is unramified at all extensions of P' and so $\beta(\chi)$ is unramified at P'. The restriction of P_c to A^n is just P, and since $F[P]/F[A^{n-1}]$ is integral, all other extensions of P are nonvertical also. Hence P_c is the unique extension of P at which $\Delta(\chi(y), y-c)$ ramifies. By Lemma 1.2, $r_P(\beta(\chi)) = \chi$.

Next let $P' \subset A^n$ be a vertical irreducible divisor that is not the restriction of an $E_{i,j}$. Since $\Delta(\chi(y), y-c)$ is unramified at all extensions of P', $\beta(\chi)$ is unramified at P' (Lemma 1.2 again). The only remaining primes are the restrictions of the $E_{i,j}$'s, which are precisely the E_i. Let $f_{i,j} = (y - \bar{c}_i)^{n_{i,j}} \in F_{i,j}(y)$ be such that $f_{i,j}^{1/n}$ defines the character $\chi_{i,j}$ of $\Delta(\chi(y), y - c)$ at $E_{i,j}$. Let $F_i = F(C_i')$ be the residue field of $C_i' \subset A^{n-1}$, so E_i has residue field $F_i(y)$. The map $E_{i,j} \to E_i$ induces an inclusion of fields $F_i(y) \subset F_{i,j}(y)$. The extensions of $C_i' \subset A^{n-1}$ to $\widetilde{P}$ consist precisely of the $C_{i,j}$ and extensions C' not over C_i where χ is then unramified. All are, of course, vertical. By 0.1, the corestriction $\mathrm{Cor}_{F_{i,j}(y)/F_i(y)}(\chi_{i,j}) \in \chi(F_i(y))$ is defined by $\mathrm{Norm}_{F_{i,j}/F_i(y)}(y - \bar{c}_{i,j})^{n_{i,j}}$ which is a power of the g_i defined in the first paragraph. By 1.2 again, $r_{E_i}(\beta(\chi))$ is defined by a power of g_i, which is of course monic. Since g_i defines C_i in E_i, $r_{E_i}(\beta(\chi))$ only ramifies at C_i. $\square$

It is now very easy to provide the last step in the proof of Theorem 1.4.

THEOREM 1.9. *The complex*

$$\mathrm{Br}(F) \longrightarrow \oplus_P \chi(P) \longrightarrow \oplus_C \mathbb{Q}/\mathbb{Z}$$

is exact.

PROOF. Suppose $P_i \subset A^n$ are irreducible divisors and $\chi_i \in \chi(P_i)$ are characters such that, for all $C \subset A^n$ of codimension 2, $\sum \mathrm{ram}_C(\chi_i) = 0$. Let $C_{i,j} \subset P_i$ be the places where the χ_i ramify. Choose a projection $\pi_L : A^n \to A^{n-1}$ such that π_L is finite on all the P_i and $\pi_L(C_{i,j}) \neq \pi_L(C_{i',j'})$ if $(i, j) \neq (i', j')$. Set $C_{i,j}' = \pi_L(C_{i,j})$ and $E_{i,j} = \pi_L^{-1}(C_{i,j}')$. Finally set $\beta = \sum \beta(\chi_i) \in \mathrm{Br}(K) = \mathrm{Br}(F(A^n))$. It suffices to show that β only ramifies along the P_i with characters χ_i.

The $E_{i,j}$ are all distinct from the P_i as π_L is finite on the P_i and not on the $E_{i,j}$. It follows from the above proposition that β ramifies along the P_i with characters χ_i. The only other places β might ramify are the $E_{i,j}$. If $\chi_{i,j}$ is the ramification of β on $E_{i,j}$, $\chi_{i,j}$ is monically defined and only ramifies along $C_{i,j}$. But $0 = \mathrm{ram}_{C_{i,j}}(\chi_{i,j}) + \sum \mathrm{ram}_{C_{i,j}}(\chi_i) = \mathrm{ram}_{C_{i,j}}(\chi_{i,j})$. By 1.7, $\chi_{i,j} = 1$. $\square$

Let us continue this section with an example of how the detailed description of $\mathrm{Br}(F(x_1, \dots, x_n))$ embodied in the above proof can be used. Consider $K = F(x_1, x_2)$ as the function field of the affine space A^2. In A^2, let E be an elliptic curve, and $\chi \in F(E)$ a character corresponding to an everywhere unramified cover of E. By 1.4 or really the much earlier [AM, p. 84], there is a Brauer group element $\alpha \in \mathrm{Br}(K)$ that ramifies only along E with character χ. There has been some curiosity about this element as it is in some ways the "simplest" element of $\mathrm{Br}(K)$ and also arises in [ATV] (see [F2]). Our machinery allows us to deduce that:

PROPOSITION 1.10. α *is a product in the Brauer group of three symbol algebras.*

PROOF. By the proof of 1.4, we can describe α by choosing a line $L \subset A^2$ such that the projection $\pi_L : A^2 \to A^2/L = A^1$ is finite on E and then we can write $\alpha = \beta(\chi)$. Clearly $\pi_L : E \to A^1$ has degree 3 generically and so the definition of $\beta(\chi)$ shows that it is the corestriction of a symbol algebra from a degree 3 extension field. The result follows from the main result of [RT]. $\square$

In one application, we will have need to consider so-called multiplicative field invariants. These are invariant fields defined by actions of finite groups on tori. For this reason, we take a moment to consider the version of 1.4 that applies to the Brauer group of a torus.

Algebraically, we are considering the localization

$$R = F[x_1, \ldots, x_n](1/(x_1 \cdots x_n)).$$

Unlike the case of the polynomial ring, R has nonzero Brauer group. In fact, by [M, p. 166], we can describe the Brauer group as follows. Set $M = \mathbb{Z} \oplus \cdots \oplus \mathbb{Z}$ (n times) which we think of as the multiplicative subgroup of R generated by the x_i's. Recall that for $x, y \in R$ units we can define the Azumaya symbol algebra $(x, y)_{n,R}$ which is generated over R by α, β subject to the relations $\alpha^n = x$, $\beta^n = y$, and $\alpha\beta = \rho(n)\beta\alpha$. There is a homomorphism $\phi : ((\wedge^2 M) \otimes (\mathbb{Q}/\mathbb{Z})) \to \mathrm{Br}(R)$ defined by setting $\phi((x \wedge y) \otimes (1/n + \mathbb{Z}))$ to be the Brauer class $[(x, y)_{n,R}] \in \mathrm{Br}(R)$. What is shown in [M] is:

LEMMA 1.11. ϕ is an isomorphism.

Now let $R \subset F(x_1, \ldots, x_n) = K$ and consider how $\mathrm{Br}(R)$ fits into the exact sequence (3). To do this, let $\oplus_{P'}\chi(P') \subset \oplus_P\chi(P)$ be the direct sum over all height one primes except those generated by $\{x_1, \ldots, x_n\}$. These can be identified with the height one primes of R. Let $\oplus_{C'}\mathbb{Q}/\mathbb{Z}$ be the direct sum over all height 2 primes of R, that is, all height 2 primes of $F[x_1, \ldots, x_n]$ that do not contain any x_i. In our proof that $\oplus_P(\chi(P)) \to \oplus_C(\mathbb{Q}/\mathbb{Z})$ is surjective we had infinitely many choices of a P containing C so that there was an $\chi \in \chi(P)$ that ramifies only at C. Thus we can choose P not to be any of the finite set of primes generated by the x_i's, and we conclude that $\oplus_{P'}\chi(P') \to \oplus_{C'}\mathbb{Q}/\mathbb{Z}$ is surjective. Also, by [H, p. 90] again the kernel of $\mathrm{Br}(K) \to \oplus_{P'}\chi(P')$ is $\mathrm{Br}(R)$. We claim:

THEOREM 1.12. There are exact sequences

$$0 \to \mathrm{Br}(K)/\mathrm{Br}(R) \to \oplus_{P'}\chi(P) \to \oplus_C\mathbb{Q}/\mathbb{Z} \to 0$$

and

$$0 \to ((\overset{2}{\wedge} M) \otimes \mathbb{Q}/\mathbb{Z}) \to \mathrm{Br}(K) \to \mathrm{Br}(K)/\mathrm{Br}(R) \to 0$$

where P' ranges over all height one primes of R and C' ranges over all height 2 primes of R.

PROOF. The only thing left unsaid in order to prove the second sequence exact is that since R is smooth, $\mathrm{Br}(R) \to \mathrm{Br}(K)$ is an injection. The exactness of the first sequence has been proved at all places except $\oplus_{P'}\chi(P')$. So suppose $\alpha \in \oplus_{P'}\chi(P')$ maps to 0 in $\oplus_{C'}\mathbb{Q}/\mathbb{Z}$. Suppose $C_1, \ldots, C_s$ are the height 2 primes in $F[x_1, \ldots, x_n]$ where α has nonzero image $r_1 + \mathbb{Z}, \ldots, r_s + \mathbb{Z} \in \mathbb{Q}/\mathbb{Z}$. Of course, each C_i contains an x_{i_j}. Since $F[x_1, \ldots, x_n]/(x_i)$ is another polynomial ring, each C_i has the form (x_{i_j}, f_i). It is clear that there are characters $\chi_i \in \chi((x_i))$ such that the sum of the χ_i in $\oplus_P \chi(P)$ (call it β) has the same image in $\oplus_C \mathbb{Q}/\mathbb{Z}$ as α. If we set $\alpha' = \alpha - \beta$, then by 1.4 α' is in $\mathrm{Br}(K)$. But then α' is the required preimage of α. $\square$

2. Cohomology results

At several junctures in this paper we will find ourselves considering the boundary map from the cohomology long exact sequence and the boundary map from the Hochschild-Serre spectral sequence. As our reference for the prerequisite group cohomology, we will use the original paper [HS] because the concreteness of its development will best suit our needs. We refer the reader to this paper for terms we leave undefined etc.

In the first part of this section we will prove a pure cohomology result concerning these two maps. To this end, let A, B, and C be G modules and N a normal subgroup. Fix a pairing $A \times B \to C$. By [HS, p. 119], cup product of cochains induces a cup product

$$H^p(G/N, H^q(N, A)) \times H^{p'}(G/N, H^{q'}(N, B)) \longrightarrow H^{p+p'}(G/N, H^{q+q'}(N, C))$$

which we write as $\alpha \times \beta \to (\alpha \cup \beta)$.

At least in a special case, this cup product relates to the more usual "single cohomology" cup product in the following way.

LEMMA 2.1. *Let $A \times B \to C$ be a pairing, and $\gamma \in H^0(G/N, H^1(N, A))$ an element in the image of $H^1(G, A)$. Let $\phi : H^{q-1}(N, B) \to H^q(N, C)$ be defined by $\phi(\alpha) = \gamma \cup \alpha$ where γ is viewed as an element of $H^1(N, A)$. Then the induced map $\phi^* : H^p(G/N, H^{q-1}(N, B)) \to H^p(G/N, H^q(N, C))$ satisfies $\phi^*(\alpha') = \gamma \cup \alpha'$.*

PROOF. In this proof we lean heavily on the paper [HS]. The assumptions on γ imply that γ is represented by a one cocycle f on N that is the restriction of a one cocycle, f', on G. Since $d(f') = 0$, γ is represented by f' as an element of $H^0(G/N, H^1(N, A))$. Following the notation of and results of [HS], suppose α' is represented by a cocycle $h \in A^{p+q+1} \cap A_p^*$ with $d(h) \in A_{p+2}^*$. Recall that by the definition of A_p^*, $h : G^{p+q+1} \to B$ only depends on the N coset in the last p places. It is immediate that $f' \cup h \in A^{p+q} \cap A_p^*$ and $d(f' \cup h) = d(f') \cup h + (-1)f' \cup d(h) = -f' \cup d(h) \in A_{p+2}^*$. Thus $\gamma \cup \alpha$ is represented by $f' \cup h$. Fix $v_1, \ldots, v_p \in G/N$ and let $h'(x_1, \ldots, x_{q-1}) = h(x_1, \ldots, x_{q-1}, v_1, \ldots, v_p)$ for $x_i \in N$. Then $h' \in C^p(G/N, Z^{q-1}(N, B))$ and the definition of ϕ shows that $\phi^*(\alpha) = f \cup h'(v_1, \ldots, v_p)$ which the cup product definition shows is the restriction of $f' \cup h$. $\square$

Next, let M be an arbitrary G module. $H^p(G/N, H^q(N, M))$ is the $E_2^{p,q}$ term in the Hochschild-Serre spectral sequence that converges to $H^{p+q}(G, M)$. Thus there is a boundary map we call $d_2 : H^p(G/N, H^q(N, M)) \to H^{p+2}(G/N, H^{q-1}(N, M))$.

PROPOSITION 2.2. *Suppose*

$$\alpha \in H^p(G/N, H^q(N, A)) \quad and \quad \beta \in H^{p'}(G/N, H^{q'}(N, B))$$

with A, B, and C as above. Then

$$d_2(\alpha \cup \beta) = d_2(\alpha) \cup \beta + (-1)^{p+q} \alpha \cup d_2(\beta) \,.$$

PROOF. Note that both the left and the right sides of the equation belong to $H^{p+p'+2}(G/N, H^{q+q'-1}(N, C))$. The result follows from the discussion at the top of [HS, p. 119] and the equation

$$d(f \cup g) = d(f) \cup g + (-1)^n f \cup d(g)$$

where f, g are cochains, and f has degree n. $\square$

In order to derive our result on the boundary of long exact sequences, we realize this boundary as a cup product in the following way. Let

(1) $$0 \to A \to B \to C \to 0$$

be an exact sequence of G modules which splits as a sequence of abelian groups. Denote by $\delta : H^{q-1}(G, C) \to H^q(G, A)$ the boundary map from the long exact sequence.

Long exact sequences like (1) that are split as abelian groups correspond to elements of $H^1(G, \text{Hom}_{\mathbb{Z}}(C, A))$ as follows. As an abelian group we write $B = A \oplus C$. For each $g \in G$ define $f_g \in \text{Hom}(C, A)$ via the relation $g(x, y) = (gx + f_g(gy), gy)$ where $(x, y) \in B = A \oplus C$. The f_g define a one cocycle and hence an element $\gamma \in H^1(G, \text{Hom}(C, A))$ Furthermore, $\text{Hom}(C, A)$ and C have the natural pairing $\text{Hom}(C, A) \times C \to A$ defined by setting $(f, c) \to f(c)$. We use this pairing to define a cup product on cohomology and observe:

PROPOSITION 2.3. *For $\alpha \in H^{q-1}(G, C)$, $\delta(\alpha) = \gamma \cup \alpha \in H^q(G, A)$.*

PROOF. Suppose $h : G^{q-1} \to C$ is a cocycle defining α, and define $\bar{h} : G^{q-1} \to B$ by $\bar{h}(g_1, \ldots, g_{q-1}) = (0, h(g_1, \ldots, g_{q-1}))$. Computing the boundary we have

$$d(\bar{h})(g_1, \ldots, g_q) = g_1 \bar{h}(g_2, \ldots, g_q)$$
$$+ \sum_i (-1)^i \bar{h}(g_1, \ldots, g_i g_{i+1}, g_{i+2}, \ldots, g_q)$$
$$+ (-1)^q \bar{h}(g_1, \ldots, g_{q-1}) \,.$$

Since h is a cocycle this equals $(f_{g_1}(g_1 \bar{h}(g_2, \dots, g_q)), 0) \in A$. But this represents both $\delta(\alpha)$ and $\gamma \cup \alpha$. $\square$

Next, let A, B, C, G, and $\gamma \in H^1(G, \mathrm{Hom}(C, A))$ be as above and assume $N \subset G$ is a normal subgroup. Restricting γ to N, call it γ_N, we have an element of $H^0(G/N, H^1(N, \mathrm{Hom}(C, A)))$. Applying Lemma 2.1 we have:

COROLLARY 2.4. *Let* $\delta^*: H^p(G/N, H^{q-1}(N, C)) \to H^p(G/N, H^q(N, A))$ *be induced by the boundary* $\delta: H^{q-1}(N, C) \to H^q(N, A)$. *Then for* $\alpha \in H^p(G/N, H^{q-1}(N, C))$, $\delta^*(\alpha) = \gamma_N \cup \alpha$.

In the case of extensions (1) that are split as extensions of abelian groups we can now easily derive our result.

THEOREM 2.5. *Suppose* $0 \to A \to B \to C \to 0$ *is an extension of G modules split as an extension of abelian groups. Let* $N \subset G$ *be a normal subgroup, and* $\delta: H^{q-1}(N, C) \to H^q(N, A)$ *the boundary map from the cohomology long exact sequence. Then the following diagram commutes up to multiplication by* -1.

$$(2) \quad \begin{array}{ccc} H^p(G/N, H^{q-1}(N, C)) & \xrightarrow{\delta^*} & H^p(G/N, H^q(N, A)) \\ \downarrow{d_2} & & \downarrow{d_2} \\ H^{p+2}(G/N, H^{q-2}(N, C)) & \xrightarrow{\delta^*} & H^{p+2}(G/N, H^{q-1}(N, A)). \end{array}$$

PROOF. We have shown that $\delta^*(\alpha) = \gamma_N \cup \alpha$. We know that $d_2(\delta^*(\alpha)) = d_2(\gamma_N \cup \alpha) = d_2(\gamma_N) \cup \alpha + (-1)\gamma_N \cup d_2(\alpha) = -(\gamma_N \cup d_2(\alpha)) = -\delta^*(d_2(\alpha))$. $\square$

Of course, we want to prove the above result for all sequences. The tool we need is:

PROPOSITION 2.6. *Suppose* $\alpha \in H^a(G/N, H^b(N, C))$. *Then there is a* $\mathbb{Z}$ *free G module* C' *and a surjection* $C' \to C$ *such that* α *is the image of some* $\alpha' \in H^a(G/N, H^b(N, C'))$.

PROOF. Recall that G/N was assumed to be finite. It follows that $H^a(G/N, H^b(N, C))$ is the direct limit of groups $H^a(G/N', H^b(N', C'))$ where N' is a finite image of N and C' is a finite generated submodule of C. Thus we may assume N is finite and C is finitely generated.

Next we observe that the "single cohomology" version of this result is easy (and well known). Let H be a finite group, define $M_0(H) = \mathbb{Z}$, and inductively define $M_b(H)$ via the exact sequence $0 \to M_b(H) \to P \to M_{b-1}(H) \to 0$ where P is a free $\mathbb{Z}[H]$ module. An easy induction argument shows that $M_b(H)$ is torsion free and if $\alpha \in H^b(H, C)$ there is an H module homomorphism $\phi: M_b(H) \to C$ with α in the image of the induced map $H^b(H, M_b(H)) \to H^b(H, C)$.

Suppose now that $b > 1$. Form an exact sequence $0 \to C \to P \to D \to 0$ where P is a free G module and so a free N module under restriction. It follows that the boundary map $H^{b-1}(N, D) \to H^b(N, C)$ is an isomorphism of G/N modules. In particular, the induced map $H^a(G/N, H^{b-1}(N, D)) \to H^a(G/N, H^b(N, C))$ is an isomorphism. If $\beta \in H^a(G/N, H^{b-1}(N, D))$ is the preimage of α, we assume by induction that there is a $\mathbb{Z}$ free D' and a map $D' \to D$ such that β is in the image of $H^a(G/N, H^{b-1}(N, D'))$. Form an exact sequence $0 \to C' \to P' \to D' \to 0$ where P' is a free $\mathbb{Z}[G]$ module. We have the diagram:

$$
\begin{array}{ccccccccc}
0 & \longrightarrow & C' & \longrightarrow & P' & \longrightarrow & D' & \longrightarrow & 0 \\
& & \downarrow & & \downarrow & & \downarrow & & \\
0 & \longrightarrow & C & \longrightarrow & P & \longrightarrow & D & \longrightarrow & 0
\end{array}
$$

from which it quickly follows that α is in the image of $H^a(G/N, H^b(N, C'))$.

Thus we may assume $b = 0$ or $b = 1$. In the first case, $\alpha \in H^b(G/N, C^N)$ and so there is a $\phi : M_b(G/N) \to C^N \subset C$ such that α is in the image of $H^b(G/N, M_b(G/N))$. Viewing $M_b(G/N)$ as a G module, we are done with this case.

Finally, now, we are reduced to the case $b = 1$. Note that we can take $M_1(N) = I(N)$, where $I(N)$ is the kernel of $\mathbb{Z}[N] \to \mathbb{Z}$. Let $mI(N)$ be the m-fold direct sum of $I(N)$, and $M(m)$ the induced module $\mathbb{Z}[G] \otimes_N mI(N)$. Note that $H^1(N, I(N)) = \mathbb{Z}/n\mathbb{Z}$ where n is the order of N. Furthermore, if D is any N module and $\beta \in H^1(N, D)$, there is a $\phi : I(N) \to D$ such that the induced map $\phi^* : \mathbb{Z}/n\mathbb{Z} = H^1(N, I(N)) \to H^1(N, D)$ satisfies $\phi^*(1+n\mathbb{Z}) = \beta$. As a G/N module, $H^1(N, M(m))$ is the free $(\mathbb{Z}/n\mathbb{Z})[G/N]$ module of rank m with basis, say, $c_1, \ldots, c_m$. Thus if C is a G module, and $\alpha_1, \ldots, \alpha_m \in H^1(N, C)$ there is a G module map $\phi : M(m) \to C$ with $\phi^*(c_i) = \alpha_i$.

Let C' be a $\mathbb{Z}$ free G module and $\alpha_1, \ldots, \alpha_m \in H^1(N, C')$. Note that $\mathrm{Ext}_G(m\mathbb{Z}[G/N], C') = \mathrm{Ext}_N(m\mathbb{Z}, C') = mH^1(N, C')$, and let

$$0 \to C' \to R \to m\mathbb{Z}[G/N] \to 0$$

correspond to $(\alpha_1, \ldots, \alpha_m)$. It quickly follows that a map $\phi : C' \to D$ extends to a map $R \to D$ if and only if $\phi^*(\alpha_i) = 0$ in $H^1(N, D)$ for $i = 1, \ldots, m$.

We are finally ready to finish the $b = 1$ case. Let $\alpha \in H^a(G/N, H^1(N, C))$ be given by a cocycle $f : (G/N)^a \to H^1(N, C)$. Form $M(m)$ as above where $H^1(N, M(m))$ is the free $\mathbb{Z}[G/N]$ module with basis the elements of $(G/N)^a$. It follows that there is a $\phi : M(m) \to C$ and a cochain $f' : (G/N)^a \to H^1(N, M(m))$ such that $\phi^*(f') = f$. Now use the construction of the last paragraph to embed $M(m) \to R$ such that we exactly impose the relations turning f' into a cocycle $f'' : (G/N)^a \to H^1(N, R)$. Since f is a cocycle ϕ extends to $\phi : R \to C$ such that $\phi^*(f'') = f$. $\square$

With this proposition we can prove 2.5 in full generality.

THEOREM 2.7. *Suppose* $0 \to A \to B \to C \to 0$ *is an extension of* G *modules. Let* $N \subset G$ *be a normal subgroup. Then the diagram* (2) *commutes up to multiplication by* (-1).

PROOF. Suppose $\alpha \in H^p(G/N, H^{q-1}(N, C))$ and choose C' and α' as in the above lemma. Taking pullbacks we have the diagram:

$$
\begin{array}{ccccccccc}
0 & \longrightarrow & A & \longrightarrow & B' & \longrightarrow & C' & \longrightarrow & 0 \\
  &   & \| &   & \downarrow & & \downarrow & & \\
0 & \longrightarrow & A & \longrightarrow & B & \longrightarrow & C & \longrightarrow & 0
\end{array}
$$

and by naturality it suffices to prove the result for the top row. Since C' is $\mathbb{Z}$ free, we are done. $\square$

Discrete valuations on a field induce completions of that field. Complete discrete valued fields with residue characteristic 0 have absolute Galois groups of the form $\hat{\mathbb{Z}} \oplus N$ where N is the Galois group of the maximum unramified extension and the $\hat{\mathbb{Z}}$ term corresponds to taking roots of the prime. It will therefore be beneficial to study the the cohomology of groups of the form $\hat{\mathbb{Z}} \oplus N$ where N is an arbitrary profinite group. In particular, we are interested in describing the cohomology groups $H^i(\hat{\mathbb{Z}} \oplus N, \mu)$ where μ can be taken to be $\mathbb{Q}/\mathbb{Z}$ with the trivial action. To study this cohomology group we define a series of modules and maps. To begin with, let the $\hat{\mathbb{Z}} \oplus N$ module $\mathbb{Q}^*$ be defined as follows. As an N module, $\mathbb{Q}^* = \mu \oplus \mathbb{Q}$. To make it a $\hat{\mathbb{Z}} \oplus N$ module, set σ to be the standard topological generator of $\hat{\mathbb{Z}}$ corresponding to $1 \in \mathbb{Z}$. We need to define an action of σ on $\mathbb{Q}^*$. In fact, set $\sigma(\epsilon, r/n) = (\epsilon + r/n + \mathbb{Z}, r/n)$. Note that this is well defined and the fixed module $(\mathbb{Q}^*)^{\hat{\mathbb{Z}}} = \mu \oplus \mathbb{Z}$. Furthermore, there is an exact sequence $0 \to \mu \to \mathbb{Q}^* \to \mathbb{Q} \to 0$ of $\hat{\mathbb{Z}} \oplus N$ modules.

We have the maps
(3)
$$
H^i(N, \mu) \oplus H^{i-1}(N, \mathbb{Q}/\mathbb{Z}) \xrightarrow{1 \oplus \delta} H^i(N, \mu \oplus \mathbb{Z}) \xrightarrow{\iota_1} H^i(\hat{\mathbb{Z}} \oplus N, \mathbb{Q}^*) \xleftarrow{\iota_2} H^i(\hat{\mathbb{Z}} \oplus N, \mu)
$$

where the maps are as follows. The map ι_1 is the inflation map and ι_2 is the map induced by the inclusion $\mu \subset \mathbb{Q}^*$. Without any fear of confusion we can identify $H^i(N, \mu \oplus \mathbb{Z})$ with $H^i(N, \mu) \oplus H^i(N, \mathbb{Z})$, so δ is the boundary map induced by the sequence $0 \to \mathbb{Z} \to \mathbb{Q} \to \mathbb{Q}/\mathbb{Z} \to 0$.

LEMMA 2.8. *For* $i \geq 1$, ι_1 *is an isomorphism. For* $i \geq 2$, δ *and* ι_2 *are isomorphisms.*

PROOF. Since $\hat{\mathbb{Z}}$ has cohomological dimension one, $H^i(\hat{\mathbb{Z}}, \mathbb{Q}^*) = H^i(\hat{\mathbb{Z}}, \mu) = 0$ for $i \geq 2$. Also, $H^1(\hat{\mathbb{Z}}, \mu) \to H^1(\hat{\mathbb{Z}}, \mathbb{Q}^*)$ is easily seen to be surjective and $\mathbb{Q}^*$ splits all the 1 cohomology of μ so $H^1(\hat{\mathbb{Z}}, \mathbb{Q}^*) = 0$. The Hochschild-Serre spectral sequence now implies that ι_1 is an isomorphism. Since $H^i(N, \mathbb{Q}) = 0 = H^i(\hat{\mathbb{Z}} \oplus N, \mathbb{Q})$ for $i \geq 1$, the long exact sequence of cohomology shows that ι_2 and δ are isomorphisms for $i \geq 2$. $\square$

In order to make stating the proposition below a little smoother, let $\phi_1 :$ $H^1(N, \mu) \to H^1(\hat{\mathbb{Z}} \oplus N, \mu)$ be the inflation map and $\phi_2 : H^0(N, \mathbb{Q}/\mathbb{Z}) \to$ $H^1(\hat{\mathbb{Z}} \oplus N, \mu)$ the map defined as follows. $H^0(N, \mathbb{Q}/\mathbb{Z}) = \mathbb{Q}/\mathbb{Z}$ so it suffices to define $\phi_2(1/n + \mathbb{Z})$ as an element of $H^1(\hat{\mathbb{Z}} \oplus N, \mu) = \mathrm{Hom}(\hat{\mathbb{Z}} \oplus N, \mu)$. We do this by setting $\phi_2(1/n + \mathbb{Z})(N) = 0$ and $\phi_2(1/n + \mathbb{Z})(\sigma) = -1/n + \mathbb{Z}$.

PROPOSITION 2.9. *Diagram* (3) *yields an isomorphism* $\tau_i : H^i(\hat{\mathbb{Z}} \oplus N, \mu) \cong$ $H^i(N, \mu) \oplus H^{i-1}(N, \mathbb{Q}/\mathbb{Z})$ *for* $i \geq 2$. *There is also an isomorphism* $\tau_1 :$ $H^1(\hat{\mathbb{Z}} \oplus N, \mu) \cong H^1(N, \mu) \oplus H^0(N, \mathbb{Q}/\mathbb{Z})$ *where* τ^{-1} *is* (ϕ_1, ϕ_2).

PROOF. The case $i = 1$ is straightforward and the case $i \geq 2$ follows from 2.8. □

From τ_i there is an induced map $\rho : H^i(\hat{\mathbb{Z}} \oplus N, \mu) \to H^{i-1}(N, \mathbb{Q}/\mathbb{Z})$ defined by projection. This map ρ is the cohomological version of the ramification map as we will easily see in section three. As such, it is very important to us. However, the peculiarities of the method of proof require we now consider the map in the opposite direction. The inverse $\tau^{-1} : H^i(N, \mu) \oplus$ $H^{i-1}(N, \mathbb{Q}/\mathbb{Z}) \cong H^i(\hat{\mathbb{Z}} \oplus N, \mu)$ must have the form (ϕ_1, ϕ_2) (extending our notation from the $i = 1$ case) where $\phi_1 : H^i(N, \mu) \to H^i(\hat{\mathbb{Z}} \oplus N, \mu)$ and $\phi_2 : H^{i-1}(N, \mathbb{Q}/\mathbb{Z}) \to H^i(\hat{\mathbb{Z}} \oplus N, \mu)$. Our definition of τ shows immediately that ϕ_1 is the inflation map so ϕ_2 is the harder object. ϕ_2 is particularly difficult because it has two distinct definitions in the cases $i = 1$ and $i > 1$. We must remedy this by giving a alternate description of ϕ_2. This description will take much of the rest of the section. For our discussion we will need to assume that N is "large enough" in the following sense. Note that the condition below is always satisfied if N is the absolute Galois group of a field containing all roots of one.

CONDITION LE. If n is any integer, $N \to N'$ a surjection to a finite group N', then this map factors as a composition of surjections $N \to N'' \to N'$ where N'' is finite of order a multiple of n.

View $\mathbb{Z}/n\mathbb{Z} \subset \mathbb{Q}/\mathbb{Z}$ by identifying $1 + n\mathbb{Z}$ with $1/n + \mathbb{Z}$. Then we have that $H^{i-1}(N, \mathbb{Q}/\mathbb{Z})$ is the direct limit of groups $H^{i-1}(N', \mathbb{Z}/n\mathbb{Z})$ where N' is a finite image of N. Since we will observe that the maps we define will commute with this directed system, it suffices to define a homomorphism $H^{i-1}(N', \mathbb{Z}/n\mathbb{Z}) \to H^i(\hat{\mathbb{Z}} \oplus N', \mu)$ where N' is finite of order a multiple of n.

To define this map we call ϕ_2', form a resolution

(4) $$0 \longrightarrow K \longrightarrow P \longrightarrow \mathbb{Z}/n\mathbb{Z} \longrightarrow 0$$

where P is a free $\mathbb{Z}[N']$ module. Let $\delta : H^{i-1}(N', \mathbb{Z}/n\mathbb{Z}) \to H^i(N', K)$ be the boundary map from the long exact sequence. We next define a $\hat{\mathbb{Z}} \oplus N'$ module P^* as follows. As an N' module, $P^* = \mu \oplus P$. If $f : P \to \mathbb{Z}/n\mathbb{Z}$ is the map from (4), and if $\sigma \in \hat{\mathbb{Z}}$ is the canonical generator as above, set $\sigma(\epsilon, x) = (\epsilon + f(x), x)$. In this way we define a $\hat{\mathbb{Z}}$ and hence a $\hat{\mathbb{Z}} \oplus N'$

action on P^*. It is easy to see that there is a commutative diagram:

$$
\begin{array}{ccccccccc}
0 & \longrightarrow & \mu & \longrightarrow & P^* & \longrightarrow & P & \longrightarrow & 0 \\
 & & \| & & \downarrow & & \downarrow & & \\
0 & \longrightarrow & \mu & \longrightarrow & Q^* & \longrightarrow & Q & \longrightarrow & 0
\end{array}
$$

with exact rows. Furthermore the rightmost vertical arrow fits into a diagram:

$$
(5) \qquad
\begin{array}{ccc}
P & \rightarrow & Q \\
\downarrow & & \downarrow \\
\mathbb{Z}/n\mathbb{Z} & \subset & \mathbb{Q}/\mathbb{Z}.
\end{array}
$$

Observe that $(P^*)^{\hat{\mathbb{Z}}} = \mu \oplus K$ and so there is a inflation map $H^i(N', \mu \oplus K) \to H^i(\hat{\mathbb{Z}} \oplus N', P^*)$. Further observe that, for $i \geq 1$, $H^i(\hat{\mathbb{Z}} \oplus N', P)$ is the direct sum of cohomology groups of the form $H^i(\hat{\mathbb{Z}}, \mathbb{Z})$ and hence $H^i(\hat{\mathbb{Z}} \oplus N', P) = 0$. Since N' has order a multiple of n, we compute that $H^0(\hat{\mathbb{Z}} \oplus N', P^*) \to H^0(\hat{\mathbb{Z}} \oplus N', P)$ is surjective. Thus for $i \geq 1$, $H^i(\hat{\mathbb{Z}} \oplus N', \mu) \cong H^i(\hat{\mathbb{Z}} \oplus N', P^*)$ via the map induced by the inclusion $\mu \subset P^*$. Finally, we define ϕ_2' to be the composition:

$$
\begin{aligned}
H^{i-1}(N', \mathbb{Z}/n\mathbb{Z}) &\to H^i(N', K) \to H^i(N', \mu \oplus K) \\
&\to H^i(\hat{\mathbb{Z}} \oplus N', P^*) \cong H^i(\hat{\mathbb{Z}} \oplus N', \mu).
\end{aligned}
$$

We next observe that ϕ_2' is independent of the the choice of resolution (4). In fact, if $0 \to K' \to P' \to \mathbb{Z}/n\mathbb{Z} \to 0$ is another resolution, we have the commutative diagrams:

$$
\begin{array}{ccccccccc}
0 & \to & K & \to & P & \to & \mathbb{Z}/n\mathbb{Z} & \to & 0 \\
 & & \downarrow & & \downarrow & & \| & & \\
0 & \to & K' & \to & P' & \to & \mathbb{Z}/n\mathbb{Z} & \to & 0
\end{array}
$$

and

$$
\begin{array}{ccccccccc}
0 & \to & \mu & \to & P^* & \to & P & \to & 0 \\
 & & \| & & \downarrow & & \downarrow & & \\
0 & \to & \mu & \to & (P')^* & \to & P' & \to & 0
\end{array}
$$

and the independence of ϕ_2' is clear. Because of this it is easy to see that ϕ_2' commutes with inflation and the natural inclusions $\mathbb{Z}/n\mathbb{Z} \subset \mathbb{Z}/mn\mathbb{Z}$.

We must show that ϕ_2' is essentially the map ϕ_2 defined above. This is the content of

PROPOSITION 2.10. *Suppose* $\phi_2 : H^{i-1}(N, \mathbb{Q}/\mathbb{Z}) \to H^i(\hat{\mathbb{Z}} \oplus N, \mu)$ *is the map defined after 2.9. Let* $N \to N'$ *be a finite image of* N *of order a multiple of* n, *let* $\iota : H^{i-1}(N', \mathbb{Z}/n\mathbb{Z}) \to H^{i-1}(N, \mathbb{Q}/\mathbb{Z})$ *be induced by inflation and the inclusion* $\mathbb{Z}/n\mathbb{Z} \subset \mathbb{Q}/\mathbb{Z}$, *and let* $\iota' : H^i(\hat{\mathbb{Z}} \oplus N', \mu) \to H^i(\hat{\mathbb{Z}} \oplus N, \mu)$ *be the inflation map. Then* $\iota' \circ \phi_2'$ *is* $\phi_2 \circ \iota$ *for all* $i \geq 1$.

PROOF. There is a commutative diagram:

$$0 \longrightarrow K \longrightarrow P \longrightarrow \mathbb{Z}/n\mathbb{Z} \longrightarrow 0$$

$$\downarrow \qquad\qquad \downarrow \qquad\qquad \downarrow$$

$$0 \longrightarrow \mathbb{Z} \longrightarrow \mathbb{Q} \longrightarrow \mathbb{Q}/\mathbb{Z} \longrightarrow 0.$$

This, the diagram (5) and the naturality of inflation shows that the diagram

$$H^{i-1}(N', \mathbb{Z}/n\mathbb{Z}) \to H^i(N', K) \to H^i(N', \mu \oplus K) \to H^i(\hat{\mathbb{Z}} \oplus N', P^*) \leftarrow H^i(\hat{\mathbb{Z}} \oplus N', \mu)$$

$$\downarrow \qquad\qquad \downarrow \qquad\qquad \downarrow \qquad\qquad \downarrow \qquad\qquad \downarrow$$

$$H^{i-1}(N, \mathbb{Q}/\mathbb{Z}) \to H^i(N, \mathbb{Z}) \to H^i(N, \mu \oplus \mathbb{Z}) \to H^i(\hat{\mathbb{Z}} \oplus N, \mathbb{Q}^*) \leftarrow H^i(\hat{\mathbb{Z}} \oplus N, \mu)$$

commutes for $i \geq 2$. This proves the proposition in those cases. The case $i = 1$ amounts to showing the following. Identify $H^0(N, \mathbb{Z}/n\mathbb{Z})$ with $\mathbb{Z}/n\mathbb{Z}$. Then we must show $\phi_2'(1 + n\mathbb{Z}) \in H^1(\hat{\mathbb{Z}} \oplus N', \mu) = \mathrm{Hom}(\hat{\mathbb{Z}} \oplus N', \mu)$ is the homomorphism with $\phi_2'(1 + n\mathbb{Z})(N) = 0$ and $\phi_2'(1 + n\mathbb{Z})(\sigma) = -1/n + \mathbb{Z}$. This last fact is a direct computation best left to the reader. $\square$

Now suppose there is an exact sequence

$$1 \longrightarrow N \longrightarrow G' \longrightarrow G \longrightarrow 1$$

of profinite groups with G finite. Then there is an induced exact sequence:

$$1 \longrightarrow \hat{\mathbb{Z}} \oplus N \longrightarrow \hat{\mathbb{Z}} \oplus G' \longrightarrow G \longrightarrow 1.$$

Associated to each of these sequences is a Hochschild-Serre spectral sequence:

$$H^p(G, H^q(N, \mu)) \Rightarrow H^{p+q}(G', \mu)$$

and

$$H^p(G, H^q(\hat{\mathbb{Z}} \oplus N, \mu)) \Rightarrow H^{p+q}(\hat{\mathbb{Z}} \oplus G', \mu) .$$

More precisely, $H^p(G, H^q(N, \mu))$ or $H^p(G, H^q(\hat{\mathbb{Z}} \oplus N, \mu))$ are the E^2 terms of the respective spectral sequences. Thus there are boundaries maps:

$$(d_2')_{p,q} : H^p(G, H^q(\hat{\mathbb{Z}} \oplus N, \mu)) \to H^{p+2}(G, H^{q-1}(\hat{\mathbb{Z}} \oplus N, \mu))$$

and

$$(d_2)_{p,q} : H^p(G, H^q(N, \mu)) \to H^{p+2}(G, H^{q-1}(N, \mu)).$$

THEOREM 2.11. *Assume N is "large enough" in the sense that N satisfies Condition LE above. For $q \geq 1$ there is a commutative diagram*:

$$H^p(G, H^q(\hat{\mathbb{Z}} \oplus N, \mu)) \quad \cong \quad H^p(G, H^q(N, \mu)) \oplus H^p(G, H^{q-1}(N, \mathbb{Q}/\mathbb{Z}))$$

$$(d_2')_{p,q} \downarrow \qquad\qquad\qquad\qquad \downarrow (d_2)_{p,q} \oplus (-1)(d_2)_{p,q-1}$$

$$H^{p+2}(G, H^{q-1}(\hat{\mathbb{Z}} \oplus N, \mu)) \cong H^{p+2}(G, H^{q-1}(N, \mu)) \oplus H^{p+2}(G, H^{q-2}(N, \mathbb{Q}/\mathbb{Z}))$$

where the horizontal isomorphisms are the composition of the isomorphism induced by τ of 2.9, and the general isomorphism $H^p(G, A \oplus B) \cong H^p(G, A) \oplus H^p(G, B)$.

PROOF. This result is an immediate consequence of the commutativity of the following two diagrams:

(6)
$$
\begin{array}{ccc}
H^p(G, H^q(N, \mu)) & \xrightarrow{\phi_1^*} & H^p(G, H^q(\hat{\mathbb{Z}} \oplus N, \mu)) \\
\downarrow{(d_2)_{p,q}} & & \downarrow{(d_2')_{p,q}} \\
H^{p+2}(G, H^{q-1}(N, \mu)) & \xrightarrow{\phi_1^*} & H^{p+2}(G, H^{q-1}(\hat{\mathbb{Z}} \oplus N, \mu))
\end{array}
$$

and

(7)
$$
\begin{array}{ccc}
H^p(G, H^{q-1}(N, \mathbb{Q}/\mathbb{Z})) & \xrightarrow{\phi_2^*} & H^p(G, H^q(\hat{\mathbb{Z}} \oplus N, \mu)) \\
\downarrow{(-1)(d_2)_{p,q-1}} & & \downarrow{(d_2')_{p,q}} \\
H^{p+2}(G, H^{q-2}(N, \mathbb{Q}/\mathbb{Z})) & \xrightarrow{\phi_2^*} & H^{p+2}(G, H^{q-1}(\hat{\mathbb{Z}} \oplus N, \mu))
\end{array}
$$

where ϕ_1 and ϕ_2 are the maps defined after 2.9. That (6) commutes is just the fact that the spectral sequence commutes with inflation. To show (7) commutes for all $q \geq 1$ we turn to the description of ϕ_2 embodied in 2.10 and note that then it suffices to show:

$$
\begin{array}{ccc}
H^p(G, H^{q-1}(N', \mathbb{Z}/n\mathbb{Z})) & \xrightarrow{\phi_2'} & H^p(G, H^q(\hat{\mathbb{Z}} \oplus N', \mu)) \\
\downarrow{(-1)(d_2)_{p,q-1}} & & \downarrow{(d_2')_{p,q}} \\
H^{p+2}(G, H^{q-2}(N', \mathbb{Z}/n\mathbb{Z})) & \xrightarrow{\phi_2'} & H^{p+2}(G, H^{q-1}(\hat{\mathbb{Z}} \oplus N', \mu))
\end{array}
$$

where N' is a finite image of N of order a multiple of n. Now ϕ_2' is the composition of a boundary map from a long exact sequence with a series of natural maps and inflation maps. Since the spectral sequence commutes with the later two, the commutativity reduces to 2.7. $\square$

It will be useful later to make a brief comment about the naturality of the isomorphism in 2.11. If $\hat{K}$ is a complete discrete valuation field of residue characteristic 0, there is a natural exact sequence $1 \to \hat{\mathbb{Z}} \to G_{\hat{K}} \to N \to 1$ where N is the Galois group of the maximum unramified extension of $\hat{K}$. This sequence splits as we said, but the splitting is not natural involving, as it does, the choice of a prime element in $\hat{K}$. Of course, $\phi_1 : H^q(N, \mu) \to H^q(G_{\hat{K}}, \mu) = H^q(\hat{\mathbb{Z}} \oplus N, \mu)$ is natural being the inflation map. On the other hand, ϕ_2 depends on the choice of splitting. We do know (easily) that the map ρ defined after 2.9. is natural. However, since the definition of ρ is complicated, it will often be convenient to simply note that the kernel of ρ is the image of the inflation map ϕ_1. In fact, we have:

LEMMA 2.12. *The kernel of*

$$
\rho^* : H^p(G/N, H^q(\hat{\mathbb{Z}} \oplus N, \mu)) \to H^p(G/N, H^{q-1}(N, \mathbb{Q}/\mathbb{Z}))
$$

is the image of $\phi_1^* : H^p(G, H^q(N, \mu)) \to H^p(G, H^q(\hat{\mathbb{Z}} \oplus N, \mu))$.

Let us end this cohomology section with a simple description of $(d_2)_{0,1}$. So let G be a profinite group and $N \subset G$ a closed normal subgroup. If C is any G module with trivial action, the Hochschild-Serre boundary map $(d_2)_{0,1} : H^1(N, C)^{G/N} \to H^2(G/N, C)$ can be described in the following terms. Suppose $\alpha \in H^1(N, C) = \text{Hom}(N, C)$ is G/N invariant. Then the extension $1 \to N \to G \to G/N \to 1$ induces via α an extension $1 \to C \to G' \to G/N \to 1$, and we set $c(\alpha)$ to be the corresponding element of $H^2(G/N, C)$.

PROPOSITION 2.13. $(d_2)_{0,1}(\alpha) = -c(\alpha)$.

PROOF. Using the spectral sequence description in [HS], we can perform this proof directly on cocycles. Pick coset representatives $\{g_i\}$ for N in G. Write $g_i g_j = g_k n(n_i, n_j)$ where g_k is the coset representative in $g_i g_j N$ and $n(g_i, g_j) \in N$. Thus $\alpha(n(g_i, g_j)) \in C$ is a two cycle defining $c(\alpha)$. Extend $\alpha : N \to C$ to a set map $\alpha' : G \to C$ by setting $\alpha'(g_i n) = \alpha(n)$. Compute the boundary $d(\alpha')(g_i n_i, g_j n_j) = \alpha'(g_j n_j) + \alpha'(g_i n_i) - \alpha'(g_i n_i g_j n_j) = \alpha(n_i) + \alpha(n_j) - \alpha'(g_k n(g_i, g_j)(g_j^{-1} n_i g_j) n_j) = -\alpha(n(g_i, g_j))$ because α is a G/N invariant homomorphism.

3. The spectral sequence and Galois cohomology

In this section we will investigate the Hochschild-Serre boundary maps in the context of Galois cohomology. We begin with a theorem about $(d_2)_{1,1}$ in the following situation. Let G be a finite group acting faithfully on a field K (via automorphisms) with $L = K^G$. As usual, for any field K we let G_K be the absolute Galois group. Assume as we always do in this paper that L contains μ, the full group of roots of one and that all fields have characteristic 0. We have the exact sequence $1 \to G_K \to G_L \to G \to 1$ and so a Hochschild-Serre boundary map $(d_2)_{1,1} : H^1(G, H^1(G_K, \mu)) \to H^3(G, \mu)$. We claim:

THEOREM 3.1. *The image of* $(d_2)_{1,1}$ *is the kernel of* $H^3(G, \mu) \to H^3(G, K^*)$.

PROOF. We proceed via an intermediate module as follows. Let K_a be the algebraic closure of K and hence of L. Set $M^* = \{x \in K_a^* | x^n \in K^* \text{ for some } n\}$. Clearly M^* is a G_L submodule of K_a^*, and $\mu \subset M^*$. Since K_a is separably closed, $M = M^*/\mu$ is a torsion free divisible module. One easily sees that M has trivial G_K action. It follows that $H^i(H, M) = (0)$ for all $i > 0$ and all subgroups $H \subset G_L$. Hence, $H^i(H, \mu) \cong H^i(H, M^*)$ for all $i \geq 2$. By Kummer theory, $H^1(G_K, \mu) \to H^1(G_K, M^*)$ is the 0 map, so

(1) $$H^1(G_K, M^*) = 0.$$

Finally, we have $(M^*)^{G_K} = K^*$.

The spectral sequence is natural, so

$$H^1(G, H^1(G_K, \mu)) \longrightarrow H^1(G, H^1(G_K, M^*))$$

$$(d_2)_{1,1} \downarrow \qquad\qquad\qquad \downarrow (d_2)_{1,1}$$

$$H^3(G, \mu) \longrightarrow H^3(G, K^*)$$

commutes. By (1), the image of $(d_2)_{1,1}$ is contained in the kernel of $H^3(G, \mu)$ $\to H^3(G, K^*)$.

Conversely, suppose $\alpha \in H^3(G, \mu)$ maps to 0 in $H^3(G, K^*)$. There is a commutative diagram

$$H^3(G_L, \mu) \quad \cong \quad H^3(G_L, M^*)$$

$$\text{inf} \uparrow \qquad\qquad\qquad \uparrow \text{inf}$$

$$H^3(G, \mu) \longrightarrow H^3(G, K^*)$$

so $0 = \inf(\alpha) \in H^3(G_L, \mu)$. If α is not in the image of $(d_2)_{1,1}$, then by the spectral sequence there is a $\beta \in H^2(G_K, \mu)^G$ with $(d_2)_{0,2}(\beta) = 0$ and a $\gamma \in H^1(G, H^1(G_K, \mu))$ with

$$(2) \qquad\qquad \alpha = (d_3)_{0,2}(\beta) + (d_2)_{1,1}(\gamma).$$

Here d_3 is the boundary map from the E_3 step of the spectral sequence. Since α is not in the image of $(d_2)_{1,1}$, β is not in the image of $H^2(G_L, \mu)$. We have a commutative diagram

$$H^2(G_L, \mu) \quad \longrightarrow \quad H^2(G_K, \mu)^G$$

$$\| \wr \qquad\qquad\qquad\qquad \| \wr$$

$$H^2(G_L, M^*) \quad \longrightarrow \quad H^2(G_K, M^*)^G.$$

If $\beta^* \in H^2(G_K, M^*)^G$ is the image of β, β^* is not in the image of $H^2(G_L, M^*)$. The naturality of the spectral sequence shows that $(d_2)_{0,2}(\beta^*)$ $= 0$. Thus $(d_3)_{0,2}(\beta^*)$ is defined and nonzero. We apply the inclusion $\mu \subset M^*$ to the equation (2). By (1), γ maps to (0). If α^* is the image of α in $H^3(G, K^*)$, (2) becomes $\alpha^* = (d_3)_{0,2}(\beta^*) \neq 0$. This contradiction proves the result. $\square$

In the rest of this section we will investigate the relationship between the boundary map $(d_2)_{0,2}$ from the spectral sequence and the ramification map on cohomology. To begin, we define the ramification map and show how it is related to the map τ of 2.9. Let K be a field of characteristic 0 containing all roots of one. Suppose $R \subset K$ is a discrete valuation ring such that $q(R) = K$, and the residue field $\bar{R}$ has characteristic 0. For any field L, let G_L be the absolute Galois group of L. Set $\hat{K}$ to be the completion of K with respect to R. Then $G_{\hat{K}} \cong \hat{\mathbb{Z}} \oplus G_{\bar{R}}$. The embedding $K \subset \hat{K}$ induces an embedding $\hat{\mathbb{Z}} \oplus G_{\bar{R}} \subset G_K$ defined up to conjugation. Thus there is a unique restriction map $H^n(G_K, \mu) \to H^n(\hat{\mathbb{Z}} \oplus G_R, \mu)$. The isomorphism $\tau : H^n(\hat{\mathbb{Z}} \oplus G_R, \mu) \cong H^n(G_R, \mu) \oplus H^{n-1}(G_R, \mathbb{Q}/\mathbb{Z})$ of 2.9 induces a projection $\rho : H^n(\hat{\mathbb{Z}} \oplus G_R, \mu) \to H^{n-1}(G_R, \mathbb{Q}/\mathbb{Z})$. We would like to give the well-known alternative description of this map as a ramification map generalizing the ramification map on the Brauer group.

Let $\hat{K}_a$ be the algebraic closure of $\hat{K}$ and $\hat{K}_u$ the maximum unramified extension of $\hat{K}$, so $\hat{K}_u/\hat{K}$ has Galois group G_R and $\hat{K}_u$ is the fixed field of $\hat{\mathbb{Z}}$. This induces the natural exact sequence $1 \to \hat{\mathbb{Z}} \to G_{\hat{K}} \to G_R \to 1$. Note that $H^1(\hat{\mathbb{Z}}, \hat{K}_a^*) = 0$ by Hilbert's Theorem 90. $H^q(\hat{\mathbb{Z}}, \hat{K}_a^*) = H^q(\hat{\mathbb{Z}}, \mu) = 0$ for $q \geq 2$ because $\hat{\mathbb{Z}}$ has cohomological dimension 1. Thus by the Hochschild-Serre spectral sequence the inflation map $H^n(G_R, \hat{K}_u^*) \to H^n(G_{\hat{K}}, \hat{K}_a^*)$ is an isomorphism. The valuation on $\hat{K}$ extends to $\hat{K}_u$ and so defines a map $v : \hat{K}_u^* \to \mathbb{Z}$ that induces a cohomology map $H^n(G_R, \hat{K}_u^*) \to H^n(G_R, \mathbb{Z}) \cong H^{n-1}(G_R, \mathbb{Q}/\mathbb{Z})$. The composition $H^n(G_K, \mu) \to H^n(G_{\hat{K}}, \mu) \cong H^n(G_{\hat{K}}, \hat{K}_a^*) \cong H^n(G_R, \hat{K}_u^*) \to H^{n-1}(G_R, \mathbb{Q}/\mathbb{Z})$ is the ramification map we call r. Note that if $n = 2$ and if we identify $\mathrm{Br}(K)$ with $H^2(G_K, \mu)$ and $H^1(G_R, \mathbb{Q}/\mathbb{Z})$ with $\chi(P)$ then r is just the ramification map associated to the discrete valuation ring R. We also can use r to denote the map $H^n(G_{\hat{K}}, \mu) \to H^{n-1}(G_R, \mu)$.

LEMMA 3.2. *The following diagram commutes for* $n \geq 2$.

$$
\begin{array}{ccccc}
H^n(G_K, \mu) & \longrightarrow & H^n(G_{\hat{K}}, \mu) & \longrightarrow & H^{n-1}(G_R, \mathbb{Q}/\mathbb{Z}) \\
\| & & \| & & \| \\
H^n(G_K, \mu) & \xrightarrow{\mathrm{res}} & H^n(\hat{\mathbb{Z}} \oplus G_R, \mu) & \xrightarrow{\rho} & H^{n-1}(G_R, \mu)
\end{array}
$$

where the top row is the map r.

PROOF. The left square commutes because opposite sides are identical. As for the right square, let $\hat{L} = (\hat{K})^{G_R}$ be the fixed field. Then $\hat{L}/\hat{K}$ has Galois group $\hat{\mathbb{Z}}$ and $\hat{L} = \bigcup \hat{K}(\pi^{1/n})$ where π is some prime of $\hat{K}$. Furthermore, we made our identifications so that $\sigma(\pi_{1/n}) = \rho(n)\pi^{1/n}$ for σ the canonical generator of $\hat{\mathbb{Z}}$. Thus μ and the $\pi^{1/n}$ generate a $\hat{\mathbb{Z}} \oplus G_R$ submodule of $\hat{K}_a^*$ isomorphic to $\mathbb{Q}^*$. Using this isomorphism identify $\mathbb{Q}^*$ with its image in $\hat{K}_a^*$. Now $\mu \oplus \mathbb{Z} = (\mathbb{Q}^*)^{\hat{\mathbb{Z}}}$ is a subset of $\hat{K}^*$ and so of $\hat{K}_u^*$. If v is as above, the diagram

$$
\begin{array}{ccc}
\hat{K}_u^* & \xrightarrow{v} & \mathbb{Z} \\
\cup & & \| \\
\mu \oplus \mathbb{Z} & \longrightarrow & \mathbb{Z}
\end{array}
$$

commutes and so the diagram

$$
\begin{array}{ccccccc}
H^n(\hat{\mathbb{Z}} \oplus G_R, \mu) & \cong & H^n(\hat{\mathbb{Z}} \oplus G_R, \hat{K}_a^*) & \xleftarrow{\mathrm{inf}} & H^n(G_R, \hat{K}_u^*) & \xrightarrow{v^*} & H^n(G_R, \mathbb{Z}) \\
\| & & \uparrow & & \uparrow & & \| \\
H^n(\hat{\mathbb{Z}} \oplus G_R, \mu) & \cong & H^n(\hat{\mathbb{Z}} \oplus G_R, \mathbb{Q}^*) & \xleftarrow{\mathrm{inf}} & H^n(G_R, \mu \oplus \mathbb{Z}) & \longrightarrow & H^n(G_R, \mathbb{Z})
\end{array}
$$

commutes. The top row, followed by $H^2(G_R, \mathbb{Z}) \cong H^1(G_R, \mu)$, is, by definition, r, and the bottom row, composed with the same isomorphism, is ρ. □

We will make particular use of the case $n = 2$ of the above lemma. In different language we have:

COROLLARY 3.3. *The following diagram commutes.*

$$
\begin{array}{ccccc}
\mathrm{Br}(K) & \longrightarrow & \mathrm{Br}(\hat{K}) & \longrightarrow & \chi(\bar{R}) \\
\| \wr & & \| \wr & & \| \\
H^2(G_K, \mu) & \xrightarrow{\mathrm{res}} & H^2(\hat{\mathbb{Z}} \oplus G_R, \mu) & \xrightarrow{\rho} & H^1(G_R, \mu) \,.
\end{array}
$$

Though it is a bit artificial, it is also convenient at this point to note that the map ρ is independent of the splitting of $G_{\hat{K}}$ as a direct sum.

COROLLARY 3.4. ρ *depends only on the sequence* $1 \to \hat{\mathbb{Z}} \to G_{\hat{K}} \to G_R \to 1$ *and not on the choice of spitting.*

PROOF. This is clear for $n \geq 2$ by 3.2 because the definition of r is independent of the splitting. We realize this argument is artificial because we could have made it in section 2 using $\mathbb{Q}^*$ instead of $\hat{K}_a$, and it does not really involve fields. It does, however, save space to make the argument here.

All that remains to show is the case $n = 1$. One can immediately compute from the definitions preceding 2.9 that the map $\rho : H^1(\hat{\mathbb{Z}} \oplus G_R, \mu) \to H^0(G_R, \mu) = \mu$ sends $f : \hat{\mathbb{Z}} \oplus G_R \to \mu$ to $f(\sigma)$ where σ is the canonical generator of $\hat{\mathbb{Z}}$. The result is now clear. □

Suppose next that a finite group G acts on K as above via field automorphisms. We will investigate the interplay of the Hochschild-Serre boundary maps and the ramification map $r : \mathrm{Br}(K) \to \chi(\bar{R})$. Suppose, then, that $H = \{g \in G | g(R) = R\}$ is the stabilizer of R. We say R is faithful if H acts faithfully on $\bar{R}$. Of course, the action of H extends to the completion $\hat{K}$. If $L = K^G$, then $\hat{K}^H = \hat{L}$ is the completion of L with respect to $S = R \cap L$. Note that R is faithful if and only if $\hat{K}/\hat{L}$ is unramified. There is an exact sequence

$$(3) \qquad\qquad 1 \to G_K \to G_L \to G \to 1$$

with an associated Hochschild-Serre boundary map

$$(d_2)_{0,2} : \mathrm{Br}(K)^G \cong H^2(G_K, \mu)^G \to H^2(G, H^1(G_K, \mu)) \,.$$

It is the relationship between $(d_2)_{0,2}$ and r we are interested in.

The extension (3) when restricted to $G_{\hat{L}}$ yields an extension

$$(4) \qquad\qquad 1 \to \hat{\mathbb{Z}} \oplus G_R \to G_{\hat{L}} \to H \to 1$$

and so a Hochschild-Serre boundary map

$$(d_2)'_{0,2} : H^2(\hat{\mathbb{Z}} \oplus G_R, \mu)^H \longrightarrow H^2(H, H^1(\hat{\mathbb{Z}} \oplus G_R, \mu)) \,.$$

Using (4) to write $\hat{\mathbb{Z}}$ as a subgroup of $G_{\hat{L}}$, we have observed that the fixed field of $\hat{\mathbb{Z}}$ is $\hat{K}_u$, the maximum unramified extension of $\hat{K}$. Since $\hat{K}_u/\hat{L}$ is normal, $\hat{\mathbb{Z}}$ is a normal subgroup of $G_{\hat{L}}$. It follows that the restriction map $H^1(\hat{\mathbb{Z}} \oplus G_R, \mu) \to H^1(\hat{\mathbb{Z}}, \mu) = \mu$ is an H morphism, and so induces the composition res : $H^2(G, H^1(G_K, \mu)) \to H^2(H, H^1(\hat{\mathbb{Z}} \oplus G_R, \mu)) \to H^2(H, \mu)$. We will describe res $\circ (d_2)_{0,2}$ in terms of the ramification map r.

Set H' to be the image of H in $\mathrm{Aut}(\bar{R})$, so $\bar{R}/\bar{S}$ is H' Galois. We have an extension

(5) $1 \to G_R \to G_S \to H' \to 1.$

If $\alpha \in \mathrm{Br}(K)^G$, then clearly $r(\alpha) \in \chi(\bar{R})$ is H, or equivalently, H' invariant. We recall the definition of $c(r(\alpha)) \in H^2(H', \mu)$ from the end of section 2. Since $r(\alpha) : G_R \to \mu$ is an H' map it and (5) induce an extension $1 \to \mu \to H^* \to H' \to 1$. We let $c(r(\alpha)) \in H^2(H', \mu)$ be the corresponding element. Our main result is:

THEOREM 3.5. *Suppose* $\alpha \in \mathrm{Br}(K)^G = H^2(G_K, \mu)^G$. *Then* $\mathrm{res} \circ (d_2)_{0,2}(\alpha)$ $\in H^2(H, \mu)$ *is the image under inflation of* $c(r(\alpha))$.

PROOF. To begin with, the naturality of both the spectral sequence and of the ramification map shows that we may assume $K = \hat{K}$ and so $H = G$. We first consider the case R is faithful or $\hat{K}/\hat{L}$ is unramified and $H = H'$. The sequence (4) now looks like

$$1 \to \hat{\mathbb{Z}} \oplus G_R \to \hat{\mathbb{Z}} \oplus G_S \to H \to 1$$

where we have taken the direct sum of (5) (with $H = H'$) and $1 \to \hat{\mathbb{Z}} \to \hat{\mathbb{Z}} \to 1 \to 1$. Note that we can write (4) in this way because $\hat{K}/\hat{L}$ is unramified. In fact, writing $G_{\hat{L}}$ or $G_{\hat{K}}$ as such a direct summand amounts to choosing a prime element. Since $\hat{K}/\hat{L}$ is unramified we can choose the same prime in both fields.

By 2.11, the diagram

$$
\begin{array}{ccc}
H^2(\hat{\mathbb{Z}} \oplus G_R, \mu)^H & \longrightarrow & H^1(G_R, \mathbb{Q}/\mathbb{Z})^H \\
\Big\downarrow {\scriptstyle (d_2)_{0,2}} & & \Big\downarrow {\scriptstyle (-1)(d_2)_{0,1}} \\
H^2(H, H^1(\hat{\mathbb{Z}} \oplus G_R, \mu)) & \longrightarrow & H^2(H, H^0(G_R, \mu)) = H^2(H, \mu)
\end{array}
$$

commutes so this case is done by 2.13.

We turn to the general case, but first we show we can make a simplifying assumption. It may happen that $\bar{S}$ does not contain an element which is not an e power for e some integer. If this is the case, we may adjoin an indeterminant to K such that G fixes it and we can extend R so that the new $\bar{S}$ is a function field in one variable over the old. Proving the result for this new situation implies it for the old. Thus we assume $\bar{S}$ contains an element which is not an e power for any e.

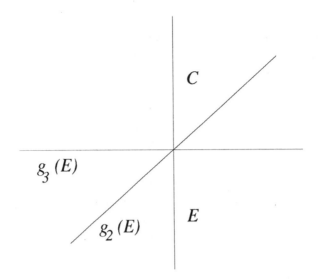

Figure 1

To continue with the general case, let N be the kernel of $H \to H'$ so that N is the ramification subgroup of $\hat{K}/\hat{L}$. By our assumptions, $\hat{K}/\hat{L}$ is tamely ramified so $N \subset H$ is a cyclic central subgroup of order, say, e. Furthermore, if $\hat{L}'' = (\hat{K})^N$, then $\hat{K} = \hat{L}''(\pi''^{1/e})$ where $\pi'' \in \hat{L}''$ is a prime. Since $\hat{L}''/\hat{L}$ is unramified, we can write $\pi'' = u\pi$ where $\pi \in \hat{L}$ is prime and u is a unit. Since we may multiply u by any unit of $\hat{L}$, we may assume that $\hat{K}'' = \hat{L}''(u^{1/e})$ has degree e over $\hat{L}''$. It quickly follows that $\hat{K}''$ is an unramified extension of $\hat{L}$ with Galois group H. $\hat{K}' = \hat{K}\hat{K}''$ is a Galois extension of $\hat{L}$ with group $N \oplus H$. Moreover, $(\pi)^{1/e} \in \hat{K}'$ and if we set $\hat{L}' = \hat{L}(\pi^{1/e})$, $\hat{K}'/\hat{L}'$ is unramified Galois with group H, and $\hat{L}'/\hat{L}$ is totally tamely ramified with group N.

We have the diagram of fields as shown in Figure 1.

In particular, $\bar{S}$ is the residue field of $\hat{L}'$. Let R' be the extension of R in $\hat{K}'$, so that $\bar{R}'$ is the residue field of $\hat{K}'$. We have the exact sequence

$$(6) \qquad\qquad 1 \to G_{R'} \to G_{\bar{S}} \to H \to 1.$$

All together, we have the commutative diagram:

$$(7) \qquad \begin{array}{ccccccccc} 1 & \to & G_{\hat{K}'} & \to & G_{\hat{L}'} & \to & H & \to & 1 \\ & & \downarrow & & \downarrow & & \| & & \\ 1 & \to & \hat{\mathbb{Z}} \oplus G_R & \to & G_{\hat{L}} & \to & H & \to & 1. \end{array}$$

Let $\eta : H^i(\hat{\mathbb{Z}} \oplus G_R, \mu) \to H^i(\hat{\mathbb{Z}} \oplus G_{R'}, \mu)$ be the restriction map. The computation of ρ in 3.4, and the fact that $\hat{K}'/\hat{K}$ is unramified, show that

$$
\begin{array}{ccc}
H^1(G_{\hat{K}}, \mu) & \xrightarrow{\eta} & H^1(G_{\hat{K}'}, \mu) \\
\downarrow & & \downarrow \\
H^0(G_R, \mu) & & H^0(G_{R'}, \mu) \\
\| & & \| \\
\mu & = & \mu
\end{array}
$$

commutes. This plus (7) and the naturality of the spectral sequence shows that

$$
\begin{array}{ccccc}
H^2(G_{\hat{K}}, \mu)^H & \xrightarrow{d_2} & H^2(H, H^1(G_{\hat{K}}, \mu)) & \longrightarrow & H^2(H, \mu) \\
\downarrow & & \downarrow & & \downarrow \\
H^2(G_{\hat{K}'}, \mu)^H & \xrightarrow{d_2} & H^2(H, H^1(G_{\hat{K}'}, \mu)) & \longrightarrow & H^2(H, \mu)
\end{array}
$$

commutes where the vertical arrows are induced by restriction and the rightmost vertical arrow is therefore the identity. Let $\alpha' \in H^2(G_{\hat{K}'}, \mu)^H$ be the restriction of α. By the faithful case, α' has image $c(r(\alpha')) \in H^2(H, \mu)$ which is also the image of α. But $r(\alpha') \in \mathrm{Hom}_c(G_{R'}, \mu)$ is the image of $r(\alpha) \in \mathrm{Hom}(G_R, \mu)$ and so $c(r(\alpha'))$ is the image under inflation of $c(r(\alpha)) \in H^2(H', \mu)$. $\square$

4. Brauer groups of invariant fields, geometrically negligible classes, and an equivariant Chow group

Let G be a finite group and V a faithful G representation over F. Form the field $F(V)$ of rational functions on V. Then G acts on $F(V)$ in the natural way. In this section we will, among other things, investigate the action of G on $\mathrm{Br}(F(V))$.

Consider the exact sequence

$$
(1) \qquad 0 \to \mathrm{Br}(F(V)) \to \oplus_P(\chi(P)) \to \oplus_C(\mathbb{Q}/\mathbb{Z}) \to 0
$$

of 1.4. G acts naturally on all these groups and this is clearly an exact sequence of G modules. Let us, however, consider in more detail how G acts on these groups. For any codimension 1 irreducible $P \subset V$, let $H_P \subset G$ be the stabilizer of P and H_P' the image of H_P in $\mathrm{Aut}(F(P))$. Then H_P acts on $\chi(P)$ in the natural way. If $\oplus_{G/H_P}(\chi(P))$ is the direct summand of $\oplus_P(\chi(P))$ corresponding to the G orbit of P, then this group is isomorphic to the induced module $\mathrm{Ind}_{H_P}^G(\chi(P))$. Set $\mathscr{P}$ to be a set of representatives of the G orbits in the set of all the codimension 1 closed irreducible $P \subset V$. Then the full group $\oplus_P(\chi(P))$ is isomorphic to $\oplus_{P \in \mathscr{P}}(\mathrm{Ind}_{H_P}^G(\chi(P)))$.

Similarly, let $C \subset V$ be irreducible of codimension 2. We set H_C to be the stabilizer of G on C. Then H_C acts trivially on $\mathbb{Q}/\mathbb{Z}$, and $\oplus_C(\mathbb{Q}/\mathbb{Z})$ is isomorphic to $\oplus_{C \in \mathscr{C}}(\mathrm{Ind}_{H_C}^G(\mathbb{Q}/\mathbb{Z}))$ where $\mathscr{C}$ is a collection of representatives of all G orbits of the C's.

We begin with trying to find some sort of description of the image of the Brauer group $\mathrm{Br}(F(V)^G)$ in $\mathrm{Br}(F(V))$. Of course, this image is in the G

fixed subgroup $\mathrm{Br}(F(V))^G$. To go any further, we invoke the Hochschild-Serre spectral sequence as follows. For convenience, set $K = F(V)$ and $L = F(V)^G = K^G$. We have an exact sequence

$$1 \to G_K \to G_L \to G \to 1$$

and so Hochschild-Serre boundary maps

$$(d_2)_{0,1} : \mathrm{Br}(K)^G \to H^2(G, H^1(G_K, \mu))$$

and

$$(d_2)_{1,1} : H^1(G, H^1(G_K, \mu)) \to H^3(G, \mu).$$

To understand our question we will have to understand these maps. With this motivation, we begin by describing $H^1(G_K, \mu)$.

Of course, we have $H^1(G_K, \mu) = \mathrm{Hom}_c(G_K, \mu) = \mathrm{Hom}_c(A, \mu)$ where $A = G_K/(G_K, G_K)$ is the Galois group of the maximum abelian extension of $K = F(V)$. To describe $H^1(G_K, \mu)$ we will describe A, which will be possible because $F[V]$ is a unique factorization domain. By Kummer theory

$$A = \varprojlim \mathrm{Hom}(K^*/(K^*)^n, \mu).$$

Since F^* is divisible, and K^*/F^* is the free abelian group on the primes of $F[V]$, it is clear that A is the direct product of $\hat{\mathbb{Z}}$'s over all the (height one) primes of $F[V]$. It will help, however, to make this description of A a bit more explicit.

Let $P \subset F[V]$ be a height one prime with generator π. Let $R_P = F[V]_P$ be the localization, a discrete valuation ring. Let $\hat{K}_P$ be the completion of K with respect to P, G_P the absolute Galois group of the residue field $F(P)$. The absolute Galois group $G_{\hat{K}_P}$ has the form $\hat{\mathbb{Z}} \oplus G_P$. We have an inclusion $\hat{\mathbb{Z}} \oplus G_P \subset G_K$ which is uniquely defined up to conjugation.

$\hat{\mathbb{Z}} \subset G_{\hat{K}_P}$ is the naturally defined ramification group. Moreover, $\hat{\mathbb{Z}}$ has a canonical generator σ_P characterized by the relation $\sigma_P(\pi'^{1/n}) = \rho(n)\pi'^{1/n}$ for any prime π' of $\hat{K}_P$. In particular this holds for π. The composition $\hat{\mathbb{Z}} \to G_{\hat{K}_P} \to G_K \to A$ is uniquely defined, and the above description of σ_P shows that this map is injective. We identify σ_P with its image in A and write $\hat{\mathbb{Z}}_P = \hat{\mathbb{Z}}\sigma_P \subset A$. We also denote by $\hat{\mathbb{Z}}_P$ a choice of image of $\hat{\mathbb{Z}}$ in G_K. Finally, we let K_a be the algebraic closure of K. Note that the fixed field $(K_a)^{\hat{\mathbb{Z}}_P}$ is isomorphic to the strict henselization of K at P. It follows that if $\pi' \in F[V]$ is a prime element not in P, $(\pi')^{1/n} \in (K_a)^{\hat{\mathbb{Z}}_P}$ for all n so $\sigma_P((\pi')^{1/n}) = (\pi')^{1/n}$. Since $F[V]$ is a UFD, elements of A are determined by their action on $(\pi')^{1/n}$ for all primes π' and all n. We immediately have:

LEMMA 4.1. (a) A *is a free profinite abelian group with basis the* σ_P *'s. If* $g \in G$, *then* $g\sigma_P g^{-1} = \sigma_{g(P)}$.

(b) $H^1(G_K, \mu) = \operatorname{Hom}_c(A, \mu) \cong \oplus_{P \in \mathscr{P}} \operatorname{Ind}_{H_P}^G(\mu)$ as G modules.

Associated to each prime P is the restriction map $\operatorname{res}_P \colon H^2(G, H^1(G_K, \mu))$ $\to H^2(H_P, \mu)$ defined prior to 3.5. Looking at the definitions, 4.1 implies that the direct sum of the res_P's yields an isomorphism $\operatorname{res} \colon H^2(G, H^1(G_K, \mu))$ $\cong \oplus_{P \in \mathscr{P}} H^2(H_P, \mu)$. Combining this with 3.5 we have:

THEOREM 4.2. *If* $\alpha \in \operatorname{Br}(F(V))^G$ *and* $r_P \colon \operatorname{Br}(F(V)) \to \chi(P)$ *is the ramification map,* $(d_2)_{0,2}(\alpha) \in H^2(G, H^1(G_K, \mu)) \cong \oplus_{P \in \mathscr{P}} H^2(H_P, \mu)$ *has* P *component equal to the image under inflation of* $c(r_P(\alpha)) \in H^2(H_P', \mu)$ *where* $c(r_P(\alpha))$ *is as defined in 2.13.*

Since the difference between H_P and H_P' plays a role in 4.2, we are led to examine what ramification is possible at primes $P \subset F[V]$. As noted before, the kernel of $H_P \to H_P'$ must be a normal central cyclic subgroup we call N. Since H_P acts faithfully on V and not on its image in $F[V]/I(P)$, $I(P) \cap V \neq (0)$. Since $I(P)$ has height one, $L = V \cap I(P)$ is a one-dimensional subspace in V. It follows that as an N module, $V = L \oplus V'$ where L is a faithful one-dimensional representation and V' is a trivial representation. It then follows that $H_P = H_P' \oplus N$. In particular, the inflation map $H^2(H_P', \mu) \to H^2(H_P, \mu)$ is injective. This fact allows a succinct description of the kernel of $(d_2)_{0,2}$. The next theorem says that α is in this kernel if and only if its characters are in the image of characters of primes restricted to $F(V)^G$.

THEOREM 4.3. *If* $\alpha \in \operatorname{Br}(F(V))^G$, *then* $(d_2)_{0,2}(\alpha) = 0$ *if and only if for all* $P \subset V$ *irreducible of codimension* 1, $r_P(\alpha) \in \chi(P)$ *is in the image of* $\chi(Q)$ *where* $Q = F[V]^G \cap P$.

PROOF. Suppose K'/L' is H' Galois, and $\chi \in \chi(K')$ is H' invariant. Let $N \subset \mu$ be the image of χ so N is the Galois group of the extension M'/K' defined by χ. The image $c(\chi) \in H^2(H, \mu)$ defines an extension, H^*, of H by μ. We have the diagram:

$$
\begin{array}{ccccccccc}
1 & \longrightarrow & N & \longrightarrow & H & \longrightarrow & H' & \longrightarrow & 1 \\
& & \downarrow & & \downarrow & & \downarrow & & \\
1 & \longrightarrow & \mu & \longrightarrow & H^* & \longrightarrow & H' & \longrightarrow & 1
\end{array}
$$

where the leftmost vertical arrow is inclusion. $c(\chi) = 0$ if and only if the bottom row splits if and only if there is a $\delta \colon H \to \mu$ such that $N \to H \to \mu$ is the inclusion. Thus if $c(\chi) = 0$, this δ can be viewed as in $\chi(L')$ and is a preimage of χ. Conversely, suppose $\delta \colon G_{L'} \to \mu$ is a preimage of χ. Since $\delta|_{G_{K'}} = \chi$, $\delta(G_{M'}) = 1$ and δ induces $\delta \colon H \to \mu$ as above. The above observation applied to the $c(r_P(\alpha))$ proves the theorem. □

Note that the conclusion of 4.3 is very natural. For $\alpha \in \operatorname{Br}(F(V))^G$ to be in the image of restriction a necessary condition is that its characters be in the image of restriction. If we view questions about characters as being easier, it is natural to focus on Brauer group elements all of whose characters are in the image of restriction and wonder whether that forces the element itself to be in the image of $\mathrm{Br}(F(V)^G)$. Thus we set $\mathrm{Br}(F(V))_0^G$ to be the set of all $\alpha \in \mathrm{Br}(F(V))^G$ such that all $r_P(\alpha) \in \chi(P)$ are in the image of $\chi(Q)$ for $Q = P \cap F(V)^G$. In other words, we define $\mathrm{Br}(F(V))_0^G$ to be the kernel of $(d_2)_{0,2}$. The rest of this section will be devoted to the quotient $\mathrm{Br}(F(V))_0^G / \mathrm{Res}(\mathrm{Br}(F(V)^G))$.

The Hochschild-Serre spectral sequence shows that $(d_3)_{0,2}$ induces an embedding $\mathrm{Br}(F(V))_0^G / \mathrm{Res}(\mathrm{Br}(F(V)^G)) \to H^3(G, \mu)/\mathrm{Im}((d_2)_{1,1})$. Moreover, the image of $(d_3)_{0,2}$ generates, modulo $\mathrm{Im}((d_2)_{1,1})$, the kernel of $H^3(G, \mu) \to H^3(G_L, \mu)$. Let us concentrate on this kernel a bit.

Inspired by a definition of Serre's, we call a $\beta \in H^n(G, \mu)$ that maps to 0 in $H^n(G_L, \mu)$ a *geometrically negligible* class. Since this concept is trivial if $n = 1$, we will always mean $n \geq 2$ when discussing these classes. We make a remark after 4.5 concerning how this definition differs from Serre's definition of negligible classes. We note below that the definition of geometrically negligible classes is independent of the choice of V as long as V is faithful over G. In fact, if V, V' are G faithful representations, then $L'' = F(V \oplus V')^G$ is rational (purely transcendental) over both $L = F(V)^G$ and $L' = F(V')^G$ (e.g. [EM, p. 16]). Thus part (b) below follows from (a).

PROPOSITION 4.4. (a) *Suppose M is a field and $M(x)$ the function field in one variable over M. Then the inflation map $H^n(G_M, \mu) \to H^n(G_{M(x)}, \mu)$ is injective.*

(b) *The group of geometrically negligible classes in $H^n(G, \mu)$ is independent of the choice of V.*

PROOF. We have to prove (a). But let $M' \supset M((x)) \supset M(x) \supset M$ be the extension of the Laurent series field $M((x))$ generated by $x^{1/n}$ for all n. Then $G_{M'} \cong G_M$, $G_{M'}$ is a subgroup of $G_{M(x)}$, and the epimorphism $G_{M(x)} \to G_M$ splits. $\square$

Because of the universality of the Galois extension $F(V)/F(V)^G$, the classes we defined as geometrically negligible are "negligible" in the following sense.

PROPOSITION 4.5. *Suppose $\alpha \in H^n(G, \mu)$ is geometrically negligible in our sense (so $n \geq 2$). Then for any G Galois extension of fields K'/L' with $F \subset L'$, α is in the kernel of the inflation map $H^n(G, \mu) \to H^n(G_{L'}, \mu)$.*

REMARK. If G is a finite group and M a G module, Serre defines an element of $\alpha \in H^n(G, M)$ to be negligible if α maps to 0 in $H^n(G_K, M)$ for every field K of characteristic 0 and every homomorphism $G_K \to G$. As we see above, our geometrically negligible classes require $M = \mu$ and K to contain the algebraically closed field F.

PROOF. Since α maps to 0 in $H^n(G_L, \mu)$, the map $G_L \to G$ factors into $G_L \to G' \to G$ where G' is finite and α maps to 0 in $H^n(G', \mu)$. Let $M \supset K \supset L$ be the extension of fields corresponding to $G_L \to G' \to G$. By, e.g., [S4, p. 253], there is a $0 \neq s \in F[V]^G$ and a G' Galois extension of commutative rings S/R such that $R = F[V]^G(1/s)$ and $M = S \otimes_R L$. Let $R \subset T \subset S$ correspond to $G' \to G$. Then $T = F[V](1/s)$ because both T and $F[V](1/s)$ are the integral closure of R in $F(V)$. By, e.g., [S4, p. 274] there is an F algebra homomorphism $\phi : R \to L'$ such that $K' \cong T \otimes_\phi L'$. Set $M' = S \otimes_\phi L'$. We have $M' \supset K' \supset L'$ and M'/L' is G Galois. Now M' may not be a field but there is a field $M'' \subset M'$ such that M''/L' is $H' \subset G'$ Galois and M' is the induced Galois extension $\text{Ind}_{H'}^{G'}(M''/L')$ (e.g. [S4, p. 253]). Since K' is a field the induced map $H' \to G$ is a surjection. But, of course, $H' \to G$ factors through G' so α maps to 0 in $H^n(H', \mu)$ and hence in $H^n(G_L', \mu)$. $\quad\square$

Among the geometrically negligible classes there are classes that are easier to describe. Let G be a finite group, $H \subset G$ a subgroup, and M a G module. Then $\text{Ext}_G(\mathbb{Z}[G/H], M) \cong H^1(G, \text{Hom}(\mathbb{Z}[G/H], M)) \cong H^1(H, M)$. Thus any $\alpha \in H^1(H, M)$ defines an extension $0 \to M \to M_\alpha \to \mathbb{Z}[G/H] \to 0$. More generally, if $\alpha_i \in H^1(H_i, M)$ are several elements, they define an extension $0 \to M \to M_{\vec{\alpha}} \to \oplus_i \mathbb{Z}[G/H_i] \to 0$ where $\vec{\alpha} = (\alpha_1, \dots, \alpha_n)$. It will be easy to see:

LEMMA 4.6. (a) *A G module morphism $\phi : M \to N$ extends to $\phi : M_{\vec{\alpha}} \to N$ if and only if $\phi^*(\alpha_i) = 0 \in H^1(H_i, N)$ for all i.*

(b) *There is an extension $0 \to \mu \to P^* \to P \to 0$ such that P is a permutation lattice, $H^1(H, \mu) \to H^1(H, P^*)$ is the zero map for all $H \subset G$, and for all $\phi : \mu \to N$, ϕ extends to P^* if and only if $H^1(H, \mu) \to H^1(H, N)$ is the zero map for all $H \subset G$. Furthermore $H^1(H, P^*) = 0$ for all $H \subset G$.*

PROOF. To prove (a), note that ϕ extends if and only if the induced extension $0 \to N \to N_{\vec{\alpha}} \to \oplus_i \mathbb{Z}[G/H_i] \to 0$ splits. The element corresponding to this extension in $\text{Ext}_G(\oplus_i \mathbb{Z}[G/H_i], N) = \oplus_i H^1(H_i, N)$ is just $(\phi^*(\alpha_1), \dots, \phi^*(\alpha_n))$ and this proves (a).

As for (b), all but the last sentence follows from (a). Since P is a permutation lattice $H^1(H, P) = 0$ for all H and the last sentence of (b) follows also. $\quad\square$

Let P^* be as in 4.6(b) and $n \geq 2$. An element in the kernel of $H^n(G, \mu) \to H^n(G, P^*)$ we call a permutation class. This name is justified as the permutation classes form the image of the coboundary map $\delta : H^{n-1}(G, P) \to H^n(G, \mu)$. In the next result we show, among other things, that the definition of permutation classes is independent of the choice of P^* and that permutation classes are geometrically negligible.

PROPOSITION 4.7. (a) *Suppose* $0 \to \mu \to Q^* \to Q \to 0$ *is an exact sequence of G modules such that* $H^1(H, Q^*) = 0$ *for all* $H \subset G$ *and* Q *is a permutation module. Then the kernel of* $H^n(G, \mu) \to H^n(G, Q^*)$ *is the group of permutation classes.*

(b) *Let* V *be a faithful* G *representation,* $K = F(V)$. *Then the kernel of* $H^n(G, \mu) \to H^n(G, K^*)$ *is exactly the set of permutation classes. In particular, permutation classes are geometrically negligible.*

(c) *The set of permutation classes form the image of* $(d_2)_{1,1}$.

PROOF. Part (c) follows from (b) by 3.1. We next show that (a) implies (b). $F[V]$ is a unique factorization domain so we have an exact sequence

$$0 \to F^* \to K^* \to Q \to 0$$

where Q is the permutation G module with basis the height one primes of $F[V]$. Now F^*/μ is uniquely divisible so $\text{Ext}_G(Q, \mu) \to \text{Ext}_G(Q, F^*)$ is surjective. It follows that we have a diagram:

$$
\begin{array}{ccccccccc}
0 & \to & F^* & \to & K^* & \to & Q & \to & 0 \\
  &   & \uparrow & & \uparrow & & \| & & \\
0 & \to & \mu & \to & Q^* & \to & Q & \to & 0
\end{array}
$$

where the vertical arrows can be assumed to be inclusions and the left square is a pushout diagram. Thus $K^*/Q^* \cong F^*/\mu$ and so $\check{H}^0(H, K^*/Q^*) = H^n(H, K^*/Q^*) = 0$ for all $H \subset G$ and $n \geq 1$. It follows that $H^1(H, Q^*) = 0$ for all $H \subset G$ and $H^n(G, Q^*) \to H^n(G, K^*)$ is injective for all $n \geq 1$. Thus the kernel of $H^n(G, \mu) \to H^n(G, P^*)$ is equal to the kernel of $H^n(G, \mu) \to H^n(G, Q^*)$ which is equal to the kernel of $H^n(G, \mu) \to H^n(G, K^*)$. To finish (b), let K_a be the algebraic closure of K and of L. Since K_a^*/μ is uniquely divisible, $H^n(G_L, \mu) \cong H^n(G_L, K_a^*)$ for $n \geq 2$. Thus $H^n(G, \mu) \to H^n(G_L, \mu)$ factors through $H^n(G, K^*)$. This proves (b).

As for (a), our construction of P^* shows that $\mu \to P^*$ extends to $P^* \to Q^*$. We also wish $\mu \to P^*$ to extend to $Q^* \to P^*$ because this will prove (a). If $\alpha \in \text{Ext}_G(Q, \mu)$ corresponds to $0 \to \mu \to Q^* \to Q \to 0$, we thus must show that α maps to 0 in $\text{Ext}_G(Q, P^*)$. But Q is a permutation module so $\text{Ext}_G(Q, P^*)$ is a direct sum of groups of the form $H^1(H, P^*) = 0$. $\square$

Let $H^3(G, \mu)_p$, respectively $H^3(G, \mu)_n$, be the group of permutation, respectively geometrically negligible, classes. The above proposition and the spectral sequence show:

COROLLARY 4.8. $(d_3)_{0,2}$ *induces an isomorphism*

$$\text{Br}(F(V))_0^G / \text{Res}(\text{Br}(F(V)^G)) \cong H^3(G, \mu)_n / H^3(G, \mu)_p .$$

Later on we will give an example where $\text{Br}(F(V))_0^G / \text{Res}(Br(F(V)^G)) \neq 0$. For now, we approach this quotient from another angle.

By the Merkuriev-Suslin theorem [MS], $\mathrm{Br}(F(V)^G)$ and $\mathrm{Res}(\mathrm{Br}(F(V)^G))$ are divisible groups. Thus the norm map $N_G(x) = \prod_{g \in G} g(x)$ is surjective when restricted to $\mathrm{Res}(\mathrm{Br}(F(V)^G))$. Of course, $\mathrm{Br}(F(V))$ is also divisible and so is $N_G(\mathrm{Br}(F(V)))$. In fact, since $\check{H}^0(G, \mathrm{Br}(F(V)))$ is $|G|$ torsion, $N_G(\mathrm{Br}(F(V)))$ is precisely the divisible part of $\mathrm{Br}(F(V))^G$. If $\mathrm{Res} : \mathrm{Br}(F(V)^G) \to \mathrm{Br}(F(V))$ and $\mathrm{Cor} : \mathrm{Br}(F(V)) \to \mathrm{Br}(F(V)^G)$ are the restriction and corestriction maps respectively, then the composition $\mathrm{Res} \circ \mathrm{Cor}$ is just N_G. Thus $N_G(\mathrm{Br}(F(V))) \subset \mathrm{Res}(\mathrm{Br}(F(V)^G))$ and altogether we have:

LEMMA 4.9. $\mathrm{Res}(\mathrm{Br}(F(V)^G)) = N_G(\mathrm{Br}(F(V)))$ is the divisible part of $\mathrm{Br}(F(V))^G$.

To analyze things further, note that 4.9 also applies to the character group $\chi(P)$ as a H_P module and hence to $\oplus_P(\chi(P))$ as a G module. If we consider the exact sequence (1), $\alpha \in \mathrm{Br}(F(V))^G$ is in $\mathrm{Br}(F(V))_0^G$ if and only if $r(\alpha) = \oplus_P r_P(\alpha) \in N_G(\oplus_P(\chi(P)))$. In other language:

PROPOSITION 4.10. $\mathrm{Br}(F(V))_0^G / \mathrm{Res}(\mathrm{Br}(F(V)^G))$ is isomorphic to the kernel of

$$\check{H}^0(G, \mathrm{Br}(F(V))) \to \check{H}^0(G, \oplus_P(\chi(P)))$$

and hence to the cokernel of

$$\check{H}^{-1}(G, \oplus_P(\chi(P))) \to \check{H}^{-1}(G, \oplus_C(\mathbb{Q}/\mathbb{Z})) .$$

The advantage of 4.10 is that it provides generators and relations for the quotient $\mathrm{Br}(F(V))_0^G / \mathrm{Res}(\mathrm{Br}(F(V)^G))$. To show this, first note that $\check{H}^{-1}(G, \oplus_C(\mathbb{Q}/\mathbb{Z}))$ is quite easy to describe. Once again, let $\mathscr{C}$ be a set of representatives of the G orbits of $C \subset V$ closed irreducible of codimension 2. Then $\oplus_C(\mathbb{Q}/\mathbb{Z}) = \oplus_{C \in \mathscr{C}}(\mathrm{Ind}_{H_C}^G(\mathbb{Q}/\mathbb{Z}))$. We have that $\check{H}^{-1}(G, \oplus_C(\mathbb{Q}/\mathbb{Z})) \cong \oplus_{\mathscr{C}}(\check{H}^{-1}(H_C, \mathbb{Q}/\mathbb{Z})) \cong \oplus_{\mathscr{C}}(\mathbb{Z}/|H_C|\mathbb{Z})$. More concretely, let $\beta'(C) \in \oplus_C(\mathbb{Q}/\mathbb{Z})$ be the element with component $1/|H_C| + \mathbb{Z}$ at C and 0 elsewhere. Let $\beta(C)$ be the image of $\beta'(C)$ in $\check{H}^{-1}(G, \oplus_C(\mathbb{Q}/\mathbb{Z}))$. Then the $\beta(C)$'s form a basis and $\beta(C)$ has order $|H_C|$.

For each C, there is an image $\alpha(C) = \delta(\beta(C)) \in \check{H}^0(G, \mathrm{Br}(F(V)))$. Let us describe this image, or more properly a representative of this image in $\mathrm{Br}(F(V))$. Choose a line $L \subset V$ such that the projection $\pi_L : V \to V/L$ is finite on C. Set $E = \pi_L^{-1}(\pi_L(C)) \supset C$. By the proof of 1.6 there is an $f \in \chi(E)$ that ramifies only at C with ramification $1/|H_C|$. Let $(f, E) \in \oplus_P(\chi(P))$ be the element with component f at E and 0 everywhere else, so (f, E) is a preimage of δ_C. Then by the definition of the coboundary map, $N_G((f, E))$ has image $\alpha(C)$ in $\check{H}^0(G, \mathrm{Br}(F(V)))$.

It will be useful if we describe $\alpha(C)$ and its ramification locus a bit more concretely. Suppose for simplicity that $H_C = G$ and $H_E = 1$. Then the

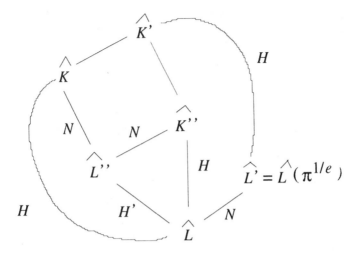

Figure 2

ramification locus of $\alpha(C)$ is as shown in Figure 2, where the characters on
each $g(E)$ ramify only at C with ramification $1/|H_C|$. If $G \not\subseteq H_C$, but
$H_E = 1$, the ramification locus of $\alpha(C)$ is the union of pictures like the
above one for each coset of H_C in G. If $H_E \neq 1$, the character on $g(E)$ is
a product of conjugates of f.

Next we turn to describing the relations among the $\alpha(C)$'s. We derive
a relation for each element of a generating set of $\check{H}^{-1}(G, \oplus_P(\chi(P))) =$
$\oplus_{P \in \mathscr{P}} \check{H}^{-1}(H_P, \chi(P))$. That is, for each $P \in \mathscr{P}$, and each $\chi \in \chi(P)$ such
that $N_{H_P}(\chi) = 0$, we get a relation among the $\alpha(C)$'s as follows. Suppose
for each $C \subset P$ of codimension 2 in V, $n_C = r_C(\chi)$ is the ramification of
χ at C. Then $\sum n_C \alpha(C) = 0$ in $\mathrm{Br}(F(V))_0^G / \mathrm{Res}(\mathrm{Br}(F(V)^G))$.

We can considerably simplify the situation by refocusing a bit. For convenience, we will make our argument about geometrically negligible classes.
One way of constructing these classes is to begin with a proper subgroup
$H \subset G$ and an element $\gamma' \in H^3(H, \mu)_n$. If we form $\gamma = \mathrm{Cor}_H^G(\gamma')$, then
is not hard to see that $\gamma \in H^3(G, \mu)_n$. In addition, the corestriction also
preserves the permutation classes. Though it might be very hard to, for example, show that $\gamma \neq 0$, in some sense the problem is purely cohomological
and can be for the moment ignored. With this philosophy in mind we will in
the next definition focus on "new" classes that are not permutation and not
corestrictions of elements of $H^3(H, \mu)_n$ for $H \subset G$ a proper subgroup.

DEFINITION. Let G be a finite group and $M \subset H^3(G, \mu)_n$ the subgroup
generated by all images $\mathrm{Cor}_H^G(H^3(H, \mu)_n)$ for all proper subgroups $H \subset G$.
Set $N^3(G) = H^3(G, \mu)_n/(M + H^3(G, \mu)_p)$.

We are about to give a geometric description of $N^3(G)$. Key to this is the following result. In order to state this result, let

$$\tau_G : \mathrm{Br}(F(V))_0^G / \mathrm{Res}(\mathrm{Br}(F(V)^G)) \cong H^3(G, \mu)_n / H^3(G, \mu)_p$$

be the isomorphism of 4.8. If C is $H \subset G$ invariant, we can restrict to H and define $\alpha_H(C) \in \mathrm{Br}(F(V))_0^H / \mathrm{Res}(\mathrm{Br}(F(V)^H))$ as above, while we set $\alpha_G(C)$ to be the $\alpha(C)$ defined for G.

LEMMA 4.11. *Suppose* $C \in \mathscr{C}$ *is* H *invariant. Then* $\mathrm{Cor}_H^G(\tau_H(\alpha_H(C))) = \tau_G(\alpha_G(C))$.

PROOF. This follows immediately from the description of corestriction in [AW, p. 104]. $\square$

Obviously, then, $N^3(G)$ is generated by the images of the $\alpha(C)$'s for all C that are G invariant. To describe the relations, first note that:

LEMMA 4.12. *If* G *is not a* p-*group for some* p, $N^3(G) = 0$. *If* G *is a* p-*group,* $pN^3(G) = 0$.

PROOF. Suppose G is not a p-group, and $\gamma \in H^3(G, \mu)_n$ has p power order. Then $r\gamma = \mathrm{Cor}_P^G \mathrm{Res}_G^P(\gamma)$ where $P \subset G$ is the p Sylow subgroup and $r = [G : P]$ is prime to p. Moreover, $\mathrm{Res}_G^P(\gamma)$ is clearly in $H^3(P, \mu)_n$. Thus if M is as above, $\gamma \in M$.

Suppose G is a p-group. Let $\gamma \in H^3(G, \mu)_n$ and choose $H \subset G$ of index p. Then $p\gamma = \mathrm{Cor}_H^G(\mathrm{Res}_G^H(\gamma))$ is in M. $\square$

With this preliminary, we turn to describing $N^3(G)$ geometrically. Because of 4.12 we assume G is a p-group. Let $F' = \oplus_{C \in \mathscr{C}}(\check{H}^{-1}(H_C, \mathbb{Q}/\mathbb{Z}))$, and view F' as the free abelian group with basis $\{\gamma'(C) | C \in \mathscr{C}\}$ such that $\gamma'(C)$ has order $|H_C|$. Let F be the $\mathbb{Z}/p\mathbb{Z}$ vector space with basis $\{\gamma(C) | C \in \mathscr{C}^G\}$ where $\mathscr{C}^G$ are the G invariant elements of $\mathscr{C}$. Define the surjection $\Phi : F' \to F$ by setting $\Phi(\gamma'(C)) = \gamma(C)$ if $C \in \mathscr{C}^{\mathscr{G}}$ and $\Phi(\gamma'(C)) = 0$ otherwise. Clearly the kernel of Φ is generated by pF' and $\{\gamma'(C) | C \notin \mathscr{C}^{\mathscr{G}}\}$. Define $R \subset F$ as follows. For each $P \subset V$, G invariant, closed, irreducible, of codimension 1, and each $f \in F(P)$ with $f^{|G|} \in F(P)^G$, write $r_C(f) = n_C/|G| + \mathbb{Z}$ for C G invariant and $n_C = 0$ otherwise. Set $r(f) = \sum (n_C)\gamma(C) \in F$. Then $R \subset F$ is generated by the $r(f)$'s for all such P and f.

THEOREM 4.13. *If* G *is not a* p-*group for some* p, $N^3(G) = 0$. *If* G *is a* p-*group,* $N^3(G) \cong F/R$.

PROOF. We have seen there is an epimorphism

$$\Psi : F' \to H^3(G, \mu)_n / H^3(G, \mu)_p$$

sending $\gamma'(C)$ to $\tau_G(\alpha_G(C))$. Composing this with the natural map

$$H^3(G, \mu)_n / H^3(G, \mu)_p \to N^3(G),$$

we have a surjection $F' \to N^3(G)$. By 4.12 pF' maps to 0, and by 4.11 $\gamma'(C)$ maps to 0 if $C \notin \mathscr{C}^{\mathscr{G}}$. Thus there is an induced surjection $F \to N^3(C)$.

Suppose $P \subset V$ and $f \in F(P)$ are as in the definition of R. Suppose $\chi \in \chi(P)$ is defined by $f^{1/n}$ for $n = |G|$. Then χ defines an element of $H^{-1}(H_P, \chi(P))$ which maps to $\sum r_C(f)(\tau_G(\alpha_G(C))) = 0$ in $H^3(G, \mu)_n / H^3(G, \mu)_p$ and hence to 0 in $N^3(G)$. It follows that there is an induced surjection $\phi : F/R \to N^3(G)$.

We must show ϕ is injective. Suppose $a \in F$ satisfies $\phi(a) = 0$. If a' is a preimage of $\phi(a)$ in $H^3(G, \mu)/H^3(G, \mu)_p$, then $a' = \sum \mathrm{Cor}_{H_i}^G(b_i)$ for some $H_i \neq G$. Now each b_i is of the form $\tau_{H_i}(\alpha_{H_i}(\sum(n_{i,j})C_{i,j}))$. By replacing each b_i by the sum of $\tau_H(\alpha_H(C_{i,j}))$ we can assume $b_i = \tau_{H_i}(\alpha_{H_i}(C_i))$. If H_i' is the stabilizer of C_i, b_i is the corestriction $\mathrm{Cor}_{H_i''}^{H_i}(b_i')$ for $H_i'' = H_i \cap H_i'$. Thus we may assume $H_i' \supset H_i$. If $H_i' \neq H_i$ then $\mathrm{Cor}_{H_i}^G(b_i)$ has the form $p\eta_i$ for some $\eta_i \in H^3(G, \mu)_n / H^3(G, \mu)_p$. Thus we can write $a' = p\eta' + \sum m_{C'}\tau_G(\alpha_G(C'))$ where all $C' \notin \mathscr{C}^{\mathscr{G}}$ and $\eta' \in H^3(G, \mu)_n / H^3(G, \mu)_p$.

Let $a'' \in F'$ be a preimage of a and $\eta'' \in F'$ a preimage of η' under Ψ. By our previous description of the kernel of Ψ, we have

$$(2) \qquad a'' = p\eta'' + \sum m_{C'}\gamma'(C') + \sum \beta(P_i, \chi_i)$$

where we define the $\beta(P_i, \chi_i)$ as follows. Suppose $P \subset V$ is codimension one closed, irreducible and $\chi \in \chi(F(P))$ has p power order and satisfies $N_{H_P}(\chi) = 0$. Set $\beta(P, \chi) = \sum r_{C_j}(\chi)\gamma'(C_j)$ where the sum is over closed irreducible codimension one subvarieties C_j of P and $r_{C_j}(\chi)$ is the ramification.

First, suppose some P is not G invariant. If $C \subset P$ is G invariant, then $|H_P|r_C(\chi) = 0$ and so $r_C(\chi)\gamma'(C) \in p(\mathbb{Z}\gamma'(C))$. It follows that $\Phi(\beta(P, \chi)) = 0 \in F$. Next, suppose P is G invariant but χ has order strictly dividing $n = |G|$. Then for any G invariant C, $r_C(\chi)\gamma'(C) \in p(\mathbb{Z}\gamma'(C))$ and again $\Phi(\beta(P, \chi)) = 0$. Finally suppose P is G invariant and χ has order, q, a multiple of $n = |G|$. If χ is defined by $f^{1/q}$ for $f \in F(P)^*$, then since $N_G(\chi) = 0$ we have $N_G(f) = x^q$ for some $x \in F(P)^*$. Since x^q is G invariant, $g(x)/x \in \mu$ for all $g \in G$ and x^n is G invariant. From this it follows that $x^q = N_G(x')$ where $x' = x^m\rho$, $\rho \in \mu$, and $m = q/n$. Thus $f = f'x'$ where $N_G(f') = 1$. If C is G invariant, we have $r_C(f') = 0$. Thus $\Phi(\beta(P, \chi)) = \Phi(\beta(P, \chi'))$ where χ' is defined by $x^{1/n}$. Finally, by definition, $\Phi(\beta(P, \chi')) \in R$.

All together we have shown that the right side of (2) above is mapped to R by Φ. Since $a = \Phi(a'')$, we are done. $\square$

At this point we can illustrate the above result with an example. Let G be a nonabelian 2 group with a normal cyclic subgroup H of index 2. H has a faithful one-dimensional representation L and we can set $V = \mathrm{Ind}_H^G(L)$.

Then V is a faithful irreducible G representation of dimension 2. The irreducible codimension 2 subvarieties are points, and it is easy to see that the origin $O \in V$ is the only G fixed point. Thus $N^3(G) = 0$ or $N^3(G) = \mathbb{Z}/2\mathbb{Z}$. We claim the second holds. It suffices to choose an image of O in $\mathrm{Br}(F(V))_0^G$ and show it is not in $\mathrm{Res}(\mathrm{Br}(F(V)^G))$.

Let L be spanned by x, choose some $g \in G$ not in H, and set $y = g(x)$. Thus x, y are a basis of V and can be viewed as elements of $F(V)$. L is defined by some injection $\rho: H \to \mu$. Since G is nonabelian, $\rho^g \neq \rho$. Note that for $h \in H$, $h(y) = \rho^g(h)y$. Let $Z \subset H$ be the subgroup $\{h \in H | \rho(h) = \rho^g(h)\}$ which is the center of G. Set $q = |Z|$. Consider the symbol algebra $(x, y)_{q, F(V)}$ and let $\alpha \in \mathrm{Br}(F(V))$ be its Brauer class. One can quickly check that $\alpha \in \mathrm{Br}(F(V))_0^G$ and has image in $\mathrm{Br}(F(V))_0^G / \mathrm{Res}(\mathrm{Br}(F(V)^G))$ equal to $\alpha(C)$ for C the point $O \in V$. Thus it suffices to show that α is not in the image of $\mathrm{Br}(F(V)^G)$. To this end blow up the point $O \in V$, and let P be the resulting exceptional divisor. An affine piece of the blow up looks like $\mathrm{Spec}(F[y, x/y])$ and P is defined by $y = 0$. Hence $F(P) = F(x/y)$ and the kernel of $G \to \mathrm{Aut}(F(P))$ is just Z. The ramification of α at P is defined by $(x/y)^{1/q}$ and the associated extension $F((y/x)^{1/q}) \supset F(y/x) \supset F(y/x)^{G/Z}$ defines the group extension

(3) $1 \to Z \to G \to G/Z \to 1$.

One calculates that if (3) corresponds to $\beta \in H^2(G/Z, Z)$, then the image of β in $H^2(G/Z, \mu)$ is not 0. Thus the ramification of α at P is not in the image of $\chi(Q)$ where Q is the prime of $F(V)$ that is the restriction of P. Thus α is not in the image of $\mathrm{Br}(F(V)^G)$.

THEOREM 4.14. *Suppose G is a nonabelian 2 group with a cyclic subgroup of index 2. Then $N^3(G) = \mathbb{Z}/2\mathbb{Z}$.*

REMARK. One can show that $H^3(A, \mu)_n/H^3(A, \mu)_p = 0$ for all abelian groups A. Then we will be able to conclude that if G is, say, dihedral of order 8 then $\mathbb{Z}/2\mathbb{Z} = \mathrm{Br}(F(V)_0^G)/\mathrm{Res}(\mathrm{Br}(F(V)^G)) = H^3(G, \mu)_n/H^3(G, \mu)_p$.

Despite the geometrical description, we do not know very much about the group $N^3(G)$. What little we do know is based on a generalization of the projection argument we used in section 1. Note that by 4.13 the problem can be roughly stated as follows. Suppose V is a G representation, and $C \subset V$ is G invariant, closed, irreducible of codimension 2. We would like to find P with $C \subset P \subset V$, P closed, irreducible, G invariant, and of codimension 1. Moreover, we need C to sit "nicely" in P.

For the rest of this section, define $\mathscr{C}^G$ to be the set of $C \subset V$ that are G invariant, closed, irreducible, and of codimension 2. For $C \in \mathscr{C}^G$, define $\delta(C) \in N^3(G)$ to be the image of $\gamma(C) \in F$. One approach to finding the needed P as above is to generalize the projection argument of section 1 as follows.

Suppose $x \in V$ is a G fixed point. Without loss of generality we may assume x is the origin O and we note that O is always G fixed. Next assume that $C \in \mathscr{C}^G$ is not a cone through O. That is, the ideal $I(C)$ is not homogeneous. As a preliminary choice for P, we consider P' to be the union of all lines through a point of C and O. This clearly is G fixed and has codimension 1. It is not hard to see that P' is irreducible, and this will follow from the discussion to come. The problem is that P' is not closed. We are about to describe its closure.

Recall the discussion of the tangent cone in [**Sh**, p.79]. For any $f \in F[V]$, let f_0 be the sum of all terms of lowest total degree and f_∞ the sum of all terms of highest total degree. Define $I_0(C)$ to be the (homogeneous) ideal generated by all f_0 for all $f \in I(C)$ and $I_\infty(C)$ the ideal generated by all f_∞ for the same f. Let C_0 be the variety on V defined by the radical of $I_0(C)$ and C_∞ the variety defined by the radical of $I_\infty(C)$. Neither C_0 nor C_∞ are necessarily irreducible, so write $C_0 = C_{0,1} \cup \cdots \cup C_{0,m}$ and $C_\infty = C_{\infty,1} \cup \cdots \cup C_{\infty,n}$ as the decompositions into irreducible components.

We can give geometric descriptions of these varieties as follows. If C does not contain O, C_0 is empty. If $O \in C$, C_0 is the tangent cone as in [**Sh**, p. 79]. To describe C_∞, embed V G equivariantly into a projective space $\mathbb{P}(E)$ where we can take $E = V \oplus F$ and let G act trivially on F. Let $\mathbb{H} = \mathbb{P}(E) - V$ be the hyperplane at infinity, and let $\bar{C} \subset \mathbb{P}(E)$ be the closure of C. Then C_∞ is the cone over $\bar{C} \cap \mathbb{H}$. Clearly both C_0 and C_∞ are cones through O, and both are G invariant. It follows that the components of C_0 and C_∞ are cones through O and are permuted by G.

We recall a bit more of the discussion of tangent cones in [**Sh**, p. 79]. Assume $0 \in C \in \mathscr{C}^G$. Let $X' \subset (V \times \mathbb{A}^1)$ be the subscheme defined as $X' = \{(x, t)|tx \in C\}$. That is, the ideal of X' is defined by the ideal J' generated by $f(tx)$ for all $f(x) \in I(C)$. Thus X' is closed and is in fact, as a set, the union of two irreducible components, namely, $V \times \{O\}$ and what we will call $\tilde{X}$. $\tilde{X}$ is the closure of $X'' = X' \cap (V \times (\mathbb{A}^1 - \{O\}))$. Note that X'' is clearly scheme theoretically isomorphic to $C \times (\mathbb{A}^1 - \{O\})$.

To continue, we describe the prime ideal $I(\tilde{X}) \subset F[V][t]$. For any $f \in F[V]$, write f as a sum of its homogeneous components $f_l + \cdots + f_h$ where $f_l = f_0$, $f_h = f_\infty$, l is the lowest degree of a term appearing in f, and h is the highest. Since $I(X')$ is generated by $f(tx)$ for all $f \in I(C)$, an exercise shows that $I(\tilde{X})$ is generated by all polynomials $f_l + f_{l+1}t + \cdots + f_h t^{h-l}$ for all $f \in I(C)$. Thus $\tilde{X} \cap (V \times \{O\})$ is scheme theoretically defined by $I_0(C)$. Let $C_0' \subset \tilde{X}$ be the associated subvariety and $C_{0,i}'$ its irreducible components. If we let G act trivially on $\mathbb{A}^1$ (or t) then $\tilde{X}$ is G invariant.

To find the closure of P' above, we must consider the closure of $\tilde{X}$ in $V \times \mathbb{P}^1$. To be precise, embed $\mathbb{A}^1$ in $\mathbb{P}^1$ such that $\mathbb{P}^1$ has homogeneous coordinates s, t and $\mathbb{A}^1$ is defined by $s \neq 0$. Let G act trivially on $\mathbb{P}^1$. The closure, $\bar{X}$, of $\tilde{X}$ in $V \times \mathbb{P}^1$ is then defined by the ideal in $F[V][s, t]$, homogeneous in s, t, with generators $s^{h-l}f_l + \cdots + t^{h-l}f_h$ for all $f \in I(C)$. Thus if $\infty \in \mathbb{P}^1 - \mathbb{A}^1$ is the point defined by $s = 0$, $i\bar{X} \cap (V \times \{\infty\})$ is defined scheme theoretically by $I_\infty(C)$ as a subscheme of $V = V \times \{\infty\}$. Let $C'_\infty \subset \bar{X}$ be the associated subvariety and $C'_{\infty,i}$ its irreducible components

Let $\pi : V \times \mathbb{P}^1 \to V$ be the projection and set $P = \pi(\bar{X})$. Since $\mathbb{P}^1$ is complete, P is closed irreducible. Since C is not a cone, P has codimension one and π restricted to $\bar{X}$ is generically finite. Since π is G invariant, so is P. Of course P is the closure of P'.

Let $a \in F(\bar{X})$ be the image of $t/s - 1 \in F(V \times \mathbb{P}^1)$. Clearly a is G invariant. $C_1 = \bar{X} \cap (V \times 1)$ is clearly isomorphic to C scheme theoretically. Since a has a zero of order 1 at C_1, and the poles of a define C'_∞ set theoretically, $(a) = [C_1] - \sum_i n_{\infty,i}[C'_{\infty,i}]$. Similarly, if $b \in F(\bar{X})$ is the image of $s/t - 1$, then b is G invariant and $(b) = [C_1] - \sum_i n_{0,i}[C'_{0,i}]$. Of course if $O \notin C$, this means that $(b) = [C_1]$. Now clearly π induces a scheme theoretic isomorphism from $C'_{0,i}, C_1, C''_{\infty,i} \subset \bar{X}$ to $C_{0,i}, C, C_{\infty,i} \subset P$. It follows that if $a' = N_{F(\bar{X})/F(P)}(a)$ and $b' = N_{F(\bar{X})/F(P)}$, then in the Chow group of P $(a') = [C] - \sum_i n_{\infty,i}[C_{\infty,i}]$ and $(b') = [C_1] - \sum_i n_{0,i}[C_{0,i}]$ (e.g. [**Fu**, p. 12]). Finally, a' and b' are clearly G invariant. We have shown:

THEOREM 4.15. *Suppose $C \in \mathscr{C}^G$ is not a cone through O. Let $\{C_{0,j} | 1 \leq j \leq m'\}$ and $\{C_{\infty,j} | 1 \leq j \leq n'\}$ be the sets of G invariant irreducible components of C_0 and C_∞ respectively. Then there are integers $n_{0,j}$ and $n_{\infty,j}$ such that, in $N^3(G)$, $\delta(C) = \sum_j n_{0,j}\delta(C_{0,j}) = \sum_j n_{\infty,j}\delta(C_{\infty,j})$. In particular, $N^3(G)$ is generated by the image of all cones and if $O \notin C$, $\delta(C) = 0$.*

We can use this result to settle one new special case.

COROLLARY 4.16. *Suppose G is a 3 group with a normal cyclic subgroup of index 3. Then $N^3(G) = 0$.*

PROOF. G has an irreducible faithful representation V of dimension 3. With the notation above, $\mathbb{H} = \mathbb{P}(E) - V$ is isomorphic to $\mathbb{P}(V)$ and so has no G fixed points. Thus V has no G invariant cones of codimension 2. $\square$

5. Unramified H^3

A field extension K/F is called *rational* if K is purely transcendental over F. If K, $K' \subset L$ are such that L/K and L/K' are rational we say K and K' are *stably isomorphic*. The nonrationality of certain fields has been shown using so-called unramified cohomology which we define below. The purpose of this section is to prove a result about the unramified cohomology of an invariant field $F(V)^G$, showing (5.3) that it is a subgroup of an image of the cohomology of G.

We begin by recalling the definition of the unramified H^3 group. Let

L/F be an extension of fields and $\mathscr{R}$ the set of all discrete valuation rings $R \subset L$ with $F \subset R$ and such that L is the field of fractions of R. To each $R \in \mathscr{R}$ there is an associated completion $\hat{L}_R$. Recall that the absolute Galois group of $\hat{L}_R$ has the form $\hat{\mathbb{Z}} \oplus G_R$ where G_R is the absolute Galois group of the residue field of R. Then to each $R \in \mathscr{R}$ there is a residue map $\rho_R : H^n(G_L, \mu) \to H^n(\mathbb{Z} \oplus G_R, \mu) \to H^{n-1}(G_R, \mu)$ where the first map is restriction and the second map is ρ as defined after 2.9 and described further in 3.2, 3.3, and 3.4.

DEFINITION. The unramified cohomology group $H_u^n(L) \subset H^n(G_L, \mu)$ is the intersection of the kernels of all ρ_R for all $R \in \mathscr{R}$.

Colliot-Thélène and Ojanguren have shown ([**CO**, p. 144]):

THEOREM 5.1. *If K and K' are stably isomorphic, $H_u^n(K) = H_u^n(K')$. In particular, if K/F is rational, $H_u^n(K) = H^n(F) = 0$ (because F was assumed algebraically closed).*

REMARK. There seems to be at least one other possible definition of ρ_R, being the map $H^n(\hat{\mathbb{Z}} \oplus G_R, \mu) \to H^{n-1}(G_R, \mu)$ arising out of the spectral sequence for $1 \to \hat{\mathbb{Z}} \to \hat{\mathbb{Z}} \oplus G_R \to G_R \to 1$. However, under any (reasonable) definition the kernel of ρ_R is the image of the inflation map $H^n(G_R, \mu) \to H^n(\hat{\mathbb{Z}} \oplus G_R, \mu)$. Thus the definition of unramified cohomology will be independent of which definition one chooses.

We are interested in $H_u^3(F(V)^G)$ where V is a faithful linear F representation of G. If V' is another faithful linear G representation over F, then by e.g. [**EM**, p. 16] $F(V \oplus V')^G$ is rational over both $F(V)^G$ and $F(V')^G$. That is, $F(V)^G$ and $F(V')^G$ are stably isomorphic. By the above theorem, $H^n(F(V)^G)$ is independent of the choice of V. For our argument we will make a special choice of V.

We would like to choose a faithful V and "double" it. That is, we would like to replace V by $V \oplus V$. We prefer $V \oplus V$ because of the following result. Recall that we called a prime P faithful if the stabilizer H_p acted faithfully on $F(P)$.

LEMMA 5.2. *Every height one prime of $F[V \oplus V]$ is faithful.*

PROOF. A nonfaithful prime Q would have the property that $L = Q \cap (V \oplus V)$ is one dimensional and G is not faithful on $(V \oplus V)/L$. Since G acts faithfully on V, this is not possible. $\square$

Our goal in this section is to prove the following theorem. In order to state it, let $K = F(V)$ for V a faithful G representation over F and set $L = F(V)^G$.

THEOREM 5.3. *$H_u^3(L)$ is a subgroup of the image of the inflation map $H^3(G, \mu) \to H^3(G_L, \mu)$.*

In order to prove this result we must explore a bit more deeply the terms of the spectral sequence associated to

(1) $$1 \to G_K \to G_L \to G \to 1.$$

One of the terms of the spectral sequence is $H^1(G, \mathrm{Br}(F(V)))$ and we study this group in the next result. Our tool is, of course, the exact sequence $0 \to \mathrm{Br}(F(V)) \to \oplus_P(\chi(P)) \to \oplus_C(\mathbb{Q}/\mathbb{Z}) \to 0$ which we have observed is an exact sequence of G modules.

LEMMA 5.4. $H^1(G, \mathrm{Br}(F(V))) \to H^1(G, \oplus_P(\chi(P)))$ is an injection.

PROOF. It suffices to show that $\check{H}^0(G, \oplus_C(\mathbb{Q}/\mathbb{Z})) = 0$. Recall from the beginning of section four that as a G module $\oplus_C(\mathbb{Q}/\mathbb{Z})$ has the form $\oplus_{C \in \mathscr{C}} \mathrm{Ind}_{H_C}^G(\mathbb{Q}/\mathbb{Z})$. Since $\check{H}^0(H_C, \mathbb{Q}/\mathbb{Z}) = 0$, the result is clear. $\square$

As a second tool in the proof of 5.3, we consider a bit more closely the map $(d_2)_{0,2} : \mathrm{Br}(F(V))^G \to H^2(G, H^1(G_K, \mu))$ defined and studied in sections three and four. To simplify the discussion, assume (as we may) that all height one primes of $F[V]$ are faithful. Recall from the discussion prior to 4.2 that $H^2(G, H^1(G_K, \mu)) \cong \oplus_{P \in \mathscr{P}} H^2(H_P, \mu))$ where the isomorphism, we call res, is the direct sum of the restriction maps res_P : $H^2(G, H^1(G_K, \mu)) \to H^2(H_P, \mu)$. Furthermore, in 4.2 we showed that for $\alpha \in \mathrm{Br}(F(V))^G$, $\mathrm{res}((d_2)_{0,2}(\alpha))$ has P component equal to $c(r_P(\alpha)) \in H^2(H_P, \mu)$. Our additional observation required here concerns how the image of $\mathrm{res} \circ (d_2)_{0,2}$ sits inside of $\oplus_{P \in \mathscr{P}} H^2(H_P, \mu)$.

PROPOSITION 5.5. $\mathrm{res}((d_2)_{0,2}(\mathrm{Br}(F(V))^G)) \subset \oplus_{P \in \mathscr{P}} H^2(H_P, \mu)$ is the direct sum of the images $\mathrm{res}_P((d_2)_{0,2}(\mathrm{Br}(F(V))^G)) \subset H^2(H_P, \mu)$.

PROOF. By 3.5 and 4.2 it suffices to prove the following. Suppose $Q \in \mathscr{P}$ and $\chi \in \chi(Q)$ is an H_Q invariant character. Then there is an $\alpha \in \mathrm{Br}(F(V))^G$ such that $c(r_Q(\alpha)) = c(\chi)$ and for any $P \in \mathscr{P}$ not equal to Q, $c(r_P(\alpha)) = 0$.

Suppose we find $\alpha' \in \mathrm{Br}(F(V))$ that is H_Q invariant and such that $c(r_Q(\alpha')) = c(\chi)$ while $c(r_P(\alpha')) = 0$ for all $P' \neq Q$. Then the G/H_Q norm $N_{G/H_Q}(\alpha')$ is the required α. In other words, we may assume $H_Q = G$. Let $r(\chi) \in \oplus_C(\mathbb{Q}/\mathbb{Z})$ be the image of χ which is therefore G fixed. By the proof of 5.4, $r(\chi) = N_G(\bar{\beta})$ for some $\bar{\beta} \in \oplus_C(\mathbb{Q}/\mathbb{Z})$. Let $\beta \in \oplus_P(\chi(P))$ be a preimage of $\bar{\beta}$ and let β_P be the P component of β for any P. If β_P is defined by $f^{1/n}$ for $f \in F(P)$, then for any $g \in G$ the $g(P)$ component of $N_G(\beta_P)$, call it $N_G(\beta_P)_{g(P)}$, is defined by $N_{H'}(g(f))^{1/n}$ where $g(f) \in F(g(P))$ corresponds to f and $H' = H_{g(P)} = gHg^{-1}$. Now $N_{H'}(g(f)) \in F(g(P))^{H'}$, so $c(N_G(\beta_P))_{g(P)} = 0$. It follows that, for any $P \in \mathscr{P}$, $c(N_G(\beta)_P) = 0$. Clearly $\chi - N_G(\beta)$ is the image of a unique $\alpha \in \mathrm{Br}(F(V))$. Since $\chi - N_G(\beta)$ is G fixed, α is G fixed and meets the requirements. $\square$

Let us introduce some notation from the Hochschild-Serre spectral sequence defined by (1) and other group extensions. If $1 \to N \to H \to M \to 1$ is an extension of groups and B is an H module then the Hochschild-Serre spectral sequence gives a filtration of $H^n(H, B)$ where we set $H^n(H, B)_{q-1}$ to be the kernel of $H^n(H, B)_q \to E_\infty^{p,q}$ and $H^n(H, B)_n = H^n(H, B)$. Call the f_q this map $H^n(H, B)_q \to E_\infty^{p,q}$.

We are ready for the proof of 5.3 which we perform in a series of steps punctuated by lemmas. Suppose for this whole argument that $\alpha \in H_u^3(F(V)^G)$.

LEMMA 5.6. $f_3(\alpha) = 0 \in H^3(G_K, \mu)$ and so α lies in $H^3(G_L, \mu)_2$, where the filtration and f_3 is defined by the extension (1).

PROOF. Of course, f_3 is just the restriction map $H^3(G_L, \mu) \to H^3(G_K, \mu)$. It was shown in [**CO**, p. 143] and is easy to check that $f_3(\alpha)$ is in $H_u^3(F(V))$. But this is 0 because $F(V)$ is rational over F.   □

By 5.6 $f_2(\alpha)$ makes sense. In fact,

LEMMA 5.7. $f_2(\alpha) = 0 \in H^1(G, \text{Br}(F(V)))$ and so α is in $H^3(G_L, \mu)_1$.

PROOF. Suppose that were not the case. Then by 5.5 $f_2(\alpha)$ has a nonzero image in $H^1(G, \oplus_P(\chi(P))) = \oplus_{P \in \mathscr{P}} H^1(H_P, \chi(P))$. In other words, there is a $P \in \mathscr{P}$ such that $r_P^*(f_2(\alpha)) \in H^1(H_P, \chi(P)) = H^1(H_P, H^1(G_P, \mu))$ is nonzero. Of course, here the map $r_P^* : H^1(G, \text{Br}(F(V))) \to H^1(H_P, \chi(P))$ is induced by the ramification map $r_P : \text{Br}(F(V)) \to \chi(P)$. If we identify $\text{Br}(F(V))$ with $H^2(G_K, \mu)$ and $\chi(P)$ with $H^1(G_P, \mu)$, then by 3.3 r_P is equal to $\rho_R : H^2(G_K, \mu) \to H^2(\mathbb{Z} \oplus G_P, \mu) \to H^1(G_P, \mu)$ defined in the beginning of this section, where $R = F[V]_P$. From now on write such a ρ_R as ρ_P.

Let Q be the restriction of P to $F(V)^G$ and G_Q the absolute Galois group of the residue field of Q. Since P is faithful we have an extension

(3) $1 \to G_Q \to G_P \to H_P \to 1$

and taking the direct sum with $\mathbb{Z}$ we get an extension

(4) $1 \to \hat{\mathbb{Z}} \oplus G_P \longrightarrow \hat{\mathbb{Z}} \oplus G_P \longrightarrow H_P \to 1$

which is the extension associated to the complete fields $\hat{L}_Q \subset \hat{K}_P$ where $L = F(V)^G$. The naturality of the spectral sequence shows that the following diagram commutes.

$$
\begin{array}{ccccc}
H^3(G_L, \mu)_2 & \xrightarrow{\;\text{res}\;} & H^3(\hat{\mathbb{Z}} \oplus G_Q, \mu)_2 & \xleftarrow{\;\text{inf}\;} & H^3(G_Q, \mu)_2 \\
\downarrow{\scriptstyle f_2} & & \downarrow{\scriptstyle f_2} & & \downarrow{\scriptstyle f_2} \\
H^1(G, H^2(G_K, \mu)) & \xrightarrow{\;\text{res}^*\;} & H^1(H_P, H^2(\hat{\mathbb{Z}} \oplus G_P, \mu)) & \xleftarrow{\;\text{inf}^*\;} & H^1(H_P, H^2(G_P, \mu)) \\
& & \downarrow{\scriptstyle \rho^*} & & \\
& & H^1(H_P, H^1(G_P, \mu)). & &
\end{array}
$$

Since α is unramified, $\rho_P(\alpha) = 0$ and so $\text{res}(\alpha) \in H^3(\hat{\mathbb{Z}} \oplus G_Q, \mu)$ is $\inf(\beta)$ for some $\beta \in \inf(H^3(G_Q, \mu))$. Since the inflation map $H^3(G_Q, \mu) \to H^3(\hat{\mathbb{Z}} \oplus G_Q, \mu)$ is injective, $f_3(\beta) = 0$ and so $\text{res}(\alpha) = \inf(\beta)$ for $\beta \in H^3(G_Q, \mu)_2$. By the above diagram, the image of α in $H^1(H_P, H^2(\hat{\mathbb{Z}} \oplus G_P, \mu))$ is in $\inf^*(H^1(H_P, H^2(G_P, \mu)))$. That is, by 2.12, $\rho^*(\text{res}^*(f_2(\alpha))) = 0 = r_P^*(f_2(\alpha))$. This contradiction proves the lemma $\square$

Of course from 5.7 we now know that $f_1(\alpha)$ is defined. Once again, we show it is zero.

LEMMA 5.8. $f_1(\alpha) = 0$.

PROOF. f_1 is defined as a map $H^3(G_L, \mu)_1 \to E_\infty^{2,1}$. From the spectral sequence we have that $E_\infty^{2,1} = H^2(G, H^1(G_K, \mu))/((d_2)_{0,2}(H^2(G_K, \mu)^G))$. By 5.5 this is equal to

$$\bigoplus_{P \in \mathscr{P}} (H^2(H_P, \mu)/\text{res}_P(d_2(H^2(G_K, \mu)^G))).$$

Let $P \in \mathscr{P}$ and let Q be the restriction of P to L. Once again by the naturality of the spectral sequence the following diagram commutes:

$$
\begin{array}{ccccc}
H^3(G_L,\mu)_1 & \xrightarrow{\ \text{res}\ } & H^3(\hat{\mathbb{Z}} \oplus G_Q,\mu)_1 & \xleftarrow{\ \text{inf}\ } & H^3(G_Q,\mu)_1 \\
\downarrow{\scriptstyle f_1} & & \downarrow{\scriptstyle f_1} & & \downarrow{\scriptstyle f_1} \\
\dfrac{H^2(G,H^1(G_K,\mu))}{\text{Im}(d_2)} & \xrightarrow{\text{res}^*} & \dfrac{H^2(H_P,H^1(\hat{\mathbb{Z}} \oplus G_P,\mu))}{\text{Im}(d_2)} & \xleftarrow{\text{inf}^*} & \dfrac{H^2(H_P,H^1(G_P,\mu))}{\text{Im}(d_2)} \\
& & \downarrow{\scriptstyle \rho^*} & & \\
& & \dfrac{H^2(H_P,\mu)}{\text{Im}(d_2)} & &
\end{array}
$$

Since α is unramified, $\text{res}(\alpha) \in H^3(\hat{\mathbb{Z}} \oplus G_Q, \mu)_1$ is the image of some $\beta \in H^3(G_Q, \mu)$. Arguing as in 5.7, $\beta \in H^3(G_Q, \mu)_2$. Since the inflation map $H^1(H_P, H^2(G_Q, \mu)) \to H^1(H_P, H^2(\hat{\mathbb{Z}} \oplus G_Q, \mu))$ is injective, once again $f_2(\beta) = 0$ and we conclude that $\text{res}(\alpha)$ is the image of some $\beta \in H^3(G_Q, \mu)_1$. From the above diagram

$$f_1(\text{res}(\alpha)) \in \frac{H^2(H_P, H^1(\hat{\mathbb{Z}} \oplus G_P, \mu))}{\text{Im}(d_2)}$$

is in the image under inflation of

$$\frac{H^2(H_P, H^1(G_P, \mu))}{\text{Im}(d_2)}$$

and so $\rho^*(f_1(\text{res}(\alpha))) = 0 \in H^2(H_P, \mu)/\text{Im}(d_2)$. But $\rho^*(f_1(\text{res}(\alpha))) = \rho^*(\text{res}^*(f_1(\alpha))) = r_P(f_1(\alpha))$. Thus $f_1(\alpha) \in H^2(G, H^1(G_K, \mu))/\text{Im}(d_2)$ is zero. $\square$

Finally, we have finished the proof of 5.3. Since $f_1(\alpha) = 0$, $\alpha \in H^3(G_L, \mu)_0$ which from the spectral sequence is the image of $H^3(G, \mu)$.

Before ending this section, we need to consider the more unsatisfactory case of multiplicative invariants. To begin, let us review a bit of multiplicative invariant theory.

Let H be a finite group and M a finitely generated H lattice. That is, M is a $\mathbb{Z}$ torsion free finitely generated $\mathbb{Z}[H]$ module. Let $\alpha \in \text{Ext}_H(M, \mu)$ correspond to the extension $0 \to \mu \to M_\alpha \to M \to 0$. Form the group algebra $F[M]$. Let $e : M \to F[M]^*$ be the natural map, where we note that as the operation of M is written additively, $e(x + y) = e(x)e(y)$. If $a_1, \dots, a_n$ are a $\mathbb{Z}$ basis for M, then as a ring $F[M] = F[x_1, \dots, x_n](1/(x_1 \cdots x_n))$ where $x_i = e(a_i)$. In [S5] it was observed that $F[M]$ has a natural action by H induced from the action on M_α. In fact, the units group $F[M]^*$ has as a subgroup $\mu \oplus M$ which is isomorphic as an abelian group to M_α. The H action on M_α extends to an action on $F[M]$ which we call the α twisted action. To emphasize that we are assuming this twisted action we will write $F[M]$ with this action as $F_\alpha[M]$. Finally, let $F_\alpha(M)$ be the field of fractions of $F_\alpha[M]$, so that the H action extends to $F_\alpha(M)$. If $\alpha = 0$ then we will drop it as a subscript.

With an eye to the specific application we have in mind, let us concentrate on special multiplicative invariant fields and prove a version of 5.3 in this case. First of all, suppose M is a H lattice and $M \subset Q$ has finite index where Q is a permutation H lattice. Form the dual $N' = \text{Hom}(Q/M, \mu)$ and note that there is a natural embedding $F(M) \subset F(Q)$. It was observed in [S, p. 233] that $F(Q)/F(M)$ is abelian Galois with Galois group N'. In fact, it was observed that $F(Q)/F(M)^H$ is Galois with group the semidirect product $G = N' \rtimes H$. In other language, the action of G on $F(Q)$ is twisted by an $\alpha \in \text{Ext}_G(\mu, Q)$ and so we will write $F(Q)$ as $F_\alpha(Q)$.

Let $x_1, \dots, x_r$ be a basis of Q that is permuted by H. Set $v_i = e(x_i) \in F(Q)$ and note that for all $g \in G$, $g(v_i) = \rho v_j$ for some j and some $\rho \in \mu$. Thus if we set $V \subset F_\alpha(Q)$ to be the F span of the v_i, then V is a G representation. It is immediate that $F(V) = F_\alpha(Q)$ and so $F(V)^G = F_\alpha(Q)^G = F(M)^H$. From 5.3 we now have:

LEMMA 5.9. $H_u^3(F(M)^H)$ is in the image of $H^3(G, \mu)$ where G is the semidirect product $N' \rtimes H$ and $N' = \text{Hom}(Q/M, \mu)$.

Let $M \subset Q$ be as above, and view both lattices as submodules of $M_\mathbb{Q} = \mathbb{Q} \otimes_\mathbb{Z} M$. Then there is an integer n such that $M \subset Q \subset (1/n)M \subset M_\mathbb{Q}$. The field extension $F((1/n)M)/F(M)^H$ has Galois group $G' = N \rtimes H$ where $N = \text{Hom}(M/nM, \mu)$. Since $F(Q) \subset F((1/n)M)$ the map $G_{F(M)^H} \to G$ factors through G' and we have:

LEMMA 5.10. $H_u^3(F(M)^H)$ is in the image of $H^3(G', \mu)$ where G' is the *semidirect product* $N \rtimes H$ *and* $N = \mathrm{Hom}(M/nM, \mu)$ *for* n *as above.*

Suppose M has the structure $M' \oplus Q$ where $Q \cong \mathbb{Z}[H/H']$ and M' is a faithful H module. Then we know from [S, p. 225] that $F(M)^H/F(M')^H$ is rational and so we would expect that the unramified H^3 would only depend on M' as the following result shows. To state the result, let n, M, and N be as above. Set $N' = \mathrm{Hom}(M'/nM', \mu)$, $G = N \rtimes H$, and $G' = N' \rtimes H$. If $S = \mathrm{Hom}(Q/nQ, \mu)$ note that $G = S \rtimes G'$.

PROPOSITION 5.11. *Suppose* $\alpha \in H^3(G, \mu)$ *has image* β *in* $H^3_u(F(M)^H)$. *Then* β *can be identified with an element* $\beta \in H^3_u(F(M')^H)$ *and as such is in the image of* $H^3(G', \mu)$.

The proof of this proposition will require two results, the first one of which is an observation that will give us some control over the transcendence base of $F(M)^H/F(M')^H$.

LEMMA 5.12. *Suppose* K/L *is* H *Galois and* V *is an* K *vector space with a semilinear action by* H. *Assume the* H *orbit of* $v \in V$ *is a basis of* V *over* K. *Let* n *be any positive integer. Then there is a basis* $\{w_1, \dots, w_r\}$ *of* V *over* K *with all* w_i G *fixed and such that if* $v = \sum k_i w_i$ *then the* k_i *are all* n^{th} *powers.*

PROOF. Denote by H' the stabilizer of v. Let $\{w'_1, \dots, w'_r\}$ be some H fixed basis. We claim that there is a $v' \in V$ which is H' fixed such that the H orbit of v' is a basis for V and if $v' = \sum k'_i w'_i$, then all the k'_i are n^{th} powers. Suppose the claim is proved. Let us show how the claim implies the lemma. Define the K linear (*not* semilinear) isomorphism $\phi : V \to V$ by setting $\phi(g(v)) = g(v')$. Note that $\phi \circ g = g \circ \phi$ for all $g \in H$. Then if we set $w_i = \phi^{-1}(w'_i)$ the w_i are a G fixed basis and $v = \sum k_i w_i$.

Thus we only must prove the claim. Let $g = g_1, \dots, g_r$ be a set of left coset representatives of H' in H. For any H' fixed k_i, $v' = \sum k_i^n w'_i$ has the property we need if and only the determinant $\det(g_i(k_j^n)) \neq 0$. We want to express this determinant in terms of polynomials over L. Suppose $\theta_1, \dots, \theta_t$ is an L basis of K where $\theta_1, \dots, \theta_s$ is an L basis for $K^{H'}$. Form the polynomial ring $K[x_{i,j} | 1 \leq i \leq r; 1 \leq j \leq s]$ where H acts trivially on the $x_{i,j}$'s. Set $y_i = (\sum x_{i,j} \theta_j)^n \in K^{H'}[x_{i,j}]$. The determinant $\det(g_i(y_j))$ can be written in the form $\sum f_k(x_{i,j})\theta_k$ for $f_k(x_{i,j}) \in L[x_{i,j}]$. Clearly it suffices to show some f_k is a nonzero polynomial. But if we extend scalars to the algebraic closure L_a of L, K becomes a direct sum of L_a's. Then all elements are nth powers and clearly there is a substitution making this determinant nonzero. This shows some f_k is nonzero and we are done. $\square$

We will use 5.12 to construct a helpful extension of $F(M)^G$, but first we study a bit the difference between G and G' (in the notation of 5.11). Recall that $G = S \rtimes G'$. Set $K' = F(M')$, $K = F(M)$, $L = K^G$, and $L' = K'^G$. Then $K = K'(x_1, \ldots, x_r)$ where $x = x_1$ is H' fixed and the x_i's are the H orbit of x. Furthermore, $F((1/n)M) = F((1/n)M')(x_1^{1/n}, \ldots, x_r^{1/n})$. Finally, therefore, S is the Galois group of $F((1/n)M')(x_1^{1/n}, \ldots, x_r^{1/n})$ over $F((1/n)(M'))(x_1, \ldots, x_r)$.

PROPOSITION 5.13. *In the setup of* 5.11 *there is a field* $L'' \supset L$ *such that the induced map* $G_{L''} \to G_L \to G_{L'}$ *is an isomorphism. Furthermore, the image of* $G_{L''} \to G_L \to G$ *is a subgroup complementary to* S *and isomorphic to* G'. *Finally under this isomorphism the given map* $G_{L'} \to G'$ *becomes the induced map* $G_{L''} \to G'$.

PROOF. One finds a transcendence base of L/L' by finding an H invariant basis of $V = \sum K'x_i$. By 5.12 we can choose this basis $\{w_1, \ldots, w_r\}$ such that if $x_1 = \sum k_i w_i$, then all the k_i are n^{th} powers. Form the iterated Laurent series field $L_1 = L((w_1))((w_2)) \cdots ((w_r))$ and let L'' be generated over L_1 by all $w_i^{1/m}$ for all m and i. Since $L = L'(w_1, \ldots, w_r)$, $L \subset L_1 \subset L''$ in a natural way. There is a canonical map (unique up to conjugation) $G_{L''} \to G_L \to G_{L'}$ which is an isomorphism. The composition $G_{L''} \to G_L \to G$ defines the Galois extension KL''. By Hensel's lemma all the x_i have n^{th} roots in $K'L''$. Thus $K'L'' = KL''$, which implies that the image of $G_{L''}$ is isomorphic to G' and complementary to S. Finally, since $G_{L'} \to G'$ and $G_{L''} \to G'$ are defined by restriction to K' and $K'L''$ respectively, we are done. □

We are finally ready to prove 5.11. Let $L'' \supset L$ be as in 5.13 and use the isomorphism to identify $G_{L'}$ and $G_{L''}$. In this way we have identified G' with a subgroup of G complementary to S. Since $\beta \in H^3(G_L, \mu)$ is unramified, it is the image of some $\beta' \in H^3(G_{L'}, \mu)$ which is also unramified. Since $G_{L''} \to G_L \to G_{L'}$ is the identity, the restriction of β to $H^3(G_{L''}, \mu)$ is β'. But then if α' is the restriction of $\alpha \in H^3(G, \mu)$ to $H^3(G', \mu)$, the inflation of α' is β' and this is precisely what is desired.

It turns out to be convenient to write G above as the image of an infinite profinite group. The idea is the following. If we consider $M_{\mathbb{Q}} = \mathbb{Q} \otimes_{\mathbb{Z}} M$ to be the union of the lattices $(1/m)M \subset M_{\mathbb{Q}}$, then we can define $F(M_{\mathbb{Q}})$ to be the union of the fields $F((1/m)M)$. What we are about to do is write down the Galois group of $F(M_{\mathbb{Q}})/F(M)^H$. Set $A = \text{Hom}(M_{\mathbb{Q}}/M, \mu)$ and note that A is a direct product of $\hat{\mathbb{Z}}$'s and is the Galois group of $F(M_{\mathbb{Q}})/F(M)$. Thus A is naturally a profinite group. Clearly, there is a natural H action on A and we can define $\tilde{G} = A \rtimes H$. We make the abelian group $\mu \oplus M_{\mathbb{Q}}$ into a $\tilde{G}$ module M^* as follows. For $a \in A$, set $a(\eta, x) = ((\eta)(a(x + M)), x)$ and $h(\eta, x) = (\eta, h(x))$ for all $\eta \in \mu$, $x \in M_{\mathbb{Q}}$, $a \in A$, and $h \in H$. Let

$\gamma \in \text{Ext}(M_{\mathbb{Q}}, \mu)$ correspond to the extension

$$(2) \qquad\qquad 0 \to \mu \to M^* \to M_{\mathbb{Q}} \to 0 .$$

Trivially generalizing the twisted multiplicative actions described above, we can define $F_\gamma(M_{\mathbb{Q}})$ which we identify with $F(M_{\mathbb{Q}})$. That is, we specify this action of $\tilde{G}$ on $F(M_{\mathbb{Q}})$. Since we have mirrored Kummer theory here, it is easy to see that $F_\gamma(M_{\mathbb{Q}})^A = F(M)$ and $F_\gamma(M_{\mathbb{Q}})^{\tilde{G}} = F(M)^H$. Of course, we have a surjection $\tilde{G} \to G$ which restricts to an H module surjection $A \to N$. In fact, it is clear that $N = A/nA$.

The purpose behind enlarging G to $\tilde{G}$ is that the cohomology simplifies. To begin with, we consider the cohomology of A.

LEMMA 5.14. *Suppose* $K = F(M)$. *Then there is a natural map* $G_K \to A$ *and the inflation map* $\iota : H^n(A, \mu) \to H^n(G_K, \mu)$ *is injective.*

PROOF. Since A is the Galois group of $F_\gamma(M_{\mathbb{Q}})/F(M)$, the map $G_K \to A$ is given by Galois theory. Let $\{m_1, \dots, m_r\}$ be a $\mathbb{Z}$ basis for M and set $x_i = e(m_i) \in F(M)$. Let $K' = F((x_1)) \cdots ((x_r))$ be the iterated Laurent series viewed as an extension of K. The algebraic closure, K_a', of K' is generated by $(x_i)^{1/q}$ for all i and q. Thus A is naturally isomorphic to the absolute Galois group of K' and viewing $G_{K'} \subset G_K$ we have that the map $G_K \to A$ splits. $\square$

Since $M_{\mathbb{Q}}$ has trivial A action and is uniquely divisible, $H^n(A, M_{\mathbb{Q}}) = (0)$ for all $n \geq 1$. Thus the exact sequence (2) implies that $H^n(A, \mu) \cong H^n(A, M^*)$ for all $n \geq 2$. The next result computes this cohomology group. To state the result, let $\mu_q \subset \mu$ be the subgroup of order q. In cohomology with μ_q coefficients there is a cup product $H^p(A, \mu_q) \times H^q(A, \mu_q) \to H^{p+q}(A, \mu_q)$. Combining this with the induced map $H^n(A, \mu_q) \to H^n(A, \mu)$ yields a series of multilinear maps

$$\phi_{q,n} : H^1(A, \mu_q) \otimes \cdots \otimes H^1(A, \mu_q) \to H^n(A, \mu).$$

Now $\text{Hom}(A, \mu_q)$ is naturally isomorphic to $((1/q)M)/M$. Thus we can rewrite the above as

$$\phi_{q,n} : ((1/q)M)/M \otimes \cdots \otimes ((1/q)M)/M \to H^n(A, \mu).$$

By 5.14 and, e.g., [**P**, Lemma 1] $\phi_{q,n}$ is an alternating map and so induces

$$\phi_{q,n} : \bigwedge^n ((1/q)M)/M \to H^n(A, \mu) .$$

Taking the union over all q we have defined

$$\phi_n : (\bigwedge^n M) \otimes_{\mathbb{Z}} \mathbb{Q}/\mathbb{Z} \to H^n(A, \mu) .$$

Our main result about the cohomology of A is:

THEOREM 5.15. ϕ_n is an isomorphism.

PROOF. To show injectivity it suffices to show that the composition $\iota \circ \phi_{q,n}$ is injective where ι is as in 5.14. By [CO, p. 144] any nonzero element of $\bigwedge^n((1/q)M)/M$ has nonzero ramification in $H^n(G_K, \mu)$ and so this is clear.

To show surjectivity it suffices to show that the set of elements of order dividing q in $(\bigwedge^n M) \otimes_{\mathbb{Z}} \mathbb{Q}/\mathbb{Z}$ and $H^n(A, \mu)$ have the same cardinality. We do this by induction. Write $A = A' \oplus \hat{\mathbb{Z}}$ and in a corresponding manner $M = M' \oplus \mathbb{Z}$. Then $H^n(A, \mu) \cong H^n(A', \mu) \oplus H^{n-1}(A', \mu)$ and $\bigwedge^n((1/q)M)/M \cong \bigwedge^n((1/q)M'/M') \oplus \bigwedge^{n-1}((1/q)M'/M')$ so the result follows by induction. $\square$

What we are aiming for is a result that would say that any element of $H_u^3(F(M)^H)$ is the image of $H^3(H, \mu \oplus M)$. Unfortunately we cannot prove this result, because there is an obstruction that as far as we can tell can actually be nonzero. What we can show is that this is the only obstruction. Recall that we have set $L = F(M)^H$ and $K = F(M)$. Below we will use the Hochschild-Serre spectral sequence with respect to the extension $1 \to A \to \tilde{G} \to H \to 1$.

PROPOSITION 5.16. Suppose $\alpha \in H^3(G, \mu)$ has image $\beta \in H^3(G_L, \mu)$ that is unramified. That is, $\beta \in H_u^3(L)$. If $\alpha' \in H^3(\tilde{G}, M^*)$ is the image of α, then $\alpha' \in H^3(\tilde{G}, M^*)_2$. If α' maps to 0 in $H^1(H, H^2(A, M^*))$, then β is in the image of $H^3(H, \mu \oplus M)$.

PROOF. Since β is unramified, its restriction to $H^3(G_K, \mu)$ must be unramified and hence 0 as K/F is rational. By 5.14, the restriction of α' to $H^3(A, \mu) = H^3(A, M^*)$ is 0, and so $\alpha' \in H^3(\tilde{G}, M^*)_2$. The result follows because $H^1(A, M^*) = 0$ and $(M^*)^A = \mu \oplus M$. $\square$

REMARK. When we apply 5.16 in the next section we will need the following observation. Suppose $\alpha \in H^3(G, \mu)$ above has order q. Then α' has order dividing q, and so does its image in $\alpha'' \in H^1(H, H^2(A, M^*))$. There is an exact sequence

$$0 \longrightarrow ((1/q)(M \bigwedge^2 M))/(M \bigwedge^2 M) \longrightarrow (\bigwedge^2 M) \otimes \mathbb{Q}/\mathbb{Z} \xrightarrow{q} (\bigwedge^2 M) \otimes \mathbb{Q}/\mathbb{Z} \longrightarrow 0$$

that shows that α'' is the image of $\alpha_1 \in H^1(H, ((1/q)(M \bigwedge M))/(M \bigwedge M))$. Of course, to satisfy the condition of 5.16 it suffices to show that $\alpha_1 = 0$.

As our final observation about $H_u^3(F(M)^H)$ we note that in some circumstances all that matters is $H^3(H, M)$.

COROLLARY 5.17. Assume that $\alpha \in H_u^3(F(M)^H)$ is in the image of $H^3(H, \mu \oplus M)$. Suppose V is a faithful H representation and $F(V)^H$ is rational over F. Then α is in the image of $H^3(H, M)$.

PROOF. Write $\alpha = \alpha_1 + \alpha_2$ where α_1 is the image of $\beta_1 \in H^3(H, \mu)$ and α_2 is the image of $\beta_2 \in H^3(H, M)$. We have fields $F(V)^H \subset F(M)(V)^H \supset$

$F(M)^H$. Now $F(M)(V)^H/F(M)^H$ is rational so the map on H^3 is injective. Similarly, $F(M)(V)^H/F(V)^H$ is unirational and so the map on H^3 is again injective. Since $F(V)^H$ is rational, the image in $H^3(F(V)^H)$ of β_1 must be either ramified or 0. In the latter case β_1 has 0 image in $H^3(F(M)^H)$ and we are done.

Suppose then that the image of β_1 is ramified. The dual lattice $\text{Hom}(M,\mathbb{Z})$ is the image of a free lattice. Taking duals, there is a free H lattice $Q \supset M$. We have an inclusion of fields $F(M)(V)^H = F(V)(M)^H \subset F(V)(Q)^H$. Now $F(V)(Q)^H/F(V)^H$ is rational so β_1 is nonzero and ramified in $H^3(F(V)(Q)^H)$. On the other hand, the image of β_2 in $H^3(F(V)(Q)^H)$ factors through $H^3(H, Q) = 0$. Thus the image of α in $H^3(F(V)(Q)^H)$ is nonzero and ramified. This contradiction proves the corollary.   □

6. PGL_n

Up until now, we have only been considering the invariant fields of finite groups acting on a vector space V. It is obvious that one would also like to consider the case G is an algebraic group acting on a V. It turns out, at least in the case G is reductive and V is "good", that there are general results that reduce the algebraic group case to the finite one. Assume then that V is good, meaning that there is an element of V with trivial stabilizer. Bogolomolov has shown that the invariant field, $F(V)^G$, is unique up to stable isomorphism. That is, if V' is another "good" G representation, $F(V)^G$ and $F(V')^G$ are stably isomorphic. In [S3], this author showed that $F(V)^G$ is stably isomorphic to a twisted multiplicative invariant field of the Weyl group W. Thus one can hope, at least in some cases, to compute the unramified H^3 using the results at the end of the last section.

Instead of working this all out in general, we consider the special case $G = PGL_n$, the projective linear group. This special case is particularly important because many fields are stably isomorphic to $F(V)^G$ for good V, including the center of the generic division algebra. In fact, writing $F(V)^G$ as the multiplicative invariants of the Weyl group S_n is much older than the general result, is due to Procesi, and appears in [Fo]. We review it now.

View S_n as the group of permutations of $\{i|1 \leq i \leq n\}$. Let X be the S_n lattice with $\mathbb{Z}$ basis $\{x_i|1 \leq i \leq n\}$ such that for $g \in S_n$, $g(x_i) = x_{g(i)}$. Clearly $X \cong \mathbb{Z}[S_n/S_{n-1}]$. Let Y' be the S_n lattice with $\mathbb{Z}$ basis $\{y_{i,j}|1 \leq i, j \leq n; i \neq j\}$ such that $g(y_{i,j}) = y_{g(i),g(j)}$ for $g \in S_n$. Clearly $Y' \cong \mathbb{Z}[S_n/S_{n-2}]$. For convenience define $y_{i,i} = 0$. There is an S_n module morphism $\phi : Y' \to X$ defined by setting $\phi(y_{i,j}) = x_i - x_j$. The image of ϕ we call I, and note that I is a sort of relative augmentation "ideal" (except X is not a ring so I is not an ideal). In fact, we have

(1) $$0 \to I \to X \to \mathbb{Z} \to 0$$

is exact where $x_i \in X$ is mapped to $1 \in \mathbb{Z}$. Looking at the other side, we define Y to be the kernel of ϕ. By e.g. [Fo] $F(V)^G$ is stably isomorphic to the multiplicative invariant field $F(X \oplus Y)^{S_n}$. By e.g. [S, p. 224–225] this field is stably isomorphic to $F(Y)^{S_n}$ because X is a permutation lattice.

In order to apply the results of section 5 we study these lattices a bit more. The extension

$$(2) \qquad\qquad 0 \longrightarrow Y \longrightarrow Y' \longrightarrow I \longrightarrow 0$$

does not split but does "split" up to a multiple of n. To be more precise, define $c(i, j, k) \in Y$ to be the element $y_{i,j} + y_{j,k} - y_{i,k}$ which one quickly sees is in the kernel of ϕ. It was observed in [Fo] that the $c(i, j, k)$'s generate Y. In [S2, p. 60] it was observed that the $c(i, j, k)$'s are a "generic" Brauer factor set. Since the cohomology element they represent can be shown to have order n (it corresponds to the generic division algebra), we have that there are $e_{i,j} \in Y$ with $nc(i, j, k) = e_{i,j} + e_{j,k} - e_{i,k}$ for all i, j, k. Define $\iota' : Y' \to Y'$ by setting $\iota'(y_{i,j}) = ny_{i,j} - e_{i,j}$. Since $\iota'(c(i, j, k)) = 0$, ι' induces a map $\iota : I \to Y'$. We note that $\phi(\iota(x_i - x_j)) = \phi(ny_{i,j} - e_{i,j}) = n(x_i - x_j)$. Thus ι is injective and we can identify I with $\iota(I) \subset Y'$. We have $Y \oplus I \subset Y'$ and quickly note that $Y'/(Y \oplus I) \cong I/nI$ has exponent n.

The extension (1) induces an extension $0 \to Y \oplus I \to Y \oplus X \to \mathbb{Z} \to 0$ which induces an extension

$$0 \to Y' \to Y'' \to \mathbb{Z} \to 0$$

and an embedding $Y \oplus X \subset Y''$. But $\text{Ext}(\mathbb{Z}, Y') = H^1(S_n, Y') = 0$ because Y' is a permutation S_n module. Thus $Y'' \cong Y' \oplus \mathbb{Z}$ is a permutation lattice. Since $Y'/(Y \oplus I) \cong Y''/(Y \oplus X)$, $Y''/(Y \oplus X)$ has exponent n. By 5.10 and 5.11 we have:

LEMMA 6.1. *Let $G = PGL_n$ and let V be a "good" G representation. Then $H_u^3(F(V)^G)$ is in the image of $H^3(H, \mu)$ where $H = \bar{Y} \rtimes S_n$ and $\bar{Y} = \text{Hom}(Y/nY, \mu)$.*

We are working towards a result on the unramified cohomology of $F(V)^G$ in the case n is odd. It will help to know that all such elements have order dividing n as follows.

PROPOSITION 6.2. *Suppose n is odd and $\alpha \in H_u^3(F(V)^G)$. Then $n\alpha = 0$.*

PROOF. It suffices to show this for $\alpha \in H_u^3(F(Y)^{S_n})$. If $S_{n-1} \subset S_n$ is the subgroup fixing 1, it then suffices to show that $F(Y)^{S_{n-1}}$ is stably rational. But as an S_{n-1} module I is permutation so $F(Y)^{S_{n-1}}$ is stably rational by (2) and [S, p. 124–125]. $\square$

We move from H to the cohomology of a certain infinite profinite group as in 5.16. To be precise, set $Y_{\mathbb{Q}} = \mathbb{Q} \otimes_{\mathbb{Z}} Y$, $A = \text{Hom}(Y_{\mathbb{Q}}/Y, \mu)$, and $H' =$

$A \rtimes S_n$. Recall that there is a H' module extension $0 \to \mu \to Y^* \to Y_{\mathbb{Q}} \to 0$ corresponding to $\gamma \in \text{Ext}(Y_{\mathbb{Q}}, \mu)$ and an action of H' on $F_\gamma(Y_{\mathbb{Q}})$ such that $F_\gamma(Y_{\mathbb{Q}})^{H'} = F(Y)^H$ and $H_u^3(F(Y)^{S_n})$ is in the image of $H^3(Y/nY \rtimes S_n, \mu)$ and hence of $H^3(H', \mu) = H^3(H', Y^*)$. If $\alpha \in H_u^3(F(Y)^{S_n})$ then α has odd order and hence is the image of an $\alpha' \in H^3(H', \mu)$ of odd order, say, q. By 5.16 $\alpha' \in H^3(H', \mu)_2$. By the remark after 5.16 the image of α' in $H^1(S_n, H^2(A, \mu))$ is in the image of $H^1(S_n, Y(q) \wedge Y(q))$ where $Y(q) = Y/qY$. We will show:

PROPOSITION 6.3. *Suppose* q *is odd. Then* $H^2(S_n, Y(q) \wedge Y(q)) = 0$. $H^3(F(V)^G)$ *is in the image of* $H^3(S_n, \mu \oplus Y)$.

Of course the second sentence of 6.3 follows from 5.16. The proof of the first sentence is a long cohomology computation we are about to engage in. Let us outline the method. For any H lattice M, write $M_2 = M \otimes_{\mathbb{Z}} \mathbb{Z}[1/2]$. Then $Y(q) \wedge Y(q) = U/qU$ where $U = Y_2 \wedge Y_2$. The exact sequence

$$(3) \qquad\qquad 0 \longrightarrow U \overset{q}{\longrightarrow} U \longrightarrow Y(q) \wedge Y(q) \longrightarrow 0$$

shows that it is sufficient to show that $H^1(S_n, U) = 0 = H^2(S_n, U)$ To show this, consider $U' = Y_2 \otimes_{\mathbb{Z}} Y_2$. There is an automorphism τ of U' that reverses the tensor product, so τ has order 2. U is the antisymmetric part of U'. Since multiplication by 2 is invertible, none of the cohomology groups with coefficients in Y_2, U, or U' has any 2 torsion. It follows that if we consider the induced action of τ on $H^n(S_n, U')$, the antisymmetric part is just $H^n(S_n, U)$. Our calculation will involve considering the action of τ on various lattices until we get our result.

To begin with, let us note that $Y \cong I \otimes_{\mathbb{Z}} I = I^2$ as S_n lattices. In fact if we tensor (1) by I we have $0 \to I^2 \to X \otimes I \to I \to 0$. Now $Y' \cong X \otimes I$ via the isomorphism ϕ induced by setting $\phi(x_i \otimes (x_i - x_j)) = y_{i,j}$. If we check that $X \otimes I \to Y' \to I$ is just the above map, we get $Y \cong I^2$. Of course, this implies that $Y_2 \cong I_2^2$. Thus Y_2 has an automorphism we call σ obtained by reversing the tensor product. We can compute that $\phi((x_i - x_j) \otimes (x_k - x_l)) = y_{i,l} - y_{i,k} + y_{j,k} - y_{j,l}$. It follows that σ extends to an automorphism of Y_2' where $\sigma(y_{i,j}) = y_{j,i}$. Thus σ induces the (-1) map on I_2. From (2) we have the exact sequence

$$(4) \qquad\qquad 0 \to Y_2 \to Y_2' \to I_2 \to 0.$$

Since the exact sequence (4) preserves the action by σ, the boundary map $H^1(S_n, I_2) \to H^2(S_n, Y_2)$ is a σ map. Since $H^2(S_n, Y_2') = H^2(S_{n-2}, \mathbb{Z}_2) = H^1(S_{n-2}, \mathbb{Q}/\mathbb{Z}_2) = 0$ by [**K**, p. 179], σ induces the (-1) map on $H^2(S_n, Y_2)$.

More generally, we consider the induced action of σ on $\text{Ext}(Y_2, M)$ for any M. Here σ acts trivially on M. Just as above, the boundary map

$\delta : \mathrm{Ext}(Y_2, M) \to \mathrm{Ext}^2(I_2, M)$ is σ preserving and of course σ acts as -1 on $\mathrm{Ext}^2(I_2, M)$. To understand the action of σ on $\mathrm{Ext}(Y_2, M)$ we would need information on $\mathrm{Ext}(Y_2', M)$. This is available in the special case that concerns us, as we show next.

The M we are particularly interested in is $M = I_2^3 = I_2 \otimes I_2 \otimes I_2$. As mentioned above we must compute $\mathrm{Ext}(Y_2', I_2^3)$. We have $\mathrm{Ext}(Y_2', I_2^3) = \mathrm{Ext}(\mathbb{Z}_2[S_n/S_{n-2}], I_2^3) = H^1(S_{n-2}, I_2^3)$.

LEMMA 6.4. $\mathrm{Ext}(Y_2', I_2^3) = H^1(S_{n-2}, I_2^3) = 0$. σ induces the -1 map on $\mathrm{Ext}(Y_2, I_2^3)$.

PROOF. The second sentence follows from the first. As for the first, we take the sequence (4) and tensor by I_2 to get

$$0 \to I_2^3 \to Y_2' \otimes I_2 \to I_2^2 \to 0.$$

We can analyze $Y_2' \otimes I_2$ as follows. Set $y_{i,j,k} = y_{i,j} \otimes (x_i - x_k) \in Y_2' \otimes I_2$ where $i \neq j$ and $i \neq k$. It is easy to verify that the $y_{i,j,k}$'s span $Y_2' \otimes I_2$. Checking ranks shows that the $y_{i,j,k}$'s form a basis for $Y_2' \otimes I_2$ over $\mathbb{Z}_2$ and this basis is permuted by S_n. Hence $H^1(S_{n-2}, Y_2' \otimes I_2) = 0$ and $H^1(S_{n-2}, I_2^3)$ is the cokernel of $(Y_2' \otimes I_2)^{S_{n-2}} \to (I_2^2)^{S_{n-2}}$. To compute these fixed groups note that $(I_2)^{S_{n-2}}$ has rank 2 spanned by $(x_1 - x_2)$ and $(\sum_{i>2} x_i) - (n-2)x_1$. $(Y_2')^{S_{n-2}}$ has rank 6 spanned by $y_{1,2}$, $\sum_{j>2} y_{1,j}$, $\sum_{j>2} y_{j,1}$, $\sum_{j>2} y_{2,j}$, $\sum_{j>2} y_{j,2}$, and $\sum_{i,j>2;i\neq j} y_{i,j}$. It follows that $(Y_2)^{S_{n-2}} = (I_2^2)^{S_{n-2}}$ has rank 4 spanned by $l_1 = \sum_{j>2}(y_{1,j} + y_{j,1})$, $l_2 = \sum_{j>2}(y_{2,j} + y_{j,2})$, $k = \sum_{j>2}(y_{1,2} + y_{2,j} - y_{1,j})$, and $d = \sum_{i,j>2;i\neq j} y_{i,j}$.

Next, we note that $y_{i,j,k} \in Y_2' \otimes I_2$ maps to $(x_i - x_j) \otimes (x_i - x_k) \in I_2^2$ which is $y_{i,k} + y_{j,i} - y_{j,k} = c(j, i, k)$ if $j \neq k$ and which is $y_{i,j} + y_{j,i}$ if $j = k$. Thus l_1 is the image of $\sum_{j>2} y_{1,j,j} \in (Y_2' \otimes I_2)^{S_{n-2}}$; l_2 is the image of $\sum_{j>2} y_{2,j,j} \in (Y_2' \otimes I_2)^{S_{n-2}}$; and k is the image of $\sum_{j>2} y_{2,1,j} \in (Y_2' \otimes I_2)^{S_{n-2}}$. Finally

$$d = \sum_{i,j>2;i\neq j} (y_{1,i} + y_{i,j} - y_{1,j}) - \sum_{i,j>2;i\neq j} y_{1,i} + \sum_{i,j>2;i\neq j} y_{1,j}$$

which equals

(5)
$$\sum_{i,j>2;i\neq j} c(1, i, j)$$

because

$$\sum_{i,j>2;i\neq j} y_{1,i} = \sum_{i,j>2;i\neq j} y_{1,j}.$$

However, (5) is the image of $\sum_{i,j>2;i\neq j} y_{1,i,j} \in (Y_2' \otimes I_2)^{S_{n-2}}$. $\square$

Define τ to be the automorphism of Y_2^2 defined by reversing the tensor product. The above lemma will help in computing the action of τ on the the cohomology of Y_2^2. Now Y_2^2 is a submodule of $Y_2'^2$ and we denote the quotient by K. We must analyze K a bit. K contains $(Y_2' \otimes Y_2)/(Y_2^2) \cong I_2^3$ and $(Y_2 \otimes Y_2')/Y_2^2 \cong I_2^3$. Thus K contains a submodule A isomorphic to $I_2^3 \oplus I_2^3$. Furthermore, $K/A \cong (Y_2'/Y_2)^2 \cong I_2^2 = Y_2$. Finally, the automorphism τ extends to an automorphism we also call τ that reverse the tensor product of $Y_2'^2$. Thus τ induces an automorphism of K (we also call τ) that preserves A and induces σ on the quotient $I_2^2 = Y_2$. Note finally that on A τ reverses the direct sum $I_2^3 \oplus I_2^3$.

The cohomology of $Y_2'^2$ is fairly easy to determine, so to compute the cohomology of Y_2^2 we must tackle the cohomology of K. To this end we begin by describing $(Y_2'^2)^{S_n}$. $Y_2'^2$ has as a $\mathbb{Z}_2$ basis the elements $y_{i,j} \otimes y_{k,l} = $ (by definition) $y_{i,j,k,l}$ where $i \neq j$ and $k \neq l$. Furthermore, if $g \in S_n$, $g(y_{i,j,k,l}) = y_{g(i),g(j),g(k),g(l)}$. On this basis S_n has 7 orbits which are the orbits of $y_{1,2,3,4}$, $y_{1,2,2,3}$, $y_{1,2,3,1}$, $y_{1,2,1,3}$, $y_{1,2,3,2}$, $y_{1,2,2,1}$, and $y_{1,2,1,2}$. It follows that $(Y_2'^2)^{S_n}$ has rank 7 with basis the sum of each of these orbits. Let us call these sums $\Sigma_1, \ldots, \Sigma_7$ respectively.

Next we consider $Y_2' \otimes I_2$ which has the basis $y_{i,j,k}$ described above. S_n has two orbits on this basis containing $y_{1,2,3}$ and $y_{1,2,2}$ respectively. Let Γ_1, Γ_2 be the sum over each of these orbits respectively, so the Γ's form a basis of $(Y_2' \otimes I_2)^{S_n}$. Finally, $(Y_2')^{S_n}$ has rank one spanned by $\Xi = \sum_{i \neq j} y_{i,j}$. Since Ξ maps to 0 in I_2 it is also the generator for $Y_2^{S_n}$. If we now consider the map $(Y_2' \otimes I_2)^{S_n} \to Y_2^{S_n}$ we compute that Γ_1 maps to $(n-2)\Xi$ and Γ_2 maps to 2Ξ. Recall that n is odd. Thus $(Y_2' \otimes I_2)^{S_n} \to Y_2^{S_n}$ is onto with kernel generated by $2\Gamma_1 - (n-2)\Gamma_2 = $ (by definition) Θ. It follows that Θ generates $(I_2^3)^{S_n}$.

Our ultimate goal is some information about $(Y_2'^2)^{S_n} \to K^{S_n}$. As a first step we consider the composition $(Y_2'^2)^{S_n} \to K^{S_n} \to Y_2^{S_n}$. We readily see that Σ_1 maps to 0, Σ_2 maps to $-(n-2)\Xi$, Σ_3 maps to $-(n-2)\Xi$, Σ_4 maps to $(n-2)\Xi$, Σ_5 maps to $(n-2)\Xi$, Σ_6 maps to $(-2)\Xi$, and Σ_7 maps to 2Ξ. In particular, $(Y_2'^2)^{S_n} \to Y_2^{S_n}$ is surjective. The natural map $Y_2'^2 \to (Y_2' \otimes I_2)$ induces a map $(Y_2'^2)^{S_n} \to (Y_2' \otimes I_2)^{S_n}$ whose composition with $(Y_2' \otimes I_2)^{S_n} \to (I_2^2)^{S_n}$ is the map we just studied. The elements Σ_1, $\Sigma_2 - \Sigma_3$, $\Sigma_4 - \Sigma_5$, $\Sigma_6 + \Sigma_7$, and $(n-2)\Sigma_7 + 2\Sigma_2$ all map to the kernel of the second map and so map to elements of $(I_2^3)^{S_n} \subset (Y_2' \otimes I_2)^{S_n}$. In fact, direct computation shows that Σ_1 and $\Sigma_6 + \Sigma_7$ map to 0 in $(I_2^3)^{S_n}$, and $\Sigma_2 - \Sigma_3$, $\Sigma_4 - \Sigma_5$, $\Sigma_2 + \Sigma_4$, and $(n-2)\Sigma_7 + 2\Sigma_2$ map to Θ.

These very same elements must then map to $A^{S_n} \subset K^{S_n}$ and what we have computed above is their image in the second component of $A^{S_n} = (I_2^3)^{S_n} \oplus (I_2^3)^{S_n}$. However, $\tau(\Sigma_2) = \Sigma_3$ and τ fixes Σ_4, Σ_5, Σ_6, and Σ_7. It follows that $\Sigma_2 - \Sigma_3$ maps to $(-\Theta, \Theta) \in A^{S_n}$, $\Sigma_4 - \Sigma_5$ maps to (Θ, Θ), and thus $\Sigma_2 + \Sigma_4$ maps to $(0, \Theta)$.

LEMMA 6.5. $H^1(S_n, Y_2^2) = 0$.

PROOF. It suffices to show $(Y_2'^2)^{S_n} \to K^{S_n}$ is surjective. But we have shown above that $(Y_2'^2)^{S_n} \to Y_2^{S_n}$ is surjective and every element of A^{S_n} is in the image of $(Y_2'^2)^{S_n}$. From this the surjectivity easily follows. □

Next we wish to analyze the action of τ on $H^2(S_n, Y_2^2)$. We have an exact sequence

$$H^1(S_n, K) \to H^2(S_n, Y_2^2) \to H^2(S_n, Y_2'^2)$$

where all maps preserve the τ action. To handle the easy case first, note that $H^2(S_n, Y_2'^2) = 0$ because it is a direct sum of cohomology groups $H^2(S_k, \mathbb{Z}_2) = \operatorname{Hom}(S_k, \mathbb{Q}/\mathbb{Z}_2) = 0$ by, e.g., [K, p. 179]. Thus we are left with handling the action of τ on $H^1(S_n, K)$.

Let us describe K in a bit more detail. The exact sequence $0 \to A \to K \to Y \to 0$ defines an element $(\alpha, \beta) \in \operatorname{Ext}(Y, I_2^3 \oplus I_2^3) = \operatorname{Ext}(Y, A)$. Since this exact sequence is τ invariant, the element (α, β) is τ invariant. But τ acts as -1 on $\operatorname{Ext}(Y, I_2^3)$ (with no action on I_2^3) and switches the direct sum in A. It follows that $\alpha = -\beta$. Furthermore $K/(I_2^3 \oplus (0)) \cong (Y_2' \otimes I_2)$ so α is $\pm\gamma$ where $\gamma \in \operatorname{Ext}(Y_2, I_2^3)$ corresponds to $Y_2' \otimes I_2$.

Now define the exact sequence $I_2^3 \to A \to I_2^3$ by setting $a \to (a, a)$ and $(b, c) \to (b/2 - c/2)$. The induced map $\operatorname{Ext}(Y_2, A) \to \operatorname{Ext}(Y, I_2^3)$ sends $(\alpha, -\alpha)$ to α. This if we view $I_2^3 \subset K$ via the first map then $K/(I_2^3) \cong (Y_2' \otimes I_2)$. But $H^1(S_n, Y_2' \otimes I_2) = 0$ because $Y_2' \otimes I_2$ is a permutation lattice over $\mathbb{Z}_2$. Thus the map $H^1(S_n, I_2^3) \to H^1(S_n, K)$ is surjective. Since τ acts as the identity on I_2^3, τ also acts as the identity on $H^1(S_n, K)$ and thus as the identity on $H^2(S_n, Y_2^2)$.

PROPOSITION 6.6. $H^2(S_n, Y_2 \wedge Y_2) = 0$.

PROOF. Consider the map $\eta : Y_2^2 \to Y_2^2$ defined by $\eta = (1/2)(1 - \tau)$. Then $\eta(Y_2^2) = Y_2 \wedge Y_2$ which is also the kernel of $1 - \eta$. It follows that as a subgroup of $H^n(S_n, Y_2^2)$ $H^n(S_n, Y_2 \wedge Y_2)$ is the image of η^* and the kernel of $1 - \eta^*$. That is, $H^n(S_n, Y_2 \wedge Y_2)$ is the antisymmetric part (under τ) of $H^n(S_n, Y_2^2)$. The proposition follows. □

As observed above, 6.5 and 6.6 proves all of 6.3. We have assembled all the pieces to show:

THEOREM 6.7. *Let* n *be odd,* V *a "good" representation, and* $G = PGL_n$. *Then* $H_u^3(F(V)^G) = (0)$. *In other language,* $H_u^3(Z(F, n, r)) = (0)$ *for* n *odd where* $Z(F, n, r)$ *is the center of the generic division algebra of degree* n *in* r *variables over* F.

PROOF. By our description of this invariant field what we must show is $H_u^3(F(Y)^{S_n}) = (0)$. Suppose $\alpha \in H^3(F(Y)^{S_n})$. It is well known (e.g. [J, p. 235]) that there is a faithful S_n representation V such that $F(V)^{S_n}$ is rational. It follows from 6.3 and 5.17 that α is in the image of $H^3(S_n, Y)$. The exact sequence (2) yields a long exact sequence

$$H^2(S_n, Y') \to H^2(S_n, I) \to H^3(S_n, Y) \to H^3(S_n, Y') .$$

Now $H^r(S_n, Y') = H^r(S_n, \mathbb{Z}[S_n/S_{n-2}]) = H^r(S_{n-2}, \mathbb{Z}) = H^{r-1}(S_{n-2}, \mathbb{Q}/\mathbb{Z})$ and so by e.g. [K, p. 179] $H^3(S_n, Y') = \mathbb{Z}/2\mathbb{Z}$ or $H^3(S_n, Y') = 0$. The exact sequence (1) shows that $H^2(S_n, I)$ is a subgroup of $H^2(S_n, X) = H^2(S_{n-1}, \mathbb{Z}) = H^1(S_{n-1}, \mathbb{Q}/\mathbb{Z}) = \mathbb{Z}/2\mathbb{Z}$. Thus $H^3(S_n, Y)$ has order dividing 4 (it actually has order 2 if $n \geq 4$) and we are done by 6.2. $\square$

REFERENCES

[AB] M. Auslander and A. Brumer, *Brauer groups of discrete valuation rings*, Nederl. Akad. Wetensch. Proc. Ser. A **71** (1968), 286–296.

[AG] M. Auslander and O. Goldman, *The Brauer group of a commutative ring*, Trans. Amer. Math. Soc. **97** (1960), 367–409.

[AM] M. Artin and D. Mumford, *Some elementary examples of unirational varieties which are not rational*, Proc. London Math. Soc. **25** (1972), 75–95.

[ATV] M. Artin, J. Tate, and M. Van den Bergh, *Modules over regular algebras of dimension 3*, Invent. Math. **106** (1991), 335–388.

[AW] M. F. Atiyah and C. T. C. Wall, *Cohomology of groups*, Algebraic Number Theory (J. W. S. Cassels, and A. Frohlich, eds.), Thompson, Washington, D.C., 1967.

[B] K. Brown, *Cohomology of groups*, Springer-Verlag,, New York/Heidelberg/Berlin, 1982.

[CO] J.-L. Colliot-Thélène and M. Ojanguren, *Variétés unirationelles nonrationelles: au-delà de l'exemple d'Artin et Mumford*, Invent. Math. **97** (1989), 141–158.

[EM] S. Endo and T. Miyata, *Invariants of finite abelian groups*, J. Math. Soc. Japan **25** (1973), 7–26.

[FS] B. Fein and M. Schacher, *Brauer groups of rational function fields over global fields*, Groupe de Brauer (M. Kervaire and M. Ojanguren, eds.), Lecture Notes in Math., vol. 844, Springer-Verlag, New York, 1981, pp. 46–74.

[FSS] B. Fein, D. J. Saltman, and M. Schacher, *Brauer-Hilbertian fields*, Trans. Amer. Math. Soc. **334** (1992), 915–928.

[F] T. Ford, *On computing the Brauer group of a localization*, Preprint.

[F2] _____, *Products of symbol algebras that ramify only on a nonsingular plane elliptic curve*, Ulam-Quart. **1** 1 (1992), 12ff.

[Fo] E. Formanek, *Polynomial identities of matrices*, Algebraists' Homage... (A. Amitsur, D. Saltman, and G. Seligman, eds.), Amer. Math. Soc., Providence, RI, 1982.

[Fu] W. Fulton, *Intersection theory*, Springer-Verlag, Berlin, 1984.

[Ha] R. Hartshorne, *Algebraic geometry*, Springer-Verlag, New York, 1977.

[H] R. Hoobler, *A cohomological interpretation of the Brauer group of rings*, Pacific J. Math. **86** 1 (1980), 89–92.

[HS] G. Hochschild and J-P. Serre, *Cohomology of group extensions*, Trans. Amer. Math. Soc. **74** (1953), 110–134.

[J]   N. Jacobson, *Basic algebra*. I, W. H. Freeman, San Francisco, CA, 1974.

[K]   G. Karpilovsky, *Projective representations of finite groups*, Marcel Dekker, New York, 1985.

[M]   A. Magid, *Brauer groups of linear algebraic groups with characters*, Proc. Amer. Math. Soc. **71** (1978), 164–168.

[MS]   A. S. Merkuriev and A. A. Suslin, *K-cohomologies of Severi-Brauer varieties and norm residue homomorphisms*, Izv. Akad. Nauk. SSSR Ser. Mat. **46** (1982), 1011–1046.

[OS]   M. Orzech and C. Small, *The Brauer group of commutative rings*, Marcel Dekker, New York, 1975.

[P]   E. Peyre, *Unramified cohomology and rationality problems*, Math.-Ann. **296** 2 (1993), 247–268.

[RT]   S. Rosset and J. Tate, *A reciprocity law for generalized traces*, Comment. Math. Helv. **58** (1983), 38–47.

[S]   D. J. Saltman, *Multiplicative field invariants*, J. Algebra **106** (1987), 221–238.

[S2]   _____, *The Brauer group and the center of generic matrices*, J. Algebra **97** (1985), 53–67.

[S3]   _____, *Invariant fields of linear groups and division algebras*, Perspectives in Ring Theory (F. van Oystaeyen and L. Le Bruyn, eds.), Kluwer, Dordrecht, 1988, pp. 279–297.

[S4]   _____, *Generic Galois extensions and problems in field theory*, Adv. Math. **43** (1982), 250–282.

[Se]   J. P. Serre, *Local fields*, Springer-Verlag, New York/Heidelberg/Berlin, 1979.

[Sh]   I. R. Shafarevich, *Basic algebraic geometry*, Springer-Verlag, Berlin, 1977.

DEPARTMENT OF MATHEMATICS, THE UNIVERSITY OF TEXAS AT AUSTIN, AUSTIN, TEXAS 78712

E-mail address: `saltman@math.utexas.edu`

Proceedings of Symposia in Pure Mathematics
Volume 58.1 (1995)

Higher Algebraic K-Theory

RICHARD G. SWAN

ABSTRACT. We give an exposition of the basic properties of Quillen's higher algebraic K-theory. These properties are then applied to compute the K-theory of Severi-Brauer varieties and of quadric hypersurfaces.

The following is intended to be an introduction to higher algebraic K-theory for those with little or no previous knowledge of this field. The proofs of the basic properties of the higher K-functors will not be given here since they require quite a bit of topology. However I will try to make them seem reasonable by discussing the classical case of the Grothendieck group K_0. I will then give a number of applications to illustrate the use of these properties. The reader interested in seeing all the details can consult the excellent book [S] of Srinivas as well as the original work of Quillen [Q, GQ].

1. Exact categories

If R is a ring, the Grothendieck group $K_0(R)$ is defined to be the abelian group with one generator $[P]$ for each finitely generated projective R-module P, and a relation $[P] = [P'] + [P'']$ for each short exact sequence $0 \to P' \to P \to P'' \to 0$ of such modules. The same definition can clearly be applied to any category $\mathscr{C}$ in which short exact sequences are defined yielding a group $K_0(\mathscr{C})$. However, for the generalization to higher K-functors $K_i(\mathscr{C})$ we must be a bit more specific about the categories to be used.

For readers unfamiliar with the notion of an abelian category it will suffice for our purposes to define such a category as either the category $\mathscr{M}od(R)$ of modules over a ring R (or, more generally, the category $\mathscr{M}od(\mathscr{O})$ of sheaves of modules over a sheaf of rings $\mathscr{O}$), or a full subcategory of such a category which is closed under kernels, cokernels, and finite direct sums. A simple example is the category $\mathscr{M}odfg(R)$ of finitely generated modules over a

1991 *Mathematics Subject Classification.* Primary 19-02; Secondary 19D06, 19D50, 19E08.
Key words and phrases. Higher K-theory, Severi-Brauer varieties, quadric hypersurfaces.
Partly supported by the National Science Foundation.
This paper is in final form and no version of it will be submitted for publication elsewhere.

noetherian ring R. The usual definition of exact sequence obviously makes sense in such a category and it is not hard to show that it is independent of the embedding in the category $\mathcal{M}od(R)$. The general theory of abelian categories is presented in [**F, Ga, Mit**].

Now the category $\mathcal{P}(R)$ of finitely generated projective R-modules is not abelian in general since a map in $\mathcal{P}(R)$ need not have a kernel or cokernel in $\mathcal{P}(R)$. In order to allow such categories we define the more general notion of an exact category following Quillen [**Q**].

DEFINITION. An exact category $\mathcal{E}$ is a full subcategory of an abelian category $\mathcal{A}$ which is closed under extensions and which contains a zero object of $\mathcal{A}$.

A sequence $0 \to E' \to E \to E'' \to 0$ in $\mathcal{E}$ is called exact if it is exact in $\mathcal{A}$. In contrast to the abelian case, this notion of exactness depends on the embedding of $\mathcal{E}$ in $\mathcal{A}$.

EXAMPLE. Let $\mathcal{E}$ be any additive category and consider the Yoneda embedding $E \mapsto Hom(-, E)$ of $\mathcal{E}$ in the functor category $\mathcal{A}b^{\mathcal{E}^{op}}$ of contravariant functors from $\mathcal{E}$ to abelian groups. It is easy to see that $0 \to E' \to E \to E'' \to 0$ maps to an exact sequence in $\mathcal{A}b^{\mathcal{E}^{op}}$ if and only if it splits. Applying this observation to $\mathcal{E} = \mathcal{M}od(\mathbb{Z})$ shows that an additive category may have more than one structure as an exact category.

DEFINITION. A functor $F : \mathcal{E} \to \mathcal{F}$ between exact categories is called exact if it is additive and preserves short exact sequences. Two exact categories $\mathcal{E}$ and $\mathcal{F}$ are called equivalent if there is an equivalence of categories $F : \mathcal{E} \to \mathcal{F}$ such that F and F^{-1} are exact.

We say that a map $i : M \to N$ in an exact category $\mathcal{E}$ is an admissible monomorphism (written $i : M \rightarrowtail N$) if there is a short exact sequence $0 \to M \xrightarrow{i} N \to L \to 0$ in $\mathcal{E}$. In other words, the cokernel of i in the abelian category $\mathcal{A}$ lies in $\mathcal{E}$. Similarly $j : M \to N$ is an admissible epimorphism (written $i : M \twoheadrightarrow N$) if there is a short exact sequence $0 \to L \to M \xrightarrow{j} N \to 0$ in $\mathcal{E}$.

The assumption that $\mathcal{E}$ is closed under extensions implies that a composition of admissible monomorphisms is an admissible monomorphism and similarly for admissible epimorphisms.

We say that a sequence $0 \to E_0 \to \cdots \to E_n \to 0$ is exact in $\mathcal{E}$ if it splits into short exact sequences $0 \to E_0 \to E_1 \to E_2' \to 0$, $0 \to E_2' \to E_2 \to E_3' \to 0$, ..., $0 \to E_{n-1}' \to E_{n-1} \to E_n \to 0$ in $\mathcal{E}$ (so that the original map $E_i \to E_{i+1}$ is the composition $E_i \to E_{i+1}' \to E_{i+1}$). This is more restrictive than asking that the sequence be exact in the abelian category $\mathcal{A}$.

EXAMPLE. Let $R = \mathbb{R}[x, y, z]/(x^2 + y^2 + z^2 - 1)$, let $\mathcal{A} = \mathcal{M}od(R)$, and let $\mathcal{E}$ be the full subcategory of finitely generated free R-modules. The Koszul complex $K.(x, y, z) = (R \xrightarrow{x} R) \otimes (R \xrightarrow{y} R) \otimes (R \xrightarrow{z} R)$ has the form

$$0 \to R \to R^3 \to R^3 \xrightarrow{x,y,z} R \to 0.$$

This is exact in $\mathscr{A}$ but not in $\mathscr{E}$ since the kernel of $R^3 \to R$ is not free and so does not lie in $\mathscr{E}$.

If $\mathscr{E}$ and $\mathscr{F}$ are exact categories, F', F, $F'' : \mathscr{E} \to \mathscr{F}$ are exact functors, and $F' \xrightarrow{i} F \xrightarrow{j} F''$ are natural transformations, we say that $0 \to F' \xrightarrow{i} F \xrightarrow{j} F'' \to 0$ is exact if $0 \to F'(E) \xrightarrow{i} F(E) \xrightarrow{j} F''(E) \to 0$ is exact for each object E of $\mathscr{E}$. More generally $0 \to F_0 \to \cdots \to F_n \to 0$ is exact if $0 \to F_0(E) \to \cdots \to F_n(E) \to 0$ is exact (in the sense defined above) for each object E of $\mathscr{E}$.

2. The Grothendieck group

If $\mathscr{E}$ is an exact category, the Grothendieck group $K_0(\mathscr{E})$ is the abelian group with one generator $[E]$ for each object E of $\mathscr{E}$ and a relation $[E] = [E'] + [E'']$ for each short exact sequence $0 \to E' \to E \to E'' \to 0$ in $\mathscr{E}$.

If $0 \to E_0 \to \cdots \to E_n \to 0$ is exact in $\mathscr{E}$, then $\sum (-1)^i [E_i] = 0$ in $K_0(\mathscr{E})$. This is clear from the definition of exactness given in §1.

If $F : \mathscr{E} \to \mathscr{F}$ is an exact functor, it induces a map $F_* : K_0(\mathscr{E}) \to K_0(\mathscr{F})$ by $F_*([E]) = [F(E)]$.

Here are some basic properties of K_0.

THEOREM 2.1.

(1) $K_0(\mathscr{E} \times \mathscr{F}) = K_0(\mathscr{E}) \times K_0(\mathscr{F})$.
(2) $K_0(\mathscr{E}^{\mathrm{op}}) = K_0(\mathscr{E})$.
(3) If F, $G : \mathscr{E} \to \mathscr{F}$ are exact and $F \approx G$, then $F_* = G_* : K_0(\mathscr{E}) \to K_0(\mathscr{F})$.
(4) If $\mathscr{E}$, $\mathscr{F}$ are equivalent exact categories, then $K_0(\mathscr{E}) = K_0(\mathscr{F})$.
(5) If an exact category $\mathscr{E}$ is a filtered colimit of exact categories $\mathscr{E}_\alpha$ then $K_0(\mathscr{E}) = \operatorname{colim} K_0(\mathscr{E}_\alpha)$.

All of these properties are immediate. The hypothesis of (5) is to be understood in the following sense: Let $\mathrm{ob}(\mathscr{E})$, $\mathrm{mor}(\mathscr{E})$, $\mathrm{ses}(\mathscr{E})$ be the sets of objects, morphisms, and short exact sequences of $\mathscr{E}$. Then $\mathrm{ob}(\mathscr{E}) = \operatorname{colim} \mathrm{ob}(\mathscr{E}_\alpha)$, $\mathrm{mor}(\mathscr{E}) = \operatorname{colim} \mathrm{mor}(\mathscr{E}_\alpha)$, and $\mathrm{ses}(\mathscr{E}) = \operatorname{colim} \mathrm{ses}(\mathscr{E}_\alpha)$.

THEOREM 2.2 (Additivity). Let $0 \to F' \to F \to F'' \to 0$ be a short exact sequence of exact functors between exact categories $\mathscr{E}$ and $\mathscr{F}$. Then $F_* = F'_* + F''_* : K_0(\mathscr{E}) \to K_0(\mathscr{F})$.

This is clear since $[F(E)] = [F'(E)] + [F''(E)]$.

COROLLARY 2.3. If $0 \to F_0 \to \cdots \to F_n \to 0$ is an exact sequence of exact functors then $\sum (-1)^i F_{i*} = 0 : K_0(\mathscr{E}) \to K_0(\mathscr{F})$.

PROOF. Let $F'_2(E)$ be the cokernel of $F_0(E) \to F_1(E)$. This lies in $\mathscr{F}$

by the definition of exactness. One verifies easily that F_2' is again an exact functor. We have $0 \to F_0 \to F_1 \to F_2' \to 0$ and $0 \to F_2' \to F_2 \to \cdots \to F_n \to 0$ and the result follows by induction on n.

COROLLARY 2.4. *Let* $F : \mathscr{E} \to \mathscr{F}$ *be an exact functor which has a filtration* $0 = F_0 \subset F_1 \subset \cdots \subset F_n = F$ *where the subfunctors* F_i *are exact and such that for each* $E \in \mathscr{E}$, $F_{i-1}(E) \to F_i(E)$ *is an admissible monomorphism in* $\mathscr{F}$. *Define* $(F_i/F_{i-1})(E)$ *to be the cokernel of* $F_{i-1}(E) \to F_i(E)$. *Then* F_i/F_{i-1} *is an exact functor and* $F_* = \sum(F_i/F_{i-1})_* : K_0(\mathscr{E}) \to K_0(\mathscr{F})$.

Let $\mathscr{M}$ be an exact category defined by a full embedding $\mathscr{M} \subset \mathscr{A}$ in an abelian category $\mathscr{A}$ such that $\mathscr{M}$ is closed under extensions. Let $\mathscr{M}'$, $\mathscr{M}'' \subset \mathscr{M}$ be full subcategories closed under extension. Let $\mathscr{E}$ be the category of all short exact sequences $0 \to M' \to M \to M'' \to 0$ in $\mathscr{M}$ with $M' \in \mathscr{M}'$ and $M'' \in \mathscr{M}''$. We regard $\mathscr{E}$ as a full subcategory of the category of chain complexes in $\mathscr{A}$ so that an exact sequence in $\mathscr{E}$ is a diagram

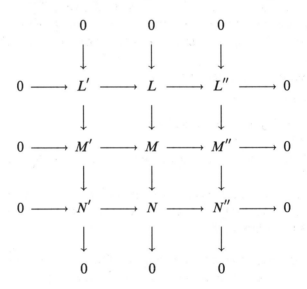

with exact rows and columns. We have exact functors $s : \mathscr{E} \to \mathscr{M}'$, $t : \mathscr{E} \to \mathscr{M}$, $q : \mathscr{E} \to \mathscr{M}''$ sending $0 \to M' \to M \to M'' \to 0$ to M', M, and M'' respectively.

COROLLARY 2.5 ([W]). $(s_*, q_*) : K_0(\mathscr{E}) \xrightarrow{\approx} K_0(\mathscr{M}') \times K_0(\mathscr{M}'')$.

PROOF. $f = (s, q) : \mathscr{E} \to \mathscr{M}' \times \mathscr{M}''$ is exact. Define $g : \mathscr{M}' \times \mathscr{M}'' \to \mathscr{E}$ by $g(M', M'') = (0 \to M' \to M' \oplus M'' \to M'' \to 0$. Clearly fg is the identity. Define $i : \mathscr{M}' \to \mathscr{E}$ by $i(M') = g(M', 0)$ and $j : \mathscr{M}'' \to \mathscr{E}$ by

$j(M'') = g(0, M'')$. If $0 \to M' \to M \to M'' \to 0$ lies in $\mathscr{E}$ we have

$$
\begin{array}{ccccccccc}
 & & 0 & & 0 & & 0 & & \\
 & & \downarrow & & \downarrow & & \downarrow & & \\
0 & \longrightarrow & M' & \longrightarrow & M' & \longrightarrow & 0 & \longrightarrow & 0 \\
 & & \downarrow & & \downarrow & & \downarrow & & \\
0 & \longrightarrow & M' & \longrightarrow & M & \longrightarrow & M'' & \longrightarrow & 0 \\
 & & \downarrow & & \downarrow & & \downarrow & & \\
0 & \longrightarrow & 0 & \longrightarrow & M'' & \longrightarrow & M'' & \longrightarrow & 0 \\
 & & \downarrow & & \downarrow & & \downarrow & & \\
 & & 0 & & 0 & & 0 & &
\end{array}
$$

showing that $0 \to is \to 1_{\mathscr{E}} \to jq \to 0$ is exact. By Theorem 2.2, $1_{K_0(\mathscr{E})} = (is)_* + (jq)_*$ showing that the composition

$$
K_0(\mathscr{E}) \xrightarrow{(s_*, q_*)} K_0(\mathscr{M}') \times K_0(\mathscr{M}'') \xrightarrow{(i_*, j_*)} K_0(\mathscr{E})
$$

is the identity. But the other composition is also the identity since $si = 1_{\mathscr{M}'}$, $sj = 0$, $qi = 0$, and $qj = 1_{\mathscr{M}''}$.

Conversely, one can deduce Theorem 2.2 from Corollary 2.5 [Q].

THEOREM 2.6 (Dévissage). *Let $\mathscr{A}$ be an abelian category and let $\mathscr{B}$ be a full subcategory closed under subobjects, quotient objects, and finite direct sums (so $\mathscr{B}$ is also abelian). Suppose each object A of $\mathscr{A}$ has a filtration $0 = A_0 \subset A_1 \subset \cdots \subset A_n = A$ with all A_i/A_{i-1} in $\mathscr{B}$. Then $K_0(\mathscr{B}) \xrightarrow{\approx} K_0(\mathscr{A})$.*

PROOF. Define $K_0(\mathscr{A}) \to K_0(\mathscr{B})$ by sending $[A]$ to $\sum[A_i/A_{i-1}]$. The Jordan-Hölder-Zassenhaus theorem shows that this is well defined and one verifies immediately that it is inverse to $K_0(\mathscr{B}) \to K_0(\mathscr{A})$.

REMARK. If (A_i), (A_i') are two filtrations of A, the proof of the Jordan-Hölder theorem involves forming $A_i \cap A_j'$. We have to assume the categories are abelian to insure that such intersections lie in $\mathscr{A}$. There does not seem to be a more general form applicable to exact categories unless one assumes the filtration is functorial so that Corollary 2.4 applies.

If R is a noetherian ring define $G_0(R) = K_0(\mathscr{M}odfg(R))$.

COROLLARY 2.7. *If R is a noetherian ring, I is a 2-sided ideal of R, and I is nilpotent then $G_0(R) = G_0(R/I)$.*

We apply the theorem to $\mathscr{A} = \mathscr{M}odfg(R)$ and $\mathscr{B} = \mathscr{M}odfg(R/I)$.

THEOREM 2.8 (Resolution). *Let $\mathcal{M}$ be an exact category and let $\mathcal{P} \subset \mathcal{M}$ be a full subcategory closed under extensions. Assume*

(a) *For every object M of $\mathcal{M}$ there is an admissible epimorphism $P \twoheadrightarrow M$ with P in $\mathcal{P}$.*

(b) *If $0 \to M' \to P \to M'' \to 0$ is a short exact sequence in $\mathcal{M}$ with P in $\mathcal{P}$, then M' also lies in $\mathcal{P}$.*

Then $K_0(\mathcal{P}) \xrightarrow{\approx} K_0(\mathcal{M})$.

Here $\mathcal{P}$ is given the obvious exact category structure; i.e. a short exact sequence in $\mathcal{P}$ is one in $\mathcal{M}$ which lies in $\mathcal{P}$.

PROOF. We try to define a map $\gamma : K_0(\mathcal{M}) \to K_0(\mathcal{P})$ as follows: Given M, find $0 \to M' \to P \to M \to 0$ by (a). By (b), M' lies in $\mathcal{P}$ and we set $\gamma(M) = [P] - [M'] \in K_0(\mathcal{P})$. If $0 \to N' \to Q \to M \to 0$ is a different choice then $P \oplus Q \to M$ is an admissible epimorphism being the composition of the admissible epimorphisms $P \oplus Q \twoheadrightarrow M \oplus M \twoheadrightarrow M$. Its kernel X lies in $\mathcal{P}$ by (b) and we have

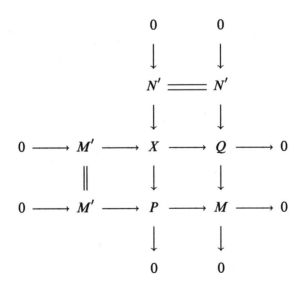

so $[X] = [P] + [N'] = [Q] + [M']$ showing that $[P] - [M'] = [Q] - [N']$.

If $0 \to M' \to M \to M'' \to 0$ choose $P' \twoheadrightarrow M'$, $P \twoheadrightarrow M$ and consider

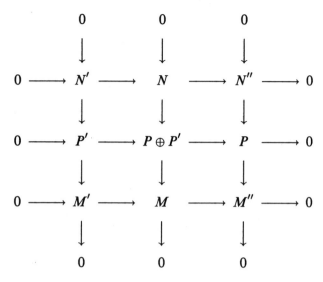

to show that $\gamma(M) = \gamma(M') + \gamma(M'')$.

It is clear that γ is inverse to $K_0(\mathscr{P}) \to K_0(\mathscr{M})$.

If R is noetherian, the inclusion $\mathscr{P}(R) \subset \mathscr{M}odfg(R)$ induces a map $K_0(R) \to G_0(R)$. We say that R is regular if it is noetherian and every finitely generated R-module has finite projective dimension.

COROLLARY 2.9. *If R is regular, $K_0(R) \xrightarrow{\approx} G_0(R)$.*

PROOF. Let $\mathscr{P}_n$ be the full subcategory of $\mathscr{M}odfg(R)$ consisting of modules of projective dimension $\leq n$. Then $\mathscr{P} = \mathscr{P}_0 \subset \mathscr{P}_1 \subset \ldots$ and $\mathscr{M}odfg(R) = \text{colim}\,\mathscr{P}_n$. If $M \in \mathscr{M}odfg(R)$ then $M \in \mathscr{P}_n$ if and only if $\text{Ext}_R^i(M, -) = 0$ for $i > n$. We can map a free module P onto M and a glance at the exact Ext-sequence shows that $\mathscr{P}_{n-1} \subset \mathscr{P}_n$ satisfies all the hypotheses of Theorem 2.8 so $K_0(R) = K_0(\mathscr{P}) \xrightarrow{\approx} K_0(\mathscr{P}_1) \xrightarrow{\approx} \ldots \xrightarrow{\approx} K_0(\mathscr{P}_n) \xrightarrow{\approx} \ldots$ and $G_0(R) = \text{colim}\, K_0(\mathscr{P}_n)$.

More general results of this sort are given in [Q].

3. Localization

A Serre subcategory $\mathscr{S}$ of an abelian category $\mathscr{A}$ is a full subcategory closed under subobjects, quotient objects, and extensions. In other words, if $0 \to M' \to M \to M'' \to 0$ is exact in $\mathscr{A}$ then M lies in $\mathscr{S}$ if and only if M' and M'' lie in $\mathscr{S}$.

One can define a quotient category $\mathscr{A}/\mathscr{S}$ and an exact functor $U : \mathscr{A} \to \mathscr{A}/\mathscr{S}$ which is universal for exact functors to abelian categories sending $\mathscr{S}$ to 0. In other words if $F : \mathscr{A} \to \mathscr{B}$ is an exact functor (with $\mathscr{B}$ abelian) and $F(S) = 0$ for all S in $\mathscr{S}$ then there is a unique G making the diagram

$$\mathscr{A} \xrightarrow{\;\;U\;\;} \mathscr{A}/\mathscr{S}$$

$$\Big\| \qquad\qquad \Big\downarrow G$$

$$\mathscr{A} \xrightarrow{\;\;F\;\;} \mathscr{B}$$

commute. G will also be exact. A proof can be found in [Ga] and also in [SK].

For our purposes we will only need the following 4 properties which are satisfied by U and which characterize it up to equivalence.

Let $T : \mathscr{A} \to \mathscr{B}$ be an additive functor between abelian categories and let $\mathscr{S}$ be a full subcategory of $\mathscr{A}$. Consider the following properties:

(1) T is exact.

(2) $T(A) = 0$ if and only if A lies in $\mathscr{S}$.

(3) Each $B \in \mathrm{ob}(\mathscr{B})$ is isomorphic to some $T(A)$.

(4) If $\varphi : T(A) \to T(A')$, there are maps $A \xleftarrow{\alpha} X \xrightarrow{\beta} A'$ in $\mathscr{A}$ such that $T(\alpha)$ is an isomorphism and

$$\begin{array}{ccc} T(X) & =\!=\!=\!= & T(X) \\ T(\alpha)\Big\downarrow & & T(\beta)\Big\downarrow \\ T(A) & \xrightarrow{\;\varphi\;} & T(A') \end{array}$$

commutes.

Note that (4) is equivalent to its dual

(4$'$) If $\varphi : T(A) \to T(A')$, there are maps $A \xrightarrow{\gamma} Y \xleftarrow{\delta} A'$ in $\mathscr{A}$ such that $T(\delta)$ is an isomorphism and

$$\begin{array}{ccc} T(A) & \xrightarrow{\;\varphi\;} & T(A') \\ T(\gamma)\Big\downarrow & & T(\delta)\Big\downarrow \\ T(Y) & =\!=\!=\!= & T(Y) \end{array}$$

commutes.

To pass from (4) to (4$'$) we need only form the pushout of the diagram

$$\begin{array}{ccc} X & \longrightarrow & A' \\ \Big\downarrow & & \\ A & & \end{array}$$

THEOREM 3.1. *If $\mathscr{S}$ is a Serre subcategory of $\mathscr{A}$ then $U : \mathscr{A} \to \mathscr{A}/\mathscr{S}$ satisfies* (1) *to* (4). *If $T : \mathscr{A} \to \mathscr{B}$ satisfies* (1) *to* (4), *the map $G : \mathscr{A}/\mathscr{S} \to \mathscr{B}$ is an equivalence of categories. If T also satisfies*

($3'$) $T : \mathrm{ob}\,\mathscr{A} \approx \mathrm{ob}\,\mathscr{B}$,

then G is an isomorphism.

This is clear from the construction of $\mathscr{A}/\mathscr{S}$. We will work directly with (1) to (4) here.

EXAMPLES.

(A) Let S be a central multiplicative set in a ring R. The localization functor $M \mapsto M_S$ from $\mathscr{M}od(R)$ to $\mathscr{M}od(R_S)$ satisfies (1) to (4).
This is trivial since we can take $A = B$ in (3) and $X = A$, $\alpha = 1_A$ in (4). We use (2) as the definition of $\mathscr{S}$.

(B) If in (A) R is noetherian then $\mathscr{M}odfg(R) \to \mathscr{M}odfg(R_S)$ satisfies (1) to (4).
This is a standard exercise in localization.

(C) If U is an open subset of a topological space X, the restriction map $\mathrm{Sh}_X \to \mathrm{Sh}_U$ sending each sheaf $\mathscr{F}$ on X to $\mathscr{F}|U$ satisfies (1) to (4).

(D) If U is an open subset of a noetherian scheme X, the same is true for coherent sheaves; i.e. $\mathrm{Coh}_X \to \mathrm{Coh}_U$ satisfies (1) to (4).

THEOREM 3.2. *If $\mathscr{A}$, $\mathscr{B}$, T, $\mathscr{S}$ satisfy (1) to (4) then $K_0(\mathscr{S}) \to K_0(\mathscr{A}) \to K_0(\mathscr{B}) \to 0$ is exact.*

PROOF. Define $\gamma : K_0(\mathscr{B}) \to \mathrm{ckr}[K_0(\mathscr{S}) \to K_0(\mathscr{A})]$ by $\gamma(B) = [A]$ where $T(A) \approx B$. If $T(A') \approx B$ find $A \xleftarrow{\alpha} X \xrightarrow{\beta} A'$ as in (4). Then $T(\alpha)$, $T(\beta)$ are isomorphisms so $\ker\alpha$, $\ker\beta$, $\mathrm{ckr}\,\alpha$, $\mathrm{ckr}\,\beta$ lie in $\mathscr{S}$. From $0 \to \ker\alpha \to X \to \mathrm{im}\,\alpha \to 0$, $0 \to \mathrm{im}\,\alpha \to A \to \mathrm{ckr}\,\alpha \to 0$ we deduce that $[A] \equiv [X] \bmod \mathrm{im}\,K_0(\mathscr{S})$ and similarly $[X] \equiv [A']$ so that $\gamma(B)$ is well-defined. If $0 \to B' \xrightarrow{i} B \xrightarrow{j} B'' \to 0$ choose $T(A) \approx B$, $T(A'') \approx B''$, and $A \xleftarrow{\alpha} X \xrightarrow{\beta} A''$ as in (4) for $\varphi = j$. Then $0 \to \ker\beta \to X \to \mathrm{im}\,\beta \to 0$ maps under T to a short exact sequence isomorphic to $0 \to B' \to B \to B'' \to 0$ and we see that $\gamma(B) = \gamma(B') + \gamma(B'')$. It is clear that γ is inverse to $K_0(\mathscr{A})/\mathrm{im}\,K_0(\mathscr{S}) \to K_0(\mathscr{B})$.

COROLLARY 3.3. *If R is a noetherian ring and S is a central multiplicative subset then $K_0(\mathscr{S}) \to G_0(R) \to G_0(R_S) \to 0$ is exact where $\mathscr{S} \subset \mathscr{M}odfg(R)$ is the full subcategory of modules M with $M_S = 0$.*

COROLLARY 3.4. *If R is a noetherian ring and s is a central element then $G_0(R/sR) \to G_0(R) \to G_0(R_s) \to 0$ is exact.*

PROOF. $\mathscr{M}odfg(R/sR) \subset \mathscr{S}$ and if $M \in \mathscr{S}$, $s^n M = 0$ for some n since M is finitely generated. We can filter M by $0 = s^n M \subset s^{n-1} M \subset \cdots \subset sM \subset M$ and the Dévissage Theorem 2.6 shows that $G_0(R/sR) = K_0(\mathscr{S})$.

The above results do not extend to projective modules since $\mathscr{P}(R)$ is usually not abelian. However, we can get a weaker result of this type provided S is regular, i.e. consists of nonzero-divisors.

If S is a central multiplicative set in R we define $\mathscr{H}_S(R)$ to be the full subcategory of $\mathscr{M}od(R)$ of modules M such that

(1) $M_S = 0$.
(2) There is a short exact sequence $0 \to Q \to P \to M \to 0$ with P, Q in $\mathscr{P}(R)$.

THEOREM 3.5. *Let S be a central multiplicative set in R consisting of regular elements. Then there is an exact sequence $K_0(\mathscr{H}_S(R)) \to K_0(R) \to K_0(R_S)$.*

In general, we cannot put a 0 on the right here. Also in the situation of Corollary 3.4, $K_0(\mathscr{H}_S(R))$ cannot be replaced by $K_0(R/sR)$. However, if R is regular we can use Corollary 2.9 to replace $K_0(R)$ and $K_0(R_S)$ by $G_0(R)$ and $G_0(R_S)$ so that $K_0(R) \to K_0(R_S)$ is onto in this case. Note that R_S is regular if R is. If R/sR is also regular we can apply Corollary 3.4 to get a similar sequence involving K_0.

PROOF. The map $K_0(\mathscr{H}_S(R)) \to K_0(R)$ sends $[M]$ to $[P] - [Q]$ where $0 \to Q \to P \to M \to 0$ with P, $Q \in \mathscr{P}(R)$. One can check directly that it is well defined but the following approach will be more useful later.

Define $\mathscr{P}_S(R)$ to be the full subcategory of $\mathscr{M}od(R)$ consisting of modules M such that

(1) $M_S \in \mathscr{P}(R_S)$.
(2) There is a short exact sequence $0 \to Q \to P \to M \to 0$ with P, $Q \in \mathscr{P}(R)$.

LEMMA 3.6. *$\mathscr{P}_S(R)$ and $\mathscr{H}_S(R)$ are closed under extensions.*

PROOF. There is no problem with (1). If $0 \to M' \to M \to M'' \to 0$ and we have $0 \to Q' \to P' \to M' \to 0$ and $0 \to Q'' \to P'' \to M'' \to 0$, lift $P'' \to M''$ to $P'' \to M$ and form

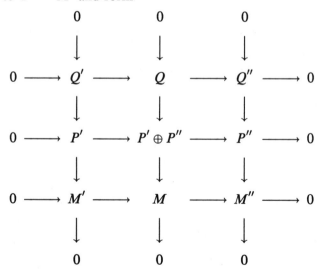

(a standard construction in homological algebra [**CE**]). Since Q', Q'' lie in $\mathscr{P}(R)$ so does Q and the middle column shows that M satisfies (2).

LEMMA 3.7. $\mathscr{P}(R) \hookrightarrow \mathscr{P}_S(R)$ satisfies the hypothesis of the Resolution Theorem 2.8.

PROOF. (a) is clear from property (2) above. If $0 \to M' \to P \to M'' \to 0$ with $P \in \mathscr{P}(R)$, $M'' \in \mathscr{P}_S(R)$ find $0 \to Q'' \to P'' \to M'' \to 0$ as in (2). Schanuel's lemma [**B**, I, 6.3] shows that $P'' \oplus M' \approx Q'' \oplus P$ so that $M' \in \mathscr{P}(R)$ as required.

We now have $\mathscr{H}_S(R) \hookrightarrow \mathscr{P}_S(R) \hookleftarrow \mathscr{P}(R)$ inducing

$$K_0(\mathscr{H}_S(R)) \to K_0(\mathscr{P}_S(R)) \xleftarrow{\approx} K_0(\mathscr{P}(R))$$

which gives the required map. If $0 \to Q \to P \to M \to 0$, $[M] = [P] - [Q]$ in $K_0(\mathscr{P}_S(R))$ so that the map has the form mentioned above. Since $M_S = 0$ for $M \in \mathscr{H}_S(R)$ we see that $P_S \approx Q_S$ showing that the composition $K_0(\mathscr{H}_S(R)) \to K_0(R) \to K_0(R_S)$ is zero.

LEMMA 3.8. Let A be any ring. Then any element of $K_0(A)$ has the form $[P] - [Q]$ with $P, Q \in \mathscr{P}(A)$. Moreover, $[P] - [Q] = 0$ if and only if P and Q are stably isomorphic; i.e. $P \oplus F \approx Q \oplus F$ for some finitely generated free F.

PROOF. The first statement is clear. For the second, express the relation $[P] - [Q] = 0$ in terms of the defining relations of $K_0(A)$ and deduce that $P \oplus G \approx Q \oplus G$ for some $G \in \mathscr{P}(A)$. Find $F \twoheadrightarrow G$. This splits giving $F = G \oplus H$ and the result follows (see [**B**, **SK**]).

Now if $[P] - [Q]$ goes to 0 in $K_0(R_S)$ we will have $P_S \oplus F_S \approx Q_S \oplus F_S$. Replace P and Q by $P \oplus F$ and $Q \oplus F$ so that $P_S \approx Q_S$. Find a map $f : Q \to P$ inducing such an isomorphism. Since S consists of regular elements, $P \to P_S$ is injective and therefore so is f. We get $0 \to Q \to P \to M \to 0$ with $M \in \mathscr{H}_S(R)$ showing that the K_0 sequence is exact.

4. Generalizations

In the topological case, Atiyah and Hirzebruch defined a sequence of K-functors in such a way that the exact sequence of Theorem 3.5 could be extended to a long exact sequence. They defined $K^0(X)$ as $K_0(\mathscr{VB}(X))$ where $\mathscr{VB}(X)$ is the category of vector bundles (real or complex) on X. They then extended this to a cohomology theory by applying K^0 to successive suspensions of X. The suspension ΣX can be defined as the union of 2 cones $C^+(X)$ and $C^-(X)$ with common base X. Vector bundles on ΣX will be trivial on $C^+(X)$ and $C^-(X)$ and so can be described by clutching functions $X \to GL_n(\mathbb{R})$ (or $X \to GL_n(\mathbb{C})$) which tell us how the two trivial bundles should be glued together over the common base X. One takes these modulo homotopy which corresponds to isomorphism of vector bundles. To get a good exact sequence, it turns out that one also has to ignore trivial summands of the vector bundle. This amounts to identifying $X \to GL_n(\mathbb{R})$ with $X \to GL_n(\mathbb{R}) \hookrightarrow GL_{n+1}(\mathbb{R})$ under the natural inclusion $GL_n(\mathbb{R}) \hookrightarrow GL_{n+1}(\mathbb{R})$ by $A \mapsto \begin{pmatrix} A & 0 \\ 0 & 1 \end{pmatrix}$.

Bass proposed a similar definition in the algebraic case. For any ring R let $GL(R) = \operatorname{colim} GL_n(R)$ under the natural inclusions. Bass replaced the subgroup of homotopically trivial elements by the subgroup $E(R)$ generated by elementary matrices $e_{ij}(r) = I + re_{ij}$, $i \neq j$, where $r \in R$ and e_{ij} has 1 in position ij and 0 elsewhere. Whitehead's lemma [**B**, V, 1.7] shows that $E(R) = [GL(R), GL(R)]$ so that $K_1(R) = GL(R)/E(R)$ is an abelian group.

THEOREM 4.1 (Bass [**B**, IX,6.3]). *In the situation of Theorem* 3.5 *we have an exact sequence*

$$K_1(R) \to K_1(R_S) \xrightarrow{\partial} K_0(\mathscr{H}_S(R)) \to K_0(R) \to K_0(R_S).$$

The map ∂ sends $\alpha \in GL_n(R_S)$ to $[\operatorname{ckr}(s\alpha)] - [\operatorname{ckr} sI_n]$ where $s \in S$ is chosen so that $s\alpha \in \mathscr{M}_n(R)$.

Bass also extended his definition to categories by defining a universal determinant functor $BK_1(\mathscr{M})$. This is the abelian group with a generator $[M, \alpha]$ for each $M \in \operatorname{ob}\mathscr{M}$ and automorphism $\alpha : M \approx M$ and with relations

(1) $[M, \alpha\beta] = [M, \alpha] + [M, \beta]$.
(2) $[M, \alpha] = [M', \alpha'] + [M'', \alpha'']$ if

$$
\begin{array}{ccccccccc}
0 & \longrightarrow & M' & \longrightarrow & M & \longrightarrow & M'' & \longrightarrow & 0 \\
& & \alpha' \downarrow \approx & & \alpha \downarrow \approx & & \alpha'' \downarrow \approx & & \\
0 & \longrightarrow & M' & \longrightarrow & M & \longrightarrow & M'' & \longrightarrow & 0
\end{array}
$$

commutes.

This satisfies the Additivity Theorem and special cases of the Dévissage and Resolution Theorems. Under special conditions it can also be used to extend the exact sequences [**B**].

A definition of $K_2(R)$ was then proposed by Milnor [**Mi**]. The elements $e_{ij}(r)$ of $E(R)$ satisfy certain relations due to Steinberg. Let $St(R)$, the Steinberg group of R, be the group generated by symbols $x_{ij}(r)$, $i \neq j$, subject to the Steinberg relations:

(1) $x_{ij}(r + s) = x_{ij}(r)x_{ij}(s)$.
(2) $[x_{ij}(r), x_{k\ell}(s)] = 1$ if $i \neq \ell$, $j \neq k$.
(3) $[x_{ij}(r), x_{jk}(s)] = x_{ik}(rs)$ if $i \neq k$.

We have a map $\varphi : St(R) \to GL(R)$ by $\varphi(x_{ij}(r)) = e_{ij}(r)$. The cokernel is $K_1(R)$ and Milnor defines $K_2(R) = \ker \varphi$ which turns out to be the center of $St(R)$.

A number of definitions were then proposed for the higher K-functors $K_n(R)$ but little could be done with them until a good definition was found by Quillen using a topological construction. He took the classifying space

$BGL(R)$ of $GL(R)$ and adjoined 2-cells to kill the subgroup $E(R)$ of $\pi_1(BGL(R)) = GL(R)$. He then showed that one could adjoin 3-cells in such a way that the homology of the resulting space $BGL^+(R)$ was restored to its original value, that of $BGL(R)$. Quillen defined $K_n(R) = \pi_n(BGL^+(R))$ for $n \geq 1$. This agrees with the definitions of Bass and Milnor for $n = 1, 2$.

Using methods he developed in his proof of the Adams conjecture, Quillen was able to compute $K_n(F)$ for a finite field F.

THEOREM 4.2 ([QC]). *Let q be a prime power. For $n \geq 1$, $K_{2n}(\mathbb{F}_q) = 0$ and $K_{2n-1}(\mathbb{F}_q) = \mathbb{Z}/(q^n - 1)\mathbb{Z}$.*

Quillen also shows that the Frobenius $x \mapsto x^p$ on $\mathbb{F}_q$ induces p^n on $K_{2n-1}(\mathbb{F}_q)$ and that $K_i(\mathbb{F}_q) \xrightarrow{\approx} K_i(\mathbb{F}_{q^d})^{\mathrm{Gal}(\mathbb{F}_{q^d}/\mathbb{F}_q)}$.

The space $BGL^+(R)$ is an H-space so one can get quite a bit of information about its homotopy $K_*(R)$ from information about its homology $H_*(BGL(R))$. By computing this homology, Borel proved the following theorem.

THEOREM 4.3 ([Bo]). *Let R be the ring of integers of a number field with r_1 real places and r_2 complex places. Then $K_0(R) = \mathbb{Z}$, $\operatorname{rk} K_1(R) = r_1 + r_2 - 1$, and, for $n \geq 2$,*

$$\operatorname{rk} K_n(R) = \begin{cases} 0 & \textit{for } n \textit{ even}, \\ r_2 & \textit{for } n \equiv 3 \bmod 4, \\ r_1 + r_2 & \textit{for } n \equiv 1 \bmod 4. \end{cases}$$

Quillen [QF] showed that the groups $K_n(R)$ are finitely generated if R is the ring of integers of a number field.

The space $BGL^+(R)$ is even an infinite loop space. There is a highly developed theory of such spaces [MaE] which can be applied to get further information about this space.

5. Homotopy theory of categories

In order to get results similar to those of §2 for the higher K_i-functors one would like to define $K_i(\mathcal{M})$ for any exact category $\mathcal{M}$ in such a way that $K_i(\mathcal{P}(R)) = K_i(R)$. Once again the correct definition was found by Quillen [Q]. This makes use of simplicial sets and classifying spaces of categories. I will give here a brief sketch of these topics. A detailed exposition of the theory of simplicial sets is given in [Ma].

Let Δ be the category whose objects are the sets $[n] = \{0, 1, \dots, n\}$ for $n \geq 0$ and whose morphisms are nondecreasing functions. A simplicial object in any category $\mathcal{C}$ is defined to be a contravariant functor $T : \Delta^{\mathrm{op}} \to \mathcal{C}$ from Δ to $\mathcal{C}$.

This notion arose from the construction of singular homology in topology. Let $\Delta^n = \{(t_0, \dots, t_n) \in \mathbb{R}^{n+1} | t_i \geq 0, \sum t_i = 1\}$. be the standard n-simplex. If X is a topological space, a singular n-simplex of X is defined to be a continuous map $T : \Delta \to X$. Let $S_n(X)$ be the set of such maps. We will define the structure of a simplicial set on $S_\bullet(X) = \coprod S_n(X)$.

We first define a covariant functor Δ from $\mathbf{\Delta}$ to topological spaces sending $[n]$ to Δ^n. Think of $[n]$ as the set of vertices of Δ^n so that i corresponds to $v_i = (0, \dots, 0, 1, 0, \dots, 0)$ with 1 in the i-th position. Each map $\alpha : [m] \to [n]$ then extends linearly to $\alpha_* : \Delta^m \to \Delta^n$. Explicitly, α_* sends $(t_0, \dots, t_m) = \sum t_i v_i$ to $\sum t_i v_{\alpha(i)}$ so that $\alpha_*(t_0, \dots, t_m) = (u_0, \dots, u_m)$ where $u_j = \sum t_i$ over i with $\alpha(i) = j$.

The composite functor $\mathbf{\Delta} \xrightarrow{\Delta} Top \xrightarrow{\mathrm{Hom}(-,X)} Sets$ takes $[n]$ to $S_n(X)$ and defines a simplicial set $S_\bullet(X)$, the singular complex of X. The functor $S_\bullet : Top \to SimplSets$ has a left adjoint, the geometric realization functor. Given a simplicial set X we take a copy of Δ^n for each $x \in X_n$ and glue these together using the maps $X(\alpha)$. Explicitly, $|X| = \coprod (X_n \times \Delta^n)/\sim$ where X_n is given the discrete topology and the equivalence relation $\sim$ is defined by $(\alpha^* x, t) \sim (x, \alpha_* t)$ where $\alpha^* = X(\alpha)$, $\alpha_* = \Delta(\alpha)$. It is easy to verify that $\mathrm{Hom}_{Top}(|X|, Y) = \mathrm{Hom}_{SimplSets}(X, S_\bullet(Y))$.

The construction of $S_\bullet(X)$ can be imitated for categories. Let $\mathscr{C}at$ be the category of small categories and define a functor $C : \mathbf{\Delta} \to \mathscr{C}at$, the analogue of Δ, by sending $[n]$ to $[n] = \{0, 1, \dots, n\}$ regarded as a category. Note that any partially ordered set P can be regarded as a category whose objects are the elements of P and where $\mathrm{Hom}(x, y)$ has one element if $x \le y$ and $\mathrm{Hom}(x, y) = \varnothing$ otherwise.

If $\mathscr{C}$ is any small category the composite functor $\mathbf{\Delta} \xrightarrow{C} \mathscr{C}at \xrightarrow{\mathrm{Hom}(-, \mathscr{C})} Sets$ defines a simplicial set, the nerve $N\mathscr{C}$ of $\mathscr{C}$. Explicitly $N_n\mathscr{C}$ is the set of all diagrams $A_0 \to A_1 \to \cdots \to A_n$ in $\mathscr{C}$. If $\alpha : [m] \to [n]$ then $\alpha_*(A_0 \to \cdots \to A_n) = (A_{\alpha(0)} \to \cdots \to A_{\alpha(n)})$ where $A_{\alpha(i)} \to A_{\alpha(i+1)}$ is 1 if $\alpha(i+1) = \alpha(i)$ and otherwise is the composition $A_{\alpha(i)} \to A_{\alpha(i)+1} \to \cdots \to A_{\alpha(i+1)}$.

DEFINITION. The classifying space $B\mathscr{C}$ of $\mathscr{C}$ is the geometric realization $|N\mathscr{C}|$ of the nerve of $\mathscr{C}$.

If $*$ is an object of $\mathscr{C}$, the diagram $*$ in $N_0\mathscr{C} = \mathrm{ob}\,\mathscr{C}$ represents a vertex of $N\mathscr{C}$ and so a point of $B\mathscr{C}$. We define the homotopy groups of $\mathscr{C}$ with basepoint $*$ to be $\pi_i(\mathscr{C}, *) = \pi_i(B\mathscr{C}, *)$. I will usually omit references to the basepoint here. We usually choose $*$ to be a zero object of $\mathscr{C}$ when this makes sense.

The classifying space construction has many nice properties. For example $B(\mathscr{A} \times \mathscr{B}) = B(\mathscr{A}) \times B(\mathscr{B})$. Also $B(\mathscr{C}^{\mathrm{op}}) = B(\mathscr{C})$ since if we reverse the arrows in $\mathscr{C}$, the elements of $N\mathscr{C}$ remain the same. The meaning of α_* changes, of course, but the patching needed to construct $B\mathscr{C}$ remains the same.

LEMMA 5.1. Let $F, G : \mathscr{A} \to \mathscr{B}$. If there is a natural transformation $\eta : F \to G$ then $BF \simeq BG : B\mathscr{A} \to B\mathscr{B}$.

PROOF. Let $\mathcal{J}$ be the ordered set $\{0, 1\}$ considered as a category. Define $H : \mathcal{J} \times \mathcal{A} \to \mathcal{B}$ by $H(0 \times A) = F(A)$, $H(1 \times A) = G(A)$, and $H((0 \to 1) \times 1_A) = \eta_A : F(A) \to G(A)$. Then if $\alpha : A \to B$, $0 \times A \xrightarrow{0 \times \alpha} 0 \times B$ is sent to $F(\alpha)$, $1 \times A \xrightarrow{1 \times \alpha} 1 \times B$ is sent to $G(\alpha)$, and the map $0 \times A \to 1 \times B$, which factors as

$$0 \times A \longrightarrow 0 \times B$$
$$\downarrow \qquad\qquad \downarrow$$
$$1 \times A \longrightarrow 1 \times B$$

is sent to either composition in the commutative diagram

$$
\begin{array}{ccc}
F(A) & \xrightarrow{F(\alpha)} & F(B) \\
\eta_A \downarrow & & \downarrow \eta_B \\
G(A) & \xrightarrow{G(\alpha)} & G(B)
\end{array}
$$

Now $N\mathcal{J}$ is just the standard (simplicial) 1-simplex so $B\mathcal{J} = \Delta^1 = I$, the usual unit interval. We get $I \times B\mathcal{A} = B(\mathcal{J} \times A) \to B\mathcal{B}$ giving the required homotopy between BF and BG.

COROLLARY 5.2. *If* $\mathcal{A} \underset{G}{\overset{F}{\rightleftarrows}} \mathcal{B}$ *are adjoint functors, they induce homotopy equivalences* $B\mathcal{A} \underset{BG}{\overset{BF}{\rightleftarrows}} B\mathcal{B}$.

In fact, the adjunction maps $FG \to 1_{\mathcal{B}}$ and $1_{\mathcal{A}} \to GF$ become homotopies $BF \circ BG \simeq 1$ and $1 \simeq BG \circ BF$.

COROLLARY 5.3. *If* $\mathcal{A}$ *and* $\mathcal{B}$ *are equivalent categories then* $B\mathcal{A} \simeq B\mathcal{B}$.

COROLLARY 5.4. *If* $\mathcal{C}$ *has an initial object or final object then* $B\mathcal{C}$ *is contractible.*

If $*$ is initial we have adjoint functors $* \rightleftarrows \mathcal{C}$ and similarly for final objects.

Deeper results about $B\mathcal{C}$ depend on the following theorems of Quillen [Q]. If $F : \mathcal{A} \to \mathcal{B}$ and b is an object of $\mathcal{B}$ we define F/b to be the category whose objects are pairs (a, u) where $a \in \mathrm{ob}\,\mathcal{A}$ and $u : F(a) \to b$. A map $(a, u) \to (a', u')$ means a map $a \to a'$ such that

$$
\begin{array}{ccc}
F(a) & \longrightarrow & F(a') \\
\downarrow & & \downarrow \\
B & =\!=\!= & B
\end{array}
$$

commutes. This category F/b serves as a sort of fiber of F.

We can also define $b \backslash F$ whose objects are (a, u) with $u : b \to F(a)$. This is the same as the previous construction applied to $F : \mathcal{A}^{\mathrm{op}} \to \mathcal{B}^{\mathrm{op}}$.

In the following theorem, "contractible" means contractible to a point and, in particular, implies that the space is nonempty.

QUILLEN'S THEOREM A. *If $B(F/b)$ is nonempty and contractible for all $b \in \text{ob} \mathscr{B}$ then $BF : B\mathscr{A} \to B\mathscr{B}$ is a homotopy equivalence.*

By considering $F : \mathscr{A}^{\text{op}} \to \mathscr{B}^{\text{op}}$ we get a similar theorem involving $B(b \backslash F)$.

In order to get long exact sequences, we must also consider fibrations. If $f : X \to Y$ is a map of topological spaces, the homotopy fiber $F_y(f)$ over a point y of Y is defined to be the set of (x, ω) in $X \times Y^I$ with $f(x) = \omega(0)$ and $\omega(1) = y$. By standard homotopy theory one gets a long exact sequence of homotopy groups.

$$\cdots \to \pi_n(F(f)) \to \pi_n(X) \to \pi_n(Y) \xrightarrow{\partial} \pi_{n-1}(F(f)) \to \cdots.$$

Now consider a functor $F : \mathscr{A} \to \mathscr{B}$. We have a functor $F/b \to \mathscr{A}$ sending (a, u) to a and two functors $F/b \to \mathscr{B}$ sending (a, u) to $F(a)$ and to b (the constant functor). The map $u : F(a) \to b$ gives a natural transformation between these two functors from F/b to $\mathscr{B}$. By Lemma 5.1 this gives us a map $I \times B(F/b) \to B\mathscr{B}$ or, equivalently, a map $B(F/b) \to B\mathscr{B}^I$. Combining this with $B(F/b) \to B\mathscr{A}$, we get a map $B(F/b) \to B\mathscr{A} \times B\mathscr{B}^I$ and it is easy to see that this is a map $B(F/b) \to F_b(BF)$.

If $\beta : b \to b'$ in $\mathscr{B}$ we get a functor $F/\beta : F/b \to F/b'$ sending (a, u) with $u : f(a) \to b$ to $(a, \beta \circ u)$.

QUILLEN'S THEOREM B. *If $B(F/\beta) : B(F/b) \xrightarrow{\simeq} B(F/b')$ for all morphisms β in $\mathscr{B}$, then for all $b \in \text{ob} \mathscr{B}$, $B(F/b) \xrightarrow{\simeq} F_b(BF)$.*

It follows that we get a long exact sequence of homotopy groups

$$\cdots \to \pi_{n+1}(\mathscr{B}) \xrightarrow{\partial} \pi_n(F/b) \to \pi_n(\mathscr{A}) \to \pi_n(\mathscr{B}) \xrightarrow{\partial} \cdots.$$

In applying this theorem we usually want to identify $B(F/b)$ with the classifying space of some other category of interest, at least up to homotopy. This is often done as follows. The categories used will have a special class of objects, the "zero objects", singled out. These will all be isomorphic and $\text{Hom}(0, 0)$ will have just one element for any zero object 0.

Suppose we are given two functors $\mathscr{A} \xrightarrow{F} \mathscr{B} \xrightarrow{G} \mathscr{C}$ which preserve zero objects and are such that $GF(\mathscr{A})$ consists of zero objects. Choose a zero object 0 in $\mathscr{C}$ and define a functor $\mathscr{A} \to G/0$ sending a to $(F(a), u)$ where $u : 0 \to GF(a)$ is the unique map between zero objects.

DEFINITION. I will say that $\mathscr{A} \xrightarrow{F} \mathscr{B} \xrightarrow{G} \mathscr{C}$ is a standard homotopy fibration if F and G preserve zero objects, $GF(\mathscr{A})$ consists of zero objects, Quillen's Theorem B applies to G, and $B\mathscr{A} \xrightarrow{\simeq} B(G/0)$.

In this case we have a long exact sequence

$$\cdots \to \pi_{n+1}(\mathscr{C}) \xrightarrow{\partial} \pi_n(\mathscr{A}) \to \pi_n(\mathscr{B}) \to \pi_n(\mathscr{C}) \xrightarrow{\partial} \cdots$$

where all basepoints are chosen to be zero objects. This sequence is natural in the following sense.

THEOREM 5.5 [SQ]. *Let $\mathscr{A} \to \mathscr{B} \to \mathscr{C}$ and $\mathscr{D} \to \mathscr{E} \to \mathscr{F}$ be standard homotopy fibrations. Let*

$$
\begin{array}{ccc}
\mathscr{A} & \longrightarrow \mathscr{B} & \longrightarrow \mathscr{C} \\
\downarrow & \downarrow & \downarrow \\
\mathscr{D} & \longrightarrow \mathscr{E} & \longrightarrow \mathscr{F}
\end{array}
$$

commute up to natural isomorphism and assume that the three vertical functors preserve zero maps. Then we have a commutative diagram

$$
\begin{array}{ccccccc}
\cdots \longrightarrow & \pi_{n+1}(\mathscr{C}) & \xrightarrow{\partial} & \pi_n(\mathscr{A}) & \longrightarrow & \pi_n(\mathscr{B}) & \longrightarrow & \pi_n(\mathscr{C}) & \xrightarrow{\partial} \cdots \\
& \downarrow & & \downarrow & & \downarrow & & \downarrow & \\
\cdots \longrightarrow & \pi_{n+1}(\mathscr{F}) & \xrightarrow{\partial} & \pi_n(\mathscr{D}) & \longrightarrow & \pi_n(\mathscr{E}) & \longrightarrow & \pi_n(\mathscr{F}) & \xrightarrow{\partial} \cdots
\end{array}
$$

I will say that $\mathscr{A} \xrightarrow{F} \mathscr{B} \xrightarrow{G} \mathscr{C}$ is a pseudo-standard homotopy fibration if it satisfies all the conditions of the above definition except that the last condition is weakened to $(B\mathscr{A})_0 \xrightarrow{\simeq} (B(G/0))_0$ where $(\ )_0$ denotes the connected component containing the zero objects. Theorem 5.5 also applies to this case, the only difference being that the long exact sequences end at $\pi_1\mathscr{C}$ and $\pi_1\mathscr{F}$.

6. Higher K-Theory

Let $\mathscr{M}$ be an exact category. Quillen [Q] defines the higher K-groups of $\mathscr{M}$ as the homotopy groups of an associated category $Q\mathscr{M}$. Note that the homotopy groups of $\mathscr{M}$ itself are all 0 since a zero object of $\mathscr{M}$ is initial and final.

DEFINITION. If $\mathscr{M}$ is an exact category define $Q\mathscr{M}$ to be the category with the same objects as $\mathscr{M}$ and with morphisms defined as follows: A morphism from M to N in $Q\mathscr{M}$ is an equivalence class of diagrams of the form

(1) $M \leftarrow X \rightarrowtail N$

in $\mathscr{M}$.

Here, as in §1, $\leftarrow$ denotes an admissible epimorphism (with kernel in $\mathscr{M}$) and $\rightarrowtail$ denotes an admissible monomorphism (with cokernel in $\mathscr{M}$). We say that $M \leftarrow X \rightarrowtail N$ and $M \leftarrow Y \rightarrowtail N$ are equivalent if there is an isomorphism $X \approx Y$ making

$$M \leftarrow X \rightarrowtail N$$

$$\| \quad \approx\downarrow \quad \|$$

$$M \leftarrow Y \rightarrowtail N$$

commutative.

To compose $L \leftarrow X \rightarrowtail M$ and $M \leftarrow Y \rightarrowtail N$ we form the pullback Z in

$$Z \rightarrowtail Y \rightarrowtail N$$
$$\downarrow \quad \downarrow$$
$$X \rightarrowtail M$$
$$\downarrow$$
$$L.$$

Then $L \leftarrow Z \rightarrowtail M$ represents the composition.

REMARK. We could equally well define a morphism from M to N in $Q\mathscr{M}$ by an equivalence class of diagrams

(2) $M \rightarrowtail Y \leftarrow N.$

In fact, forming the pushout of (1) or the pullback of (2) gives

$$X \rightarrowtail N$$
$$\downarrow \quad \downarrow$$
$$M \rightarrowtail Y$$

showing that either diagram determines the other up to equivalence. In particular, $(Q\mathscr{M})^{op} \approx Q(\mathscr{M}^{op})$.

DEFINITION ([Q]). If $\mathscr{M}$ is an exact category define $K_i(\mathscr{M}) = \pi_{i+1}(BQ\mathscr{M})$ for $i \geq 0$.

THEOREM 6.1 ([Q]). *These functors have all the properties stated for K_0 in Theorems* 2.1, 2.2, 2.6, *and* 2.8.

In other words, aside from the elementary properties of Theorem 2.1, these functors satisfy the Additivity Theorem, the Dévissage Theorem, and the Resolution Theorem. It follows that these functors also have the properties given in Corollaries 2.3, 2.4, 2.5, 2.7, and 2.9. The proofs given in §2 apply without any change.

THEOREM 6.2 ([Q]). *If $\mathscr{S}$ is a Serre subcategory of an abelian category $\mathscr{A}$ then $Q\mathscr{S} \to Q\mathscr{A} \to Q(\mathscr{A}/\mathscr{S})$ is a standard homotopy fibration.*

DEFINITION. I will say that $\mathscr{A} \xrightarrow{F} \mathscr{B} \xrightarrow{G} \mathscr{C}$ is a K-fibration if $\mathscr{A}$, $\mathscr{B}$, $\mathscr{C}$ are exact categories, F and G are exact functors, $GF(\mathscr{A})$ consists of zero objects, and $Q\mathscr{A} \to Q\mathscr{B} \to Q\mathscr{C}$ is a standard homotopy fibration.

For example $\mathscr{S} \to \mathscr{A} \to \mathscr{A}/\mathscr{S}$ in Theorem 6.2 is a K-fibration. Any K-fibration gives rise to a long exact sequence

$$\cdots \to K_{n+1}(\mathscr{C}) \xrightarrow{\partial} K_n(\mathscr{A}) \to K_n(\mathscr{B}) \to K_n(\mathscr{C}) \xrightarrow{\partial} \cdots$$
$$\to K_0(\mathscr{B}) \to K_0(\mathscr{C}) \to 0.$$

Moreover if $\mathscr{A} \to \mathscr{B} \to \mathscr{C}$ and $\mathscr{D} \to \mathscr{E} \to \mathscr{F}$ are K-fibrations and if

$$\begin{array}{ccccc}
\mathscr{A} & \longrightarrow & \mathscr{B} & \longrightarrow & \mathscr{C} \\
\downarrow & & \downarrow & & \downarrow \\
\mathscr{D} & \longrightarrow & \mathscr{E} & \longrightarrow & \mathscr{F}
\end{array}$$

commutes up to natural isomorphism with the vertical functors exact, then we have a commutative ladder

$$\begin{array}{ccccccccc}
\cdots & \longrightarrow & K_{n+1}(\mathscr{C}) & \xrightarrow{\partial} & K_n(\mathscr{A}) & \longrightarrow & K_n(\mathscr{B}) & \longrightarrow & K_n(\mathscr{C}) & \xrightarrow{\partial} & \cdots \\
& & \downarrow & & \downarrow & & \downarrow & & \downarrow \\
\cdots & \longrightarrow & K_{n+1}(\mathscr{F}) & \xrightarrow{\partial} & K_n(\mathscr{D}) & \longrightarrow & K_n(\mathscr{E}) & \longrightarrow & K_n(\mathscr{F}) & \xrightarrow{\partial} & \cdots
\end{array}$$

I will say that $\mathscr{A} \xrightarrow{F} \mathscr{B} \xrightarrow{G} \mathscr{C}$ is a pseudo-K-fibration if $Q\mathscr{A} \to Q\mathscr{B} \to Q\mathscr{C}$ is a pseudo-standard homotopy fibration. The above results also hold in this case except that the long exact sequences end at $K_0(\mathscr{C})$ and $K_0(\mathscr{F})$ with no 0 at the right.

The notations $\mathscr{H}_S(R)$ and $\mathscr{P}_S(R)$ will have the same meaning as in Lemma 3.6.

THEOREM 6.3 ([GQ]). *Let S be a central multiplicative set consisting of regular elements of R. Then $\mathscr{H}_S(R) \to \mathscr{P}_S(R) \to \mathscr{P}(R_S)$ is a pseudo K-fibration.*

COROLLARY 6.4. *Under the hypothesis of Theorem 6.3 there is a long exact sequence*

$$\cdots \to K_{n+1}(R_S) \xrightarrow{\partial} K_n(\mathscr{H}_S(R)) \to K_n(R) \to K_n(R_S) \xrightarrow{\partial} \cdots$$
$$\to K_0(\mathscr{H}_S(R)) \to K_0(R) \to K_0(R_S).$$

We can replace $K_n(\mathscr{P}_S(R))$ by $K_n(R)$ by Lemma 3.7.

REMARK. The version of Theorem 6.3 given in [SQ, Prop. A13] makes use of an auxiliary category $\mathscr{V}$ defined in [GQ] having the property that $K_q(\mathscr{V}) = K_q(R_S)$ for $q \geq 1$ and $K_0(\mathscr{V}) = \operatorname{im}[K_0(R) \to K_0(R_S)]$. This yields an actual K-fibration $\mathscr{H}_S(R) \to \mathscr{P}_S(R) \to \mathscr{V}$.

In general, $K_0(R) \to K_0(R_S)$ is not onto. For example, Lindel has shown that $K_0(R) = \mathbb{Z}$ for $R = \mathbb{C}[w, x, y, z]/(wx - yz)$. In fact all finitely generated projective R-modules are free [SG, Th. 3.1]. Let $u = \frac{1}{2}(w + z)$, $v = \frac{1}{2}(w - z)$. Then $R = \mathbb{C}[u, v, x, y]/(u^2 - v^2 - xy)$ and $R_u = A[u, u^{-1}]$ where $A = \mathbb{C}[\xi, \eta, \zeta]/(\xi\eta + \zeta^2 - 1)$ with $\xi = x/u$, $\eta = y/u$, $\zeta = v/u$.

Therefore $K_0(R_u) = K_0(A) = \mathbb{Z} \oplus \mathbb{Z}$ by [SM, Cor.10.7] and [SM, Lemma 8.10(i)].

There is also an analogue of Theorem 6.3 for sheaves. Let X be a quasi-projective scheme. Let $Y \subset X$ be an effective Cartier divisor, i.e. a sub-scheme defined locally by a single equation $s = 0$ where s is a nonzero-divisor. Let $U = X - Y$ and assume that U is affine. Let $\mathscr{P}(X)$ be the category of locally free sheaves on X and let $\mathscr{P}_Y(X)$ be the category of quasicoherent sheaves $\mathscr{M}$ on X such that

  (1) $\mathscr{M}|U$ is locally free.
  (2) There is a resolution $0 \to \mathscr{Q} \to \mathscr{P} \to \mathscr{M} \to 0$ with $\mathscr{P}$ and $\mathscr{Q}$ locally free.

Let $\mathscr{H}_Y(X)$ be the full subcategory of $\mathscr{P}_Y(X)$ of all $\mathscr{M}$ with $\mathscr{M}|U = 0$. As in Lemma 3.7, $\mathscr{P}_Y(X) \hookrightarrow \mathscr{P}(X)$ induces an isomorphism $K_i(\mathscr{P}_Y(X)) \xrightarrow{\approx} K_i(\mathscr{P}(X)) = K_i(X)$.

THEOREM 6.5 ([GQ, SQ]). $\mathscr{H}_Y(X) \to \mathscr{P}_Y(X) \to \mathscr{P}(X)$ is a pseudo K-fibration.

See also Theorem 9.14 for a variant of this theorem applicable to the "noncommutative projective line".

COROLLARY 6.6. There is a long exact sequence

$$\cdots \to K_{n+1}(U) \xrightarrow{\partial} K_n(\mathscr{H}_Y(X)) \to K_n(X) \to K_n(U) \xrightarrow{\partial} \cdots \to K_0(X) \to K_0(U).$$

The final theorem of this section shows that the two definitions of higher K-theory given by Quillen are compatible.

THEOREM 6.7 ([GQ]). The loop space $\Omega BQ\mathscr{P}(R)$ has the homotopy type of $BGL^+(R) \times K_0(R)$.

COROLLARY 6.8. $K_i(\mathscr{P}(R)) = K_i(R)$ for all $i \geq 0$.

For the proofs of all the above results we refer the reader to Quillen's original papers [Q, GQ], or to [S].

7. Calculation of K_0

It is not at all obvious that the K_0 of Quillen's theory agrees with that of Grothendieck. The proofs of this given in [Q] and [S] make use of covering space theory. I will give a more direct proof here.

Recall that a tree is a connected graph containing no cycles. It is easy to see that any connected CW-complex X (such as the geometric realization of a connected simplicial set) contains a maximal tree T composed of edges. This will contain all vertices since a vertex not in T could be joined to T by a path along edges and T could be enlarged by adding the last edge in this path to T.

I will say that a 2-cell c of X is attached regularly if its boundary is an exact union of a finite number of edges, say $e_1, \ldots, e_n$, in order. Orient the edges and c and let $\varepsilon_i = 1$ if the orientation of e_i agrees with that of the boundary of c. Let $\varepsilon_i = -1$ if the orientations are opposite.

The following is the classical algorithm for computing the fundamental group of X [ST].

THEOREM 7.1. *Let X be a connected CW-complex with all 2-cells attached regularly. Let T be a maximal tree of X and let v be a vertex of X. Orient all edges and 2-cells of X. Then $\pi_1(X, v)$ is generated by symbols $[e]$ corresponding to the edges e of X with the following relations:*

(1) $[e] = 1$ *if e lies in T.*
(2) $[\partial c] := [e_1]^{\varepsilon_1} \ldots [e_n]^{\varepsilon_n} = 1$ *for each 2-cell c of X.*

In (2) the e_i and ε_i are as given above. We can also allow 2-cells which are attached trivially; i.e. the boundary is a point. Such cells do not affect $\pi_1(X, v)$ and we set $[\partial c] = 1$ for them.

We will now apply this theorem to compute $\pi_1(BQ\mathscr{C})$ for any exact category $\mathscr{C}$. If $\mathscr{A}$ is any category, the vertices of $B\mathscr{A}$ correspond to the objects of $\mathscr{A}$, the edges correspond to the morphisms in $\mathscr{A}$, and the 2-cells are triangles corresponding to diagrams $A \xrightarrow{f} B \xrightarrow{g} C$ in $\mathscr{A}$, the sides being f, g, and the composition. It is conventional to define the product in π_1 so that the path $a \xrightarrow{f} b \xrightarrow{g} c$ is written as fg. In order to preserve this convention I will adopt, for this section only, the convention that morphisms are written on the right so that the composition $A \xrightarrow{f} B \xrightarrow{g} C$ will be written fg.

COROLLARY 7.2. *Let $\mathscr{A}$ be a category such that $B\mathscr{A}$ is connected. Choose a maximal tree T in $B\mathscr{A}$. Let a be an object of $\mathscr{A}$ regarded as a vertex of $B\mathscr{A}$. Then $\pi_1(B\mathscr{A}, a)$ is generated by symbols $[f]$ for all morphisms f of $\mathscr{A}$ with relations $[fg] = [f][g]$ and with $[f] = 1$ if f lies in T or if f is an identity map.*

Note that identity maps correspond to degenerate edges of $B\mathscr{A}$ which are reduced to a point.

To apply this to $Q\mathscr{C}$, we first introduce (with apologies to Dirac) some notation for maps in $Q\mathscr{C}$. If $m : A \rightarrowtail B$ is an admissible monomorphism of $\mathscr{C}$, let $|m\rangle$ be the morphism $\xleftarrow{=} A \xrightarrow{m} B$ of $Q\mathscr{C}$. If $e : A \twoheadrightarrow B$ is an admissible epimorphism, let $\langle e|$ be $B \xleftarrow{e} A \xrightarrow{=} A$ in $Q\mathscr{C}$. Any morphism $A \xleftarrow{e} X \xrightarrow{m} B$ in $Q\mathscr{C}$ can be written $\langle e| \circ |m\rangle$ or $\langle e|m\rangle$ for short. This follows easily from the definition of composition in $Q\mathscr{C}$ as do the relations $|m\rangle|m'\rangle = |mm'\rangle$ and $\langle e|\langle e'| = \langle e'e|$.

Fix a zero object 0 of $\mathscr{C}$ and for each object A of $\mathscr{C}$ consider the maps $b_A : 0 \rightarrowtail A$ and $t_A : A \twoheadrightarrow 0$. In $Q\mathscr{C}$ let $\beta_A = |b_A\rangle$ and $\tau_A = \langle t_A|$ be the corresponding maps of 0 to A. The notation is intended to suggest that the 0 subquotient of A occurs at the bottom or top of A.

We choose the set of all β_A as our maximal tree T. Note that β_A is an identity map if A is the chosen zero object 0 and so is a degenerate edge not a loop. By Corollary 7.2, $\pi_1 BQ\mathscr{C}$ is generated by all $[f]$, $f \in \mathrm{mor}(Q\mathscr{C})$, with relations $[fg] = [f][g]$, $[\beta_A] = 1$ for all A, and $[1_A] = 1$ for all A.

If $m : A \rightarrowtail B$ is an admissible monomorphism in $\mathscr{C}$ then $b_A \circ m = b_B$ so $\beta_A \circ |m\rangle = \beta_B$ showing that $[|m\rangle] = 1$ in π_1. If $e : A \twoheadrightarrow B$ is an admissible epimorphism in $\mathscr{C}$ then $e \circ t_B = t_A$ so $\tau_B \circ \langle e| = \tau_A$ showing that $[\langle e|] = [\tau_B]^{-1}[\tau_A]$. Therefore π_1 is generated by the elements $[\tau_A]$ which I will write as $[A]$.

LEMMA 7.3. *If* $0 \to A \overset{i}{\rightarrowtail} B \overset{j}{\twoheadrightarrow} C \to 0$ *is a short exact sequence in* $\mathscr{C}$, *then* $\beta_C \circ \langle j| = \tau_A \circ |i\rangle$ *in* $Q\mathscr{C}$

PROOF. The left side is computed by forming the pullback diagram

This clearly gives the composition on the right.

It follows that $[\langle j|] = [\tau_A] = [A]$ but $[\langle j|] = [C]^{-1}[B]$ so we have the relation $[C][A] = [B]$ in π_1. In particular $[C][A] = [A \oplus C] = [C \oplus A] = [A][C]$ showing that π_1 is commutative and we have a surjection $K_0(\mathscr{C}) \twoheadrightarrow \pi_1$.

Define a map $\pi_1 \to K_0(\mathscr{C})$ by sending $[f]$ with f represented by $A \overset{e}{\twoheadleftarrow} X \overset{m}{\rightarrowtail} B$ to $[\ker e]$. The relations $[1_A] = 1$ and $[\beta_A] = 1$ are clearly preserved. To see that $[fg] = [f][g]$ is preserved, consider the diagram defining fg where g is represented by $B \overset{e'}{\twoheadleftarrow} Y \overset{m'}{\rightarrowtail} C$. We get

so $[fg] \mapsto [\ker(e''e)]$. The exact sequence $0 \to \ker e'' \to \ker(e''e) \to \ker e \to 0$ shows that the relation $[fg] = [f][g]$ maps to the relation $[\ker(e''e)] = [\ker e] + [\ker e']$ in $K_0(\mathscr{C})$.

It is now trivial to check that the maps $K_0 \mapsto \pi_1$ and $\pi_1 \mapsto K_0$ are inverse isomorphisms.

THEOREM 7.4 ([**Q**]). *For any exact category* $\mathscr{C}$, $\pi_1(BQ\mathscr{C})$ *is the Grothendieck group* $K_0(\mathscr{C})$.

8. Products

In [**W**], Waldhausen constructed a delooping $QQ\mathscr{C}$ of the Q construction. This has the property that $K_i(\mathscr{C}) = \pi_{i+2}(BQQ\mathscr{C})$. If $\mathscr{A}$, $\mathscr{B}$, $\mathscr{C}$ are exact categories and $T : \mathscr{A} \times \mathscr{B} \to \mathscr{C}$ is a biexact functor (exact in each variable), Waldhausen constructs a product map $BQ\mathscr{A} \wedge BQ\mathscr{B} \to \mathscr{B}QQ\mathscr{C}$ inducing a bilinear product $K_p(\mathscr{A}) \times K_q(\mathscr{B}) \to K_{p+q}(\mathscr{C})$. Further discussion of these products may be found in [**Gr, W, We**].

The maps $K_0(\mathscr{A}) \times K_q(\mathscr{B}) \to K_q(\mathscr{C})$ and $K_p(\mathscr{A}) \times K_0(\mathscr{B}) \to K_p(\mathscr{C})$ agree with those obtained from the additivity theorem. In other words, $K_0(\mathscr{A}) \times K_q(\mathscr{B}) \to K_q(\mathscr{C})$ sends $([A], \xi)$ to the image of ξ in $K_q(\mathscr{C})$ under the map induced by $T(A, -) : \mathscr{B} \to \mathscr{C}$ and similarly for $K_p(\mathscr{A}) \times K_0(\mathscr{B}) \to K_p(\mathscr{C})$. The additivity theorem shows directly that these maps are well defined.

These products have many naturality properties. Here are some copied from Grayson's paper [**Gr**].

(1) NATURALITY. If $F : \mathscr{A} \to \mathscr{A}'$, $G : \mathscr{B} \to \mathscr{B}'$, $H : \mathscr{C} \to \mathscr{C}'$ are exact, T, T' are biexact, and

$$
\begin{array}{ccc}
\mathscr{A} \times \mathscr{B} & \xrightarrow{\ T\ } & \mathscr{C} \\
{\scriptstyle F \times G}\downarrow & & \downarrow{\scriptstyle H} \\
\mathscr{A}' \times \mathscr{B}' & \xrightarrow{\ T'\ } & \mathscr{C}'
\end{array}
$$

commutes up to natural isomorphism, then

$$
\begin{array}{ccc}
K_p(\mathscr{A}) \times K_q(\mathscr{B}) & \longrightarrow & K_{p+q}(\mathscr{C}) \\
\downarrow & & \downarrow \\
K_p(\mathscr{A}') \times K_q(\mathscr{B}') & \longrightarrow & K_{p+q}(\mathscr{C}')
\end{array}
$$

commutes.

(2) ASSOCIATIVITY. Suppose $f : \mathscr{A} \times \mathscr{B} \to \mathscr{D}$, $g : \mathscr{B} \times \mathscr{C} \to \mathscr{E}$, $h : \mathscr{D} \times \mathscr{C} \to \mathscr{F}$. $k : \mathscr{A} \times \mathscr{E} \to \mathscr{F}$ are biexact. If

$$
\begin{array}{ccc}
\mathscr{A} \times \mathscr{B} \times \mathscr{C} & \xrightarrow{\ f \times 1\ } & \mathscr{D} \times \mathscr{C} \\
{\scriptstyle 1 \times g}\downarrow & & \downarrow{\scriptstyle h} \\
\mathscr{A} \times \mathscr{E} & \xrightarrow{\ k\ } & \mathscr{F}
\end{array}
$$

commutes up to natural isomorphism, then

$$
\begin{array}{ccc}
K_p(\mathscr{A}) \times K_q(\mathscr{B}) \times K_r(\mathscr{C}) & \longrightarrow & K_{p+q}(\mathscr{D}) \times K_r(\mathscr{C}) \\
\downarrow & & \downarrow \\
K_p(\mathscr{A}) \times K_{q+r}(\mathscr{E}) & \longrightarrow & K_{p+q+r}(\mathscr{F})
\end{array}
$$

commutes.

(3) COMMUTATIVITY. If $T : \mathscr{A} \times \mathscr{B} \to \mathscr{C}$ is biexact and $\tilde{T} : \mathscr{B} \times \mathscr{A} \to \mathscr{C}$ by $\tilde{T}(B, A) = T(A, B)$ then

$$
\begin{array}{ccc}
K_p(\mathscr{A}) \times K_q(\mathscr{B}) & \longrightarrow & K_{p+q}(\mathscr{C}) \\
\downarrow & & \| \\
K_q(\mathscr{B}) \times K_p(\mathscr{A}) & \longrightarrow & K_{p+q}(\mathscr{C})
\end{array}
$$

commutes up to a sign $(-1)^{pq}$.

(4) PRESERVATION OF ∂. Let $\mathscr{A} \xrightarrow{F} \mathscr{B} \xrightarrow{G} \mathscr{C}$ and $\mathscr{A}' \xrightarrow{F'} \mathscr{B}' \xrightarrow{G'} \mathscr{C}'$ be (pseudo) K-fibrations. Let $\mathscr{A} \times \mathscr{E} \to \mathscr{A}'$, $\mathscr{B} \times \mathscr{E} \to \mathscr{B}'$, $\mathscr{C} \times \mathscr{E} \to \mathscr{C}'$ be biexact such that

$$
\begin{array}{ccccc}
\mathscr{A} \times \mathscr{E} & \xrightarrow{F \times 1} & \mathscr{B} \times \mathscr{E} & \xrightarrow{G \times 1} & \mathscr{C} \times \mathscr{E} \\
\downarrow & & \downarrow & & \downarrow \\
\mathscr{A}' & \longrightarrow & \mathscr{B}' & \longrightarrow & \mathscr{C}'
\end{array}
$$

commutes up to natural isomorphism. Then

$$
\begin{array}{ccccccccc}
\cdots \to K_p(\mathscr{A}) \times K_q(\mathscr{E}) & \to & K_p(\mathscr{B}) \times K_q(\mathscr{E}) & \to & K_p(\mathscr{C}) \times K_q(\mathscr{E}) & \xrightarrow{\partial \times 1} & K_{p-1}(\mathscr{A}) \times K_q(\mathscr{E}) & \to & \cdots \\
\downarrow & & \downarrow & & \downarrow & & \downarrow & & \\
\cdots \to \quad K_{p+q}(\mathscr{A}') & \to & K_{p+q}(\mathscr{B}') & \to & K_{p+q}(\mathscr{C}') & \xrightarrow{\partial} & K_{p+q-1}(\mathscr{A}') & \to & \cdots
\end{array}
$$

commutes.

If instead we have $\mathscr{E} \times \mathscr{A} \to \mathscr{A}'$, $\mathscr{E} \times \mathscr{B} \to \mathscr{B}'$, $\mathscr{E} \times \mathscr{C} \to \mathscr{C}'$, and

$$
\begin{array}{ccccc}
\mathscr{E} \times \mathscr{A} & \longrightarrow & \mathscr{E} \times \mathscr{B} & \longrightarrow & \mathscr{E} \times \mathscr{C} \\
\downarrow & & \downarrow & & \downarrow \\
\mathscr{A}' & \longrightarrow & \mathscr{B}' & \longrightarrow & \mathscr{C}'
\end{array}
$$

commutes up to natural isomorphism, then

$$
\begin{array}{ccccccccc}
\cdots \to K_p(\mathscr{E}) \times K_q(\mathscr{A}) & \to & K_p(\mathscr{E}) \times K_q(\mathscr{B}) & \to & K_p(\mathscr{E}) \times K_q(\mathscr{C}) & \xrightarrow{(-1)^p \times \partial} & K_p(\mathscr{E}) \times K_{q-1}(\mathscr{A}) & \to & \cdots \\
\downarrow & & \downarrow & & \downarrow & & \downarrow & & \\
\cdots \to \quad K_{p+q}(\mathscr{A}') & \to & K_{p+q}(\mathscr{B}') & \to & K_{p+q}(\mathscr{C}') & \xrightarrow{\partial} & K_{p+q-1}(\mathscr{A}') & \to & \cdots
\end{array}
$$

commutes.

Note that the second assertion of (4) follows from the first together with (2).

It is also known that Waldhausen's product agrees with the products defined by Loday and May using the BGL^+ construction [We].

9. The projective line

In [Q], Quillen computes the K-theory of a projective space $\mathbb{P}_A^r$ over a commutative ring A or, more generally, over a quasicompact ground scheme. He observes that, if $r = 1$, the calculation works even if A is not commutative. In this case, instead of trying to define a "scheme" $\mathbb{P}_A^1$, Quillen works directly with a category which, when A is commutative, is equivalent to the category of quasicoherent sheaves on $\mathbb{P}_A^1$. I will give a detailed exposition of this case here since it requires no knowledge of algebraic geometry (except as motivation for the definitions).

DEFINITION. If A is any ring (not necessarily commutative) let $\mathcal{M}od(\mathbb{P}_A^1)$ be the category whose objects are triples (M^+, M^-, θ) where M^+ is an $A[t]$-module, M^- is an $A[t^{-1}]$-module, and θ is an isomorphism θ : $A[t, t^{-1}] \otimes_{A[t]} M^+ \approx A[t, t^{-1}] \otimes_{A[t^{-1}]} M^-$.

Here t is an indeterminate.

A morphism $(M^+, M^-, \theta) \to (N^+, N^-, \varphi)$ is a pair (f^+, f^-) where $f^+ : M^+ \to N^+$ over $A[t]$, $f^- : M^- \to N^-$ over $A[t^{-1}]$, and

$$
\begin{array}{ccc}
A[t, t^{-1}] \otimes_{A[t]} M^+ & \xrightarrow[\approx]{\theta} & A[t, t^{-1}] \otimes_{A[t^{-1}]} M^- \\
{\scriptstyle 1 \otimes f^+} \downarrow & & \downarrow {\scriptstyle 1 \otimes f^-} \\
A[t, t^{-1}] \otimes_{A[t]} N^+ & \xrightarrow[\approx]{\varphi} & A[t, t^{-1}] \otimes_{A[t^{-1}]} N^-
\end{array}
$$

commutes.

Since $A[t, t^{-1}]$ is flat over $A[t]$ and over $A[t^{-1}]$, it is easy to check that $\ker(f^+, f^-) = (\ker f^+, \ker f^-, \theta')$ and $\mathrm{ckr}(f^+, f^-) = (\mathrm{ckr}\, f^+, \mathrm{ckr}\, f^-, \varphi')$ where θ' and φ' are induced by θ and φ. From this it follows easily that $\mathcal{M}od(\mathbb{P}_A^1)$ is an abelian category. We can also embed $\mathcal{M}od(\mathbb{P}_A^1)$ as a full subcategory of modules over the ring

$$
\begin{pmatrix}
A[t, t^{-1}] & A[t, t^{-1}] & A[t, t^{-1}] \\
0 & A[t] & 0 \\
0 & 0 & A[t^{-1}]
\end{pmatrix}
$$

A module over this ring can be written as

$$
\begin{pmatrix}
M^{\pm} \\
M^+ \\
M^-
\end{pmatrix}
$$

where $M^{\pm}$ is an $A[t, t^{-1}]$-module, M^+ is an $A[t]$-module, and M^- is an $A[t^{-1}]$-module. It is specified by giving $M^{\pm}$, M^+, M^-, and homomorphisms $M^+ \to M^{\pm}$, $M^- \to M^{\pm}$ defined by the matrix units e_{12} and e_{23}. $\mathcal{M}od(\mathbb{P}_A^1)$ is equivalent to the full subcategory of modules for which $A[t, t^{-1}] \otimes_{A[t]} M^+ \to M^{\pm}$ and $A[t, t^{-1}] \otimes_{A[t^{-1}]} M^- \to M^{\pm}$ are isomorphisms.

If $\mathcal{M} = (M^+, M^-, \theta)$, then we define the Serre twist $\mathcal{M}(n)$ by $\mathcal{M}(n) = (M^+, M^-, t^{-n}\theta)$. Let $\mathcal{O} = (A[t], A[t^{-1}], 1)$. If N is an A-module let $\mathcal{O}(n) \otimes_A N = (A[t] \otimes_A N, A[t^{-1}] \otimes_A N, t^{-n}) = (N[t], N[t^{-1}], t^{-n})$.

If $\mathscr{M} = (M^+, M^-, \theta)$, we define $H^i(\mathscr{M})$ to be the cohomology of the complex $C^\bullet(\mathscr{M})$ defined by

$$0 \to M^+ \oplus M^- \xrightarrow{d} A[t, t^{-1}] \otimes_{A[t^{-1}]} M^- \to 0$$

where $d(x, y) = \theta(1 \otimes x) - 1 \otimes y$. We have $H^i(\mathscr{M}) = 0$ for $i \neq 0, 1$. Since $C^\bullet$ is an exact functor, a short exact sequence $0 \to \mathscr{M}' \to \mathscr{M} \to \mathscr{M}'' \to 0$ yields a long exact sequence
(1)
$$0 \to H^0(\mathscr{M}') \to H^0(\mathscr{M}) \to H^0(\mathscr{M}'') \to H^1(\mathscr{M}') \to H^1(\mathscr{M}) \to H^1(\mathscr{M}'') \to 0.$$

Note that $H^0(\mathscr{M}) = \operatorname{Hom}(\mathscr{O}, \mathscr{M})$.

LEMMA 9.1. *If N is any A-module,*

$$H^0(\mathscr{O}(n) \otimes_A N) = \begin{cases} 0 & \text{if } n \leq -1, \\ N^{n+1} & \text{if } n \geq -1, \end{cases}$$

and

$$H^1(\mathscr{O}(n) \otimes_A N) = \begin{cases} 0 & \text{if } n \geq -1, \\ N^{-n-1} & \text{if } n \leq -1. \end{cases}$$

This follows directly from the definition.

LEMMA 9.2. *Let $\mathscr{M} = (M^+, M^-, \theta)$. If $x \in M^+$ there is an n^+ such that for $n \geq n^+$ there is an element $(x, y) \in H^0(\mathscr{M}(n)) = \{(a, b) \in M^+ \oplus M^- \mid t^{-n}\theta(1 \otimes a) = 1 \otimes b\}$ and similarly for $y \in M^-$.*

PROOF. $\theta(1 \otimes x)$ lies in $A[t, t^{-1}] \otimes_{A[t]} M^-$ and it is clear that, for sufficiently large n, $t^{-n}\theta(1 \otimes x)$ lies in $1 \otimes M^-$.

COROLLARY 9.3. *If $f : \mathscr{M} \to \mathscr{N}$ induces an epimorphism $H^0(\mathscr{M}(n)) \twoheadrightarrow H^0(\mathscr{N}(n))$ for all large n, then f is an epimorphism.*

PROOF. Let $x \in N^+$. For large n we can find $(x, y) \in H^0(\mathscr{N}(n))$ and lift it to $(x', y') \in H^0(\mathscr{M}(n))$ showing that $M^+ \to N^+$ is onto. A similar argument shows that $M^- \to N^-$ is onto.

We say that $\mathscr{M} = (M^+, M^-, \theta)$ is of finite type if M^+ and M^- are finitely generated modules over $A[t]$ and $A[t^{-1}]$.

COROLLARY 9.4. *If $\mathscr{M}$ is of finite type there is an n_0 such that for all $n \geq n_0$ there is an epimorphism $\mathscr{O}(-n)^k \twoheadrightarrow \mathscr{M}$.*

PROOF. Let $a_1, \ldots, a_r$ generate M^+ and let $b_1, \ldots, b_s$ generate M^-. Let $k = r + s$. If n is large we can find maps $\mathscr{O} \to \mathscr{M}(n)$ sending $(1, 1)$ to (a_i, c_i) and maps sending $(1, 1)$ to (d_i, b_i). Clearly $\mathscr{O}^k \to \mathscr{M}(n)$ is onto and we twist it by $-n$.

COROLLARY 9.5. *If $\mathcal{M}$ is of finite type then $H^1(\mathcal{M})$ is a finitely generated A-module and $H^1(\mathcal{M}(n)) = 0$ for large n.*

PROOF. The cohomology sequence (1) applied to $0 \to \mathcal{N} \to \mathcal{O}(-n)^k \to \mathcal{M} \to 0$ gives $H^1(\mathcal{O}(-n)^k) \twoheadrightarrow H^1(\mathcal{M})$ and we apply Lemma 9.1. Also $H^1(\mathcal{O}(m-n)^k) \twoheadrightarrow H^1(\mathcal{M}(m))$ and $H^1(\mathcal{O}(m-n)^k) = 0$ for large m by Lemma 9.1.

REMARK. If A is noetherian, $H^0(\mathcal{M})$ will also be finitely generated over A. This follows from Corollary 9.5 applied to $\mathcal{N}$ and $H^0(\mathcal{O}(-n)^k) \to H^0(\mathcal{M}) \to H^1(\mathcal{N})$.

LEMMA 9.6 (Koszul sequence). *For any $\mathcal{M} \in \mathcal{M}od(\mathscr{P}_A^1)$ we have an exact sequence*

$$0 \to \mathcal{M}(n-2) \to \mathcal{M}(n-1)^2 \to \mathcal{M}(n) \to 0$$

for all n.

PROOF. Define $\xi : \mathcal{M}(n-1) \to \mathcal{M}(n)$ by $\xi = (1, t^{-1})$; i.e. $\xi^+ : M^+ \xrightarrow{1} M^+$ and $\xi^- : M^- \xrightarrow{t^{-1}} M^-$. Define $\eta : \mathcal{M}(n-1) \to \mathcal{M}(n)$ by $\eta = (t, 1)$. It is easy to check the exactness of

$$0 \to \mathcal{M}(n-2) \xrightarrow{(\eta, -\xi)} \mathcal{M}(n-1)^2 \xrightarrow{(\xi, \eta)} \mathcal{M}(n) \to 0.$$

DEFINITION. We say that $\mathcal{M}$ is regular if $H^1(\mathcal{M}(-1)) = 0$.

For example, $\mathcal{O}(n)$ is regular for $n \geq 0$ by Lemma 9.1.

COROLLARY 9.7. *If $\mathcal{M}$ is regular so is $\mathcal{M}(1)$.*

PROOF. The cohomology sequence of $0 \to \mathcal{M}(-2) \to \mathcal{M}(-1)^2 \to \mathcal{M} \to 0$ shows that $0 \to H^1(\mathcal{M}(-1))^2 \to H^1(\mathcal{M}) \to 0$ is exact

LEMMA 9.8. *If $\mathcal{M}$ is regular then $\mathcal{O} \otimes_A H^0(\mathcal{M}) \to \mathcal{M}$ is an epimorphism.*

PROOF. The map is given by evaluation after identifying $H^0(\mathcal{M})$ with $\text{Hom}(\mathcal{O}, \mathcal{M})$. Let $\mathcal{N} = \mathcal{O} \otimes_A H^0(\mathcal{M})$. By Lemma 9.1, $\mathcal{N}$ is regular. It is clear that $H^0(\mathcal{N}) \to H^0(\mathcal{M})$ is onto (even an isomorphism). Suppose $H^0(\mathcal{N}(n-1)) \to H^0(\mathcal{M}(n-1))$ is onto for some $n \geq 1$. Since $H^1(\mathcal{M}(n-2)) = H^1(\mathcal{N}(n-2)) = 0$, the cohomology of

$$
\begin{array}{ccccccccc}
0 & \longrightarrow & \mathcal{N}(n-2) & \longrightarrow & \mathcal{N}(n-1)^2 & \longrightarrow & \mathcal{N}(n) & \longrightarrow & 0 \\
& & \downarrow & & \downarrow & & \downarrow & & \\
0 & \longrightarrow & \mathcal{M}(n-2) & \longrightarrow & \mathcal{M}(n-1)^2 & \longrightarrow & \mathcal{M}(n) & \longrightarrow & 0
\end{array}
$$

is

$$
\begin{array}{ccccccccc}
0 & \longrightarrow & H^0(\mathcal{N}(n-2)) & \longrightarrow & H^0(\mathcal{N}(n-1))^2 & \longrightarrow & H^0(\mathcal{N}(n)) & \longrightarrow & 0 \\
& & \downarrow & & \downarrow & & \downarrow & & \\
0 & \longrightarrow & H^0(\mathcal{M}(n-2)) & \longrightarrow & H^0(\mathcal{M}(n-1))^2 & \longrightarrow & H^0(\mathcal{M}(n)) & \longrightarrow & 0
\end{array}
$$

and it follows that $H^0(\mathcal{N}(n)) \to H^0(\mathcal{M}(n))$ is also onto. So $H^0(\mathcal{N}(n)) \to H^0(\mathcal{M}(n))$ is onto for $n \geq 0$ and we can apply Corollary 9.3.

If $\mathcal{M}$ is regular, let $T_0(\mathcal{M}) = H^0(\mathcal{M})$ and consider the exact sequence

(2) $$0 \to \mathcal{Z}(\mathcal{M}) \to \mathcal{O} \otimes_A T_0(\mathcal{M}) \to \mathcal{M} \to 0.$$

Since $\mathcal{O} \otimes_A T_0(\mathcal{M})$ and $\mathcal{M}$ are regular and $H^0(\mathcal{O} \otimes_A T_0(\mathcal{M})) \xrightarrow{\approx} H^0(\mathcal{M})$, the cohomology sequence shows that $H^0(\mathcal{Z}) = H^1(\mathcal{Z}) = 0$. In particular $\mathcal{Z}(1)$ is regular. We can therefore repeat the construction getting $0 \to \mathcal{W} \to \mathcal{O} \otimes_A T_1(\mathcal{M}) \to \mathcal{Z}(1) \to 0$ where $T_1(\mathcal{M}) = T_0(\mathcal{Z}(\mathcal{M})(1))$. Again $H^0(\mathcal{W}) = H^1(\mathcal{W}) = 0$. Now $0 \to W(-1) \to \mathcal{O}(-1) \otimes_A T_1(\mathcal{M}) \to \mathcal{Z} \to 0$ and $H^1(\mathcal{O}(-1) \otimes_A T_1(\mathcal{M})) = 0$ by Lemma 9.1. Since $H^0(\mathcal{Z}) = 0$, the cohomology sequence shows that $H^1(\mathcal{W}(-1)) = 0$ so $\mathcal{W}$ is regular. But $H^0(\mathcal{W}) = 0$ so $\mathcal{W} = 0$ by Lemma 9.8.

LEMMA 9.9. *For $\mathcal{M}$ regular we have a canonical resolution*

$$0 \to \mathcal{O}(-1) \otimes_A T_1(\mathcal{M}) \to \mathcal{O} \otimes_A T_0(\mathcal{M}) \to \mathcal{M} \to 0.$$

Moreover, T_0 and T_1 are exact functors on the category of regular $\mathcal{M}$.

PROOF. Since $\mathcal{W} = 0$ above, $\mathcal{Z}(1) = \mathcal{O} \otimes_A T_1(\mathcal{M})$ so that $\mathcal{Z} = \mathcal{O}(-1) \otimes_A T_1(\mathcal{M})$. Since $\mathcal{M}$ is regular $H^1(\mathcal{M}) = 0$ and the cohomology sequence shows that T_0 is exact. Since $\mathcal{O}$ is flat over A, $\mathcal{M} \mapsto \mathcal{O} \otimes_A T_0(\mathcal{M})$ is exact and (2) shows that $\mathcal{M} \mapsto \mathcal{Z}(\mathcal{M})$ is exact. Since $\mathcal{Z}(\mathcal{M})(1)$ is regular, $T_1(\mathcal{M}) = H^0(\mathcal{Z}(\mathcal{M})(1))$ is also an exact functor.

DEFINITION. $\mathcal{P}(\mathbb{P}_A^1)$ is the full subcategory of $\mathcal{M}od(\mathbb{P}_A^1)$ of all $\mathcal{M} = (M^+, M^-, \theta)$ such that M^+ and M^- are finitely generated projective modules over $A[t]$ and $A[t^{-1}]$. Define $K_i(\mathbb{P}_A^1) = K_i(\mathcal{P}(\mathbb{P}_A^1))$.

LEMMA 9.10. *If $\mathcal{M} \in \mathcal{P}(\mathbb{P}_A^1)$ and $H^1(\mathcal{M}) = 0$ then $H^0(\mathcal{M}) \in \mathcal{P}(A)$*

PROOF. By Corollary 9.4 find $0 \to \mathcal{N} \to \mathcal{F} \to \mathcal{M} \to 0$ with $\mathcal{F} = \mathcal{O}(-n)^k$. Note that $\mathcal{N} \in \mathcal{P}(\mathbb{P}_A^1)$ also so that $H^1(\mathcal{N})$ is finitely generated by Corollary 9.5. The cohomology sequence

$$H^0(\mathcal{F}) \to H^0(\mathcal{M}) \to H^1(\mathcal{N}) \to H^1(\mathcal{F}) \to H^1(\mathcal{M}) = 0$$

shows that $H^1(\mathcal{N}) \to H^1(\mathcal{F})$ is onto. Since $H^1(\mathcal{F})$ is free by Lemma 9.1, this map splits showing that $\ker[H^1(\mathcal{N}) \to H^1(\mathcal{F})]$ is finitely generated. Since $H^0(\mathcal{F})$ is finitely generated by Lemma 9.1, it follows that $H^0(\mathcal{M})$ is finitely generated. Since $H^1(\mathcal{M}) = 0$,

$$0 \to H^0(\mathcal{M}) \to M^+ \oplus M^- \to A[t, t^{-1}] \otimes_{A[t^{-1}]} M^- \to 0$$

is exact. All terms but $H^0(\mathcal{M})$ are projective over A and it follows that $H^0(\mathcal{M})$ is also projective.

The calculation of $K_i(\mathbb{P}^1_A)$ works more generally for any functor K with the following properties.

(1) K is a functor from the category of small exact categories and exact functors to abelian groups.

(2) If $\mathscr{E}$ is the union of full subcategories $\mathscr{E}_0 \subset \mathscr{E}_1 \subset \ldots$ closed under extension then $K(\mathscr{E}) = \operatorname{colim} K(\mathscr{E}_i)$.

(3) K satisfies the Additivity Theorem 2.2.

For example Bass's functor BK_1 has these properties as well as Quillen's K_i.

Let $u_n : \mathscr{P}(A) \to \mathscr{P}(\mathbb{P}^1_A)$ by $u_n(P) = \mathscr{O}(-n) \otimes_A P$.

THEOREM 9.11. *If K satisfies* (1), (2), (3) *above then*

$$(u_{0*}, u_{1*}) : K(\mathscr{P}(A))^2 \xrightarrow{\approx} K(\mathscr{P}(\mathbb{P}^1_A)).$$

To prove this consider the full subcategory $\mathscr{P}_n$ of $\mathscr{P}(\mathbb{P}^1_A)$ of all $\mathscr{M}$ such that $H^1(\mathscr{M}(k)) = 0$ for all $k \geq n$. Let $i : \mathscr{P}_n \hookrightarrow \mathscr{P}_{n+1}$.

LEMMA 9.12. $i_* : K(\mathscr{P}_n) \xrightarrow{\approx} K(\mathscr{P}_{n+1})$.

PROOF. Define $g : \mathscr{P}_{n+1} \to \mathscr{P}_n$ by $g(\mathscr{M}) = \mathscr{M}(1)^2$ and $h : \mathscr{P}_{n+1} \to \mathscr{P}_n$ by $h(\mathscr{M}) = \mathscr{M}(2)$. By Lemma 9.6 we have exact sequences $0 \to \mathscr{M} \to ig\mathscr{M} \to ih\mathscr{M} \to 0$ of functors $\mathscr{P}_{n+1} \to \mathscr{P}_{n+1}$ and $0 \to \mathscr{M} \to gi\mathscr{M} \to hi\mathscr{M} \to 0$ of functors $\mathscr{P}_n \to \mathscr{P}_n$. The additivity property (3) now shows that $g_* - h_* : K(\mathscr{P}_{n+1}) \to K(\mathscr{P}_n)$ is inverse to i_*.

COROLLARY 9.13. $K(\mathscr{P}_n) \xrightarrow{\approx} K(\mathscr{P}(\mathbb{P}^1_A))$ *for all n.*

This follows from (2) since Corollary 9.5 shows that $\mathscr{P}(\mathbb{P}^1_A) = \bigcup \mathscr{P}_n$.

PROOF OF THEOREM 9.11. Note that $u_0, u_1 : \mathscr{P}(A) \to \mathscr{P}_0$ by Lemma 9.1. Define $v_n : \mathscr{P}_0 \to \mathscr{P}(A)$ by $v_n(\mathscr{M}) = H^0(\mathscr{M}(n))$. This is well defined and exact for $n \geq 0$ by Lemma 9.10 and the cohomology sequence.

Now

$$v_n u_m(P) = H^0(\mathscr{O}(n-m) \otimes_A P) = \begin{cases} P & \text{if } n = m, \\ 0 & \text{if } n < m \end{cases}$$

by Lemma 9.1. Therefore the composition

$$K(\mathscr{P}(A))^2 \xrightarrow{u_{0*}, u_{1*}} K(\mathscr{P}_0) \xrightarrow{v_{0*}, v_{1*}} K(\mathscr{P}(A))^2$$

is given by a matrix $\begin{pmatrix} 1 & * \\ 0 & 1 \end{pmatrix}$ and so is an isomorphism.

Note that $\mathscr{P}_{-1}$ is the subcategory of regular elements of $\mathscr{P}(\mathbb{P}^1_A)$. Let $T_0, T_1 : \mathscr{P}_{-1} \to \mathscr{P}(A)$ be as in Lemma 9.9. These are exact and Lemma 9.9 gives us an exact sequence $0 \to u_1 T_1(\mathscr{M}) \to u_0 T_0 \mathscr{M}) \to i\mathscr{M} \to 0$ of functors $\mathscr{P}_{-1} \to \mathscr{P}_0$. By the additivity property (3), $i_* : K(\mathscr{P}_{-1}) \xrightarrow{\approx} K(\mathscr{P}_0)$ has the form $i_* = u_{0*} T_{0*} - u_{1*} T_{1*}$ showing that $(u_{0*}, u_{1*}) : K(\mathscr{P}(A))^2 \to K(\mathscr{P}(\mathbb{P}^1_A))$ is onto. This proves the theorem.

In §10 we will also need the following analogue of Theorem 6.5. Let $\mathscr{P}^-(\mathbb{P}_A^1)$ be the full subcategory of $\mathscr{M}od(\mathbb{P}_A^1)$ of all $\mathscr{M} = (M^+, M^-, \theta)$ such that M^- is a finitely generated projective $A[t^{-1}]$-module and such that there is a resolution $0 \to \mathscr{Q} \to \mathscr{P} \to \mathscr{M} \to 0$ with $\mathscr{P}, \mathscr{Q} \in \mathscr{P}(\mathbb{P}_A^1)$. Let $\mathscr{H}^-(\mathbb{P}_A^1)$ be the full subcategory of $\mathscr{M}$ in $\mathscr{P}^-(\mathbb{P}_A^1)$ with $M^- = 0$.

THEOREM 9.14. $\mathscr{H}^-(\mathbb{P}_A^1) \to \mathscr{P}^-(\mathbb{P}_A^1) \to \mathscr{P}(A[t^{-1}])$ is a pseudo-K-fibration and $\mathscr{P}(\mathbb{P}_A^1) \hookrightarrow \mathscr{P}^-(\mathbb{P}_A^1)$ induces an isomorphism of K_q for all q.

This is just a special case of Theorem 6.5 if A is commutative. The proof is essentially the same as that of Theorem 6.5.

10. Fundamental Theorem

For any ring R let $\mathscr{N}il(R)$ be the category with objects (P, η) where P is a finitely generated projective R-module and $\eta : P \to P$ is a nilpotent endomorphism. A morphism $(P, \eta) \to (P', \eta')$ is an R-homomorphism $f : P \to P$ such that $\eta' f = f\eta$. We can embed $\mathscr{N}il(R)$ in $\mathscr{M}od(R[T])$ by letting T act as η and it is clear that $\mathscr{N}il(R)$ is closed under extensions.

There are exact functors $\mathscr{N}il(R) \to \mathscr{P}(R)$ sending (P, f) to P and $\mathscr{P}(R) \to \mathscr{N}il(R)$ by $P \mapsto (P, 0)$. It follows that $K_q(R)$ is a direct summand of $K_q(\mathscr{N}il(R))$ and we define $\mathrm{Nil}_q(R) = K_q(\mathscr{N}il(R))/K_q(R)$ so that $K_q(\mathscr{N}il(R)) = K_q(R) \oplus \mathrm{Nil}_q(R)$.

The following result is known as the Fundamental Theorem of K-Theory [**B, GQ**].

THEOREM 10.1 [**GQ**]. *Let T be an indeterminate. Then for $q \geq 1$,*

(1) $K_q(R[T]) = K_q(R) \oplus \mathrm{Nil}_{q-1}(R)$.

(2) $K_q(R[T, T^{-1}]) = K_q(R) \oplus K_{q-1}(R) \oplus \mathrm{Nil}_{q-1}(R). \oplus \mathrm{Nil}_{q-1}(R)$.

The isomorphism in (2) can be given in a more precise form as follows.

COROLLARY 10.2. *There is a naturally split exact sequence for $q \geq 1$,*

$$0 \to K_q(R) \to K_q(R[T]) \oplus K_q(R[T^{-1}]) \to K_q(R[T, T^{-1}]) \to K_{q-1}(R) \to 0.$$

All maps here but the last are the obvious ones except for sign as we will see from the proof of the theorem.

REMARK. In [**B**] Bass considers any functor F from rings to abelian groups and defines $LF(R)$ to be the cokernel of

$$F(R[T]) \oplus F(R[T^{-1}]) \to F(R[T, T^{-1}]).$$

By Corollary 10.2 we have $K_{q-1} = LK_q$ for $q \geq 1$. Bass defines K_q for $q < 0$ by this relation. Similarly he defines $\mathrm{Nil}_{q-1} = L\mathrm{Nil}_q$. He shows that with these definitions the fundamental theorem continues to hold for $q \leq 0$.

The fundamental theorem takes a particularly simple form if R is regular.

THEOREM 10.3. *If R is regular then* $\mathrm{Nil}_q(R) = 0$ *for all* q.

To prove this define a category $\mathcal{N}il\mathcal{M}odfg(R)$ in the same way as $\mathcal{N}il(R)$ but using objects (M, η) where M is any finitely generated R-module and η is a nilpotent endomorphism.

LEMMA 10.4. *If* $(M, \eta) \in \mathcal{N}il\mathcal{M}odfg(R)$, *there is an epimorphism* (P, ξ) $\twoheadrightarrow (M, \eta)$ *with* $(P, \xi) \in \mathcal{N}il(R)$

PROOF. Let $f : Q \twoheadrightarrow M$ with $Q \in \mathcal{P}(R)$. If $\eta^n = 0$ let $P = Q^n$ and define ξ by $\xi(p_0, \ldots, p_{n-1}) = (0, p_0, \ldots, p_{n-2})$. Let $g : P \to M$ by $g(p_0, \ldots, p_{n-1}) = \sum \eta^i f(p_i)$.

Using this lemma we can apply the resolution theorem just as in the proof of Corollary 2.9 to show that $K_q(\mathcal{N}il(R)) = K_q(\mathcal{N}il\mathcal{M}odfg(R))$. Now $\mathcal{N}il\mathcal{M}odfg(R)$ is abelian and (M, η) has a filtration $M \supset \eta M \supset \cdots \supset \eta^n M = 0$ with quotients of the form $(N_i, 0)$. The Dévissage Theorem now implies that $K_q(\mathcal{N}il\mathcal{M}odfg(R)) = K_q(\mathcal{M}odfg(R)) = K_q(R)$ using Corollary 2.9.

COROLLARY 10.5. *If* R *is regular then*

$$K_q(R[T]) = K_q(R) \quad and \quad K_q(R[T, T^{-1}]) = K_q(R) \oplus K_{q-1}(R).$$

A different proof of this was given by Quillen [Q]. Bass [B] has shown that if R is regular then $K_q(R) = 0$ for $q < 0$.

Theorem 10.1 will be proved by comparing two localization sequences [GQ]. We begin with the one for $R[T] \to R[T, T^{-1}]$. By Corollary 6.4 this has the form

$$\cdots \to K_q(\mathcal{H}_S(R)) \to K_q(R[T]) \to K_q(R[T, T^{-1}]) \xrightarrow{\partial} K_{q-1}(R) \to \cdots$$

where $\mathcal{H}$ is the category of $R[T]$-modules M such that $M_T = 0$ and such that there is a resolution $0 \to Q \to P \to M \to 0$ with $P, Q \in \mathcal{P}(R[T])$.

LEMMA 10.6. *If* $M \in \mathcal{H}$ *then* M *is a finitely generated projective R-module.*

PROOF. Since M is finitely generated over $R[T]$, $T^n M = 0$ for some n so M is finitely generated over $R[T]/(T^n)$ and so over R. Since $M \approx P/Q \approx T^n P/T^n Q$ we have an exact sequence $0 \to M \to Q/T^n Q \to Q/T^n P \to 0$. Now P and Q are projective over R so $Q/T^n P$ has projective dimension ≤ 1 over R. Also $Q/T^n Q$ is projective over $R[T]/(T^n)$ and so over R. It follows that M is projective over R.

LEMMA 10.7. $\mathcal{H} \approx \mathcal{N}il(R)$.

PROOF. By Lemma 10.6, $\mathcal{H} \to \mathcal{N}il(R)$ by $M \mapsto (M, T)$ is well defined. If $(N, \eta) \in \mathcal{N}il(R)$, make N into an $R[T]$-module by letting T act as η. The characteristic sequence [B] $0 \to N[T] \xrightarrow{T-\eta} N[T] \to N \to 0$ gives the required resolution showing that $N \in \mathcal{H}$. The resulting functor $\mathcal{N}il(R) \to \mathcal{H}$ is clearly inverse to $\mathcal{H} \to \mathcal{N}il(R)$.

The other localization sequence is that of Theorem 9.14. The functor $\mathcal{M}od(\mathbb{P}_R^1) \to \mathcal{M}od(R[T])$ by $(M^+, M^-, \theta) \mapsto M^+$ induces a map between the pseudo-K-fibrations of Theorems 9.14 and 6.3 and so gives rise to an exact ladder
(1)

$$
\begin{array}{ccccccccc}
\cdots \to & K_q(\mathcal{H}^-(\mathbb{P}_R^1)) & \to & K_q(\mathbb{P}_R^1) & \xrightarrow{\alpha} & K_q(R[T^{-1}]) & \xrightarrow{\partial} & K_{q-1}(\mathcal{H}^-(\mathbb{P}_R^1)) & \to \cdots \\
& \downarrow & & \beta \downarrow & & \downarrow & & \downarrow & \\
\cdots \to & K_q(\mathcal{H}) & \to & K_q(R[T]) & \to & K_q(R[T, T^{-1}]) & \xrightarrow{\partial} & K_{q-1}(\mathcal{H}) & \to \cdots
\end{array}
$$

LEMMA 10.8. $\mathcal{H}^-(\mathbb{P}_R^1) \xrightarrow{\approx} \mathcal{H}$

PROOF. Define a functor $\mathcal{N}il(R) \to \mathcal{H}^-(\mathbb{P}_R^1)$ by sending (M, η) to $(M, 0, 0)$ where M is regarded as an $R[T]$- module by letting T act as η. To see that this is well defined we have to construct a resolution $0 \to \mathcal{Q}' \to \mathcal{Q} \to (M, 0, 0) \to 0$ with $\mathcal{Q}, \mathcal{Q}' \in \mathcal{P}(\mathbb{P}_R^1)$. For M we choose the characteristic sequence $0 \to M[T] \xrightarrow{T-\eta} M[T] \to M \to 0$. Localizing this with respect to T gives an isomorphism $M[T, T^{-1}] \xrightarrow[\approx]{T-\eta} M[T, T^{-1}]$. The diagram

$$
\begin{array}{ccc}
M[T, T^{-1}] & \xrightarrow{T-\eta} & M[T, T^{-1}] \\
T \downarrow & & \downarrow 1 \\
M[T, T^{-1}] & \xrightarrow{1-T^{-1}\eta} & M[T, T^{-1}]
\end{array}
$$

shows that this corresponds to $M[T^{-1}] \xrightarrow{1-T^{-1}\eta} M[T^{-1}]$ which is an isomorphism since η is nilpotent. The required resolution is

$$
0 \to (M[T], M[T^{-1}], T) \xrightarrow{(T-\eta, 1-T^{-1}\eta)} (M[T], M[T^{-1}], 1)
$$
(2)
$$
\to (M, 0, 0) \to 0.
$$

There is no problem in checking that $\mathcal{N}il(R) \to \mathcal{H}^-$ preserves Hom sets and that it gives an inverse to the functor of Lemma 10.8.

Consider first the upper sequence in (1). By Theorem 9.11, $(u_{0*}, u_{1*}) : K_q(R)^2 \xrightarrow{\approx} K_q(\mathbb{P}_R^1)$ where

$$
u_0(P) = (P[T], P[T^{-1}], 1), \qquad u_1(P) = (P[T], P[T^{-1}], T).
$$

The compositions $\mathcal{P}(R) \xrightarrow{u_i} \mathcal{P}(\mathbb{P}_R^1) \xrightarrow{\alpha} \mathcal{P}(R[T])$ are both given by $\alpha u_i(P) = P[T]$. They induce the canonical map $K_q(R) \to K_q(R[T])$ which is a split monomorphism split by $R[T] \to R$ sending T to 0. The same remarks apply to $\mathcal{P}(R) \xrightarrow{u_i} \mathcal{P}(\mathbb{P}_R^1) \xrightarrow{\beta} \mathcal{P}(R[T^{-1}])$. It follows that the upper sequence in (1) reduces to

(3) $0 \to K_q(R) \leftrightarrows K_q(R[T^{-1}]) \to K_{q-1}(\mathcal{N}il(R)) \to K_{q-1}(R) \to 0.$

The last term is $(u_{0*} - u_{1*})K_{q-1}(R) \subset K_{q-1}(\mathbb{P}_R^1)$. The last map is induced by $\mathcal{H}^-(\mathbb{P}_R^1) \hookrightarrow \mathcal{P}^-(\mathbb{P}_R^1)$ sending (M, η) to $(M, 0, 0)$. By (2) and the Additivity Theorem 2.2 this map is the same as $K_{q-1}(\mathcal{N}il(R)) \to K_{q-1}(R) \overset{u_{0*} - u_{1*}}{\mapsto} K_{q-1}(\mathbb{P}_R^1)$. Therefore the last map in (3) is the canonical map $K_{q-1}(\mathcal{N}il(R)) \to K_{q-1}(R)$ showing that (3) reduces to the split exact sequence

$$0 \to K_q(R) \leftrightarrows K_q(R[T^{-1}]) \to \mathrm{Nil}_{q-1}(R) \to 0.$$

This proves the first part of Theorem 10.1.

By Lemmas 10.7 and 10.8, the vertical maps $K_q(\mathcal{H}^-(\mathbb{P}_R^1)) \to K_q(\mathcal{H})$ in (1) are isomorphisms. A standard diagram chase with (1) then yields a Mayer-Vietoris sequence

$$\cdots \to K_{q+1}(R[T, T^{-1}]) \overset{\partial}{\to} K_q(\mathbb{P}_R^1) \to K_q(R[T]) \oplus K_q(R[T^{-1}]) -$$
$$\to K_q(R[T, T^{-1}]) \overset{\partial}{\to} K_{q-1}(\mathbb{P}_R^1) \to \cdots$$

As we saw above, the maps $K_q(\mathbb{P}_R^1) \to K_q(R[T])$ and $K_q(\mathbb{P}_R^1) \to K_q(R[T^{-1}])$ have the same kernel $(u_{0*} - u_{1*})K_q(R)$ and factor through the same summand $K_q(R)$ of $K_q(\mathbb{P}_R^1)$. Therefore the Mayer-Vietoris sequence reduces to
(4)
$$0 \to K_q(R) \to K_q(R[T]) \oplus K_q(R[T^{-1}]) \to K_q(R[T, T^{-1}]) \overset{\partial}{\to} K_{q-1}(R) \to 0.$$

The maps, except for ∂, are the canonical ones up to sign and $K_q(R) \to K_q(R[T]) \oplus K_q(R[T^{-1}])$ is a split monomorphism. If we can show that ∂ is a split epimorphism, Corollary 10.2 will follow and (2) of Theorem 10.1 is clearly a consequence of this.

The map splitting ∂ will be defined using products as in §8. We first consider the case $R = \mathbb{Z}$. Easy calculations show that $K_0(\mathbb{Z}) = \mathbb{Z}$ and $K_1(\mathbb{Z}) = \{\pm 1\}$ [B]. Since $\mathbb{Z}$ is regular, $K_1(\mathbb{Z}[T]) = \{\pm 1\}$ and $K_1(\mathbb{Z}[T, T^{-1}]) = K_1(\mathbb{Z}) \oplus K_0(\mathbb{Z}) = \{\pm 1\} \oplus \mathbb{Z}$. Since the group of units $U(\mathbb{Z}[T, T^{-1}]) = \{\pm 1\} \times \langle T \rangle$ is a summand of $K_1(\mathbb{Z}[T, T^{-1}])$ via $U = GL_1$ and $\det : GL_n \to U$ [B], we see that $K_1(\mathbb{Z}[T, T^{-1}]) = U(\mathbb{Z}[T, T^{-1}]) = \{\pm 1\} \times \langle T \rangle$. The sequence (4) for $\mathbb{Z}$ reduces to $0 \to \{\pm 1\} \to \{\pm 1\} \times \langle T \rangle \overset{\partial}{\to} \mathbb{Z} \to 0$ with $\partial T = 1$ when $q = 1$. (If $\partial T = -1$ the argument is unchanged.)

The construction of the Mayer-Vietoris sequence shows $\partial : K_q(R[T, T^{-1}]) \to K_{q-1}(R)$ in (4) is given by the composition

$$K_q(R[T, T^{-1}]) \overset{\partial}{\to} K_{q-1}(\mathcal{H}) \overset{\approx}{\leftarrow} K_{q-1}(\mathcal{H}^-(\mathbb{P}_R^1)) \to K_{q-1}(\mathbb{P}_R^1)$$

and so is just the composition $K_q(R[T, T^{-1}]) \overset{\partial}{\to} K_{q-1}(\mathcal{H}) \to K_{q-1}(R)$ where the second map is just the canonical map $K_{q-1}(\mathcal{N}il(R)) \to K_{q-1}(R)$.

LEMMA 10.9. *If* $M \in \mathscr{P}_T(R[T])$ *then* M *is projective as an* R-*module.*

PROOF. We have $0 \to Q \to P \to M \to 0$ with $P, Q \in \mathscr{P}(R[T])$ and $M_T \in \mathscr{P}(R[T, T^{-1}])$. Therefore $Q_T \oplus M_T \approx P_T$. Now $Q \oplus M \in \mathscr{P}_T(R[T])$ and it will suffice to prove the lemma for $Q \oplus M$ so we can assume that $M_T \approx L_T$ for some $L \in \mathscr{P}(R[T])$. Let $L \to M$ induce such an isomorphism and lift this map to $L \to P$. Since $L \to L_T$ is injective, so is $L \to M$ and it follows that $Q \oplus L$ injects into P. We have $0 \to Q \oplus L \to P \to N \to 0$ where $N = M/L$. Since $N_T = 0$, $N \in \mathscr{P}(R)$ by Lemma 10.6 and the sequence $0 \to L \to M \to N \to 0$ shows that M is projective over R.

Now consider the localization sequences for $R[T] \to R[T, T^{-1}]$ and $\mathbb{Z}[T] \to \mathbb{Z}[T, T^{-1}]$. Let $\mathscr{P}_T(\mathbb{Z}[T]) \times \mathscr{P}(R) \to \mathscr{P}_T(R[T])$ be given by $(M, P) \mapsto P \otimes_\mathbb{Z} M$. This is biexact by Lemma 10.9 and induces the vertical maps in the diagram

$$
\begin{array}{ccccc}
\mathscr{H}_T(\mathbb{Z}[T]) \times \mathscr{P}(R) & \longrightarrow & \mathscr{P}_T(\mathbb{Z}[T]) \times \mathscr{P}(R) & \longrightarrow & \mathscr{P}(\mathbb{Z}[T, T^{-1}]) \times \mathscr{P}(R) \\
\downarrow & & \downarrow & & \downarrow \\
\mathscr{H}_T(R[T]) & \longrightarrow & \mathscr{P}_T(R[T]) & \longrightarrow & \mathscr{P}(R[T, T^{-1}]).
\end{array}
$$

By §8, (4) we get a commutative diagram

$$
\begin{array}{ccc}
K_1(\mathbb{Z}[T, T^{-1}]) \times K_{q-1}(R) & \xrightarrow{\partial \times 1} & K_0(\mathscr{H}_T(\mathbb{Z}[T])) \times K_{q-1}(R) \\
\downarrow & & \downarrow \\
K_q(R[T, T^{-1}]) & \xrightarrow{\partial} & K_{q-1}(\mathscr{H}_T(R[T])).
\end{array}
$$

By §8, (1) we get, using Lemma 10.7,

$$
\begin{array}{ccc}
K_0(\mathscr{H}_T(\mathbb{Z}[T])) \times K_{q-1}(R) & \longrightarrow & K_0(\mathbb{Z}) \times K_{q-1}(R) \\
\downarrow & & \downarrow \\
K_{q-1}(\mathscr{H}_T(R[T])) & \longrightarrow & K_{q-1}(R)
\end{array}
$$

and combining this with the previous diagram gives

$$
\begin{array}{ccc}
K_1(\mathbb{Z}[T, T^{-1}]) \times K_{q-1}(R) & \xrightarrow{\partial \times 1} & K_0(\mathbb{Z}[T]) \times K_{q-1}(R) \\
\cup\downarrow & & \downarrow\cup \\
K_q(R[T, T^{-1}]) & \xrightarrow{\partial} & K_{q-1}(R).
\end{array}
$$

If $x \in K_{q-1}(R)$, then $(\partial \times 1)(T, x) = (1, x)$ which maps to x in $K_{q-1}(R)$. In the other direction this is sent to $\partial(T \cup x)$ (where $\cup$ denotes the product) so $\partial: K_q(R[T, T^{-1}]) \to K_{q-1}(R)$ is split by $x \mapsto T \cup x$.

11. Projective space

I will show here how the calculations in §9 can be extended to projective spaces of any dimension following [**Q**]. I will consider only the case of projective space over a field for simplicity. The general case of projective space over a scheme is discussed in detail in [Q] and also in [S]. I will assume that the reader is familiar with basic facts of algebraic geometry given in [H], in particular with Lemmas 11.1 and 11.2 below.

Let k be a field and let $\mathbb{P}^r_k = \operatorname{Proj} S$ where $S = k[x_0, \ldots, x_r]$, graded so that $\deg x_i = 1$.

LEMMA 11.1 (Serre).

$$H^i(\mathbb{P}^r, \mathscr{O}_{\mathbb{P}}(n)) = \begin{cases} S_n & \text{if } i = 0, \\ 0 & \text{if } i \neq 0, r, \\ S^*_{-r-1-n} & \text{if } i = r. \end{cases}$$

Here S_n consists of all forms of degree n in S and $M^* = \operatorname{Hom}_k(M, k)$.

LEMMA 11.2 (Serre). *If $\mathscr{F}$ is a coherent sheaf on $\mathbb{P}^r$ then*

(1) $H^i(\mathbb{P}^r, \mathscr{F})$ *is finite dimensional over k for all i.*
(2) $H^i(\mathbb{P}^r, \mathscr{F}) = 0$ *for $i > r$.*
(3) $H^i(\mathbb{P}^r, \mathscr{F}(n)) = 0$ *for $n \gg 0$, $i \neq 0$.*

Let $K(x_0, \ldots, x_r)$ be the Koszul complex over S defined by $x_0, \ldots, x_r$. This is the tensor product of the complexes $(S \xrightarrow{x_i} S)$. In order to keep the boundary maps of degree 0 we will write this as $(S(-1) \xrightarrow{x_i} S)$ where, as usual, $S(n)_i = S_{n+i}$. Since $x_0, \ldots, x_r$ is a regular sequence on S, $K(x_0, \ldots, x_r)$ is a free resolution of $k = S/(x_0, \ldots, x_r)$ and takes the form

$$0 \to S(-r-1) \to S(-r)^{r+1} \to \cdots \to S(-p)^{\binom{r+1}{p}} \to \cdots \to S(-1)^{r+1}$$
$$\to S \to k \to 0.$$

Passing to the associated sheaves on $\mathbb{P}^r$ and noting that k defines the zero sheaf we get the Koszul sequence

(1) $\qquad 0 \to \mathscr{O}(-r-1) \to \mathscr{O}(-r)^{r+1} \to \cdots \to \mathscr{O}(-1)^{r+1} \to \mathscr{O} \to 0$.

This is locally split since $\mathscr{O}$ is locally free. Therefore if $\mathscr{F}$ is any $\mathscr{O}$-module, tensoring it with (1) gives

(2) $\qquad 0 \to \mathscr{F}(-r-1) \to \mathscr{F}(-r)^{r+1} \to \cdots \to \mathscr{F}(-1)^{r+1} \to \mathscr{F} \to 0$.

In particular, if X is a closed subscheme of $\mathbb{P}^r$ and $\mathscr{F}$ is a sheaf of $\mathscr{O}_X$-modules then (2) will also be a sequence of sheaves of $\mathscr{O}_X$-modules.

DEFINITION. Let X be a closed subscheme of $\mathbb{P}^r$ and let $\mathscr{F}$ be a coherent sheaf on X. We say that $\mathscr{F}$ is regular if $H^p(X, \mathscr{F}(-p)) = 0$ for all $p \geq 1$.

LEMMA 11.3. *If $0 \to \mathscr{H} \to \mathscr{G} \to \mathscr{F} \to 0$ is exact and if $\mathscr{G}$ and $\mathscr{H}(1)$ are regular, then $\mathscr{F}$ is regular.*

PROOF. For $p \geq 1$ we have $0 = H^p(X, \mathscr{G}(-p)) \to H^p(X, \mathscr{F}(-p)) \to H^{p+1}(X, \mathscr{H}(-p)) = 0$.

COROLLARY 11.4. *If* $0 \to \mathscr{G}_n \to \mathscr{G}_{n-1} \to \cdots \to \mathscr{G}_0 \to \mathscr{F} \to 0$ *is exact and if* $\mathscr{G}_i(i)$ *is regular for all* i*, then* $\mathscr{F}$ *is regular.*

PROOF. By induction on n, $0 \to \mathscr{G}_n \to \cdots \to \mathscr{G}_1 \to \mathscr{H} \to 0$ implies that $\mathscr{H}(1)$ is regular so we can apply Lemma 11.3 to $0 \to \mathscr{H} \to \mathscr{G}_0 \to \mathscr{F} \to 0$.

COROLLARY 11.5. *If* $\mathscr{F}$ *is regular so is* $\mathscr{F}(1)$*.*

PROOF. Apply Corollary 11.4 to the Koszul sequence (2).

LEMMA 11.6. *If* $\mathscr{F}$ *is regular then* $\mathscr{F}$ *is generated by global sections; i.e.* $\mathscr{O} \otimes \Gamma\mathscr{F} \to \mathscr{F}$ *is an epimorphism.*

PROOF. Break up the Koszul resolution as in the proof of Corollary 11.4 getting

$$(3) \qquad 0 \to \mathscr{F}(-r-1) \to \cdots \to \mathscr{F}(-2)^{\binom{r+1}{2}} \to \mathscr{G} \to 0$$

and

$$(4) \qquad 0 \to \mathscr{G} \to \mathscr{F}(-1)^{r+1} \to \mathscr{F} \to 0.$$

Corollary 11.4 applied to (3) shows that $\mathscr{G}(2)$ is regular. By Corollary 11.5, $\mathscr{G}(n)$ is regular for $n \geq 2$. From (4) we get $\Gamma\mathscr{F}(n-1)^{r+1} \to \Gamma\mathscr{F}(n) \to H^1(X, \mathscr{G}(n)) = 0$ for $n \geq 1$ so that $S_1 \otimes \Gamma\mathscr{F}(n-1) \to \Gamma\mathscr{F}(n)$ is onto for $n \geq 1$. It follows that $S \otimes \Gamma\mathscr{F} \to \bigoplus_{n \geq 0} \Gamma\mathscr{F}(n)$ is onto and taking associated sheaves gives $\mathscr{O} \otimes \Gamma\mathscr{F} \twoheadrightarrow \mathscr{F}$.

Note that $\mathscr{O}_{\mathbb{P}^r}$ is regular by Lemma 11.1. We will also need the following two results in §13.

Let X be a hypersurface of degree d in $\mathbb{P}_k^r$. If X is defined by $f = 0$ then $X = \operatorname{Proj} A$ where $A = S/Sf$.

LEMMA 11.7. *If* X *is a hypersurface of degree* d *in* $\mathbb{P}_k^r$ *then*

$$H^i(X, \mathscr{O}_X(n)) = \begin{cases} A_n & \text{for } i = 0, \\ 0 & \text{for } i \neq 0, r-1 \\ A_{d-r-1-n}^* & \text{for } i = r-1, \end{cases}$$

PROOF. We have $0 \to S(-d) \xrightarrow{f} S \to A \to 0$ giving $0 \to \mathscr{O}_{\mathbb{P}}(-d) \to \mathscr{O}_{\mathbb{P}} \to \mathscr{O}_X \to 0$ on $\mathbb{P}^r$. The cohomology sequence of $0 \to \mathscr{O}_{\mathbb{P}}(n-d) \to \mathscr{O}_{\mathbb{P}}(n) \to \mathscr{O}_X(n) \to 0$ is

$$0 \to S_{n-d} \xrightarrow{f} S_n \to H^0(X, \mathscr{O}_X(n)) \to 0 \to \cdots \to 0 \to H^{r-1}(X, \mathscr{O}_X(n))$$
$$\to S_{-r-1-n+d}^* \xrightarrow{\varphi} S_{-r-1-n}^* \to H^r(X, \mathscr{O}_X(n)) \to 0.$$

The map φ is dual to $S_{-r-1-n} \xrightarrow{f} S_{-r-1-n+d}$ which is injective with cokernel $A_{-r-1-n+d}$.

COROLLARY 11.8. *If X is a hypersurface of degree d in $\mathbb{P}^r$ then $\mathscr{O}_X(d-1)$ is regular.*

Suppose now that $\mathscr{F}$ is regular. Let $T_0(\mathscr{F}) = \Gamma\mathscr{F}$. This is an exact functor on regular sheaves since $H^1(X, \mathscr{F}) = 0$ for $\mathscr{F}$ regular by Corollary 11.5. Define $\mathscr{Z}_0 = \mathscr{Z}_0(\mathscr{F})$ by $0 \to \mathscr{Z}_0 \to \mathscr{O}_X \otimes T_0(\mathscr{F}) \to \mathscr{F} \to 0$. Then $\mathscr{Z}_0$ is also an exact functor on regular sheaves since the other two terms are. Note that $\Gamma(\mathscr{O} \otimes T_0(\mathscr{F})) = T_0(\mathscr{F}) \xrightarrow{\approx} \Gamma(\mathscr{F})$ so that $\Gamma\mathscr{Z}_0 = 0$.

LEMMA 11.9. *$\mathscr{Z}_0(1)$ is regular if $\mathscr{O}_X(1)$ is regular.*

PROOF.

$$0 = H^{q-1}(X, \mathscr{F}(1-q)) \to H^q(X, \mathscr{Z}_0(1-q))$$
$$\to H^q(X, \mathscr{O}_X(1-q) \otimes T_0(\mathscr{F})) = H^q(X, \mathscr{O}_X(1-q)) \otimes T_0(\mathscr{F}) = 0.$$

Therefore if $\mathscr{O}_X(1)$ is regular, we can iterate the above construction. Let $T_1(\mathscr{F}) = \Gamma(\mathscr{Z}_0(1))$ and define $\mathscr{Z}_1$ by $0 \to \mathscr{Z}_1(1) \to \mathscr{O}_X \otimes T_1(\mathscr{F}) \to \mathscr{Z}_0(1) \to 0$, etc. By patching these sequences together we obtain the following result.

LEMMA 11.10. *If $\mathscr{O}_X(1)$ is regular, every regular $\mathscr{F}$ on X has a canonical resolution*

$$\cdots \to \mathscr{O}_X(-p) \otimes T_p(\mathscr{F}) \to \cdots \to \mathscr{O}_X \otimes T_0(\mathscr{F}) \to \mathscr{F} \to 0.$$

The T_p are exact functors on regular sheaves.

In general, this resolution is of infinite length, but it stops in the case $X = \mathbb{P}^r$.

LEMMA 11.11. *For $\mathscr{F}$ regular on $\mathbb{P}^r$ we have $\mathscr{Z}_r(\mathscr{F}) = 0$.*

PROOF. We use the fact that $\mathscr{O}_{\mathbb{P}^r}$ is regular. The cohomology sequence of

$$0 \to \mathscr{Z}_{n+q}(n) \to \mathscr{O}(-q) \otimes T_{n+q}(\mathscr{F}) \to \mathscr{Z}_{n+q-1}(n) \to 0$$

gives

$$H^{q-1}(\mathbb{P}^r, \mathscr{Z}_{n+q-1}(n)) \xrightarrow{\delta} H^q(\mathbb{P}^r, \mathscr{Z}_{n+q}(n)) \to H^q(\mathbb{P}^r, \mathscr{O}_{\mathbb{P}^r}(-q)) \otimes T_{n+q}(\mathscr{F}) = 0$$

so δ is onto. Since $H^0(\mathscr{Z}_n(n)) = 0$ we see that $H^1(\mathscr{Z}_{n+1}(n)) = 0, \ldots,$ $H^q(\mathscr{Z}_{n+q}(n)) = 0$. Therefore $H^q(\mathscr{Z}_r(r-q)) = 0$ for $q \le r$ (and also for $q > r$ by Lemma 11.2) showing that $\mathscr{Z}_r(r)$ is regular. Since $\Gamma(\mathscr{Z}_r(r)) = 0$, Lemma 11.6 shows that $\mathscr{Z}_r(r) = 0$.

Let K be a functor having the properties considered in §9, namely

(1) K is a functor from the category of small exact categories and exact functors to abelian groups.

(2) If $\mathscr{E}$ is the union of full subcategories $\mathscr{E}_0 \subset \mathscr{E}_1 \subset \ldots$ closed under extension then $K(\mathscr{E}) = \operatorname{colim} K(\mathscr{E}_i)$.

(3) K satisfies the Additivity Theorem 2.2.

Let X be a closed subscheme of $\mathbb{P}_k^r$ and let $\mathscr{P}(X)$ be the category of vector bundles on X, i.e. the full subcategory of locally free sheaves on X. Let $\mathscr{P}_n$ be the full subcategory of $\mathscr{P}(X)$ of all $\mathscr{F}$ such that $H^i(X, \mathscr{F}(p)) = 0$ for $p \geq n$, $i \neq 0$. Let $\mathscr{R}_n$ be the full subcategory of $\mathscr{P}(X)$ of all $\mathscr{F}$ such that $\mathscr{F}(n)$ is regular. Note that $\mathscr{R}_n \subset \mathscr{P}_n$ by Corollary 11.5. By Lemma 11.2(3), $\mathscr{P}(X) = \bigcup \mathscr{P}_n = \bigcup \mathscr{R}_n$.

LEMMA 11.12. $K(\mathscr{P}_n) \to K(\mathscr{P}_{n+1})$ and $K(\mathscr{R}_n) \to K(\mathscr{R}_{n+1})$ are isomorphisms so $K(\mathscr{P}_n) \xrightarrow{\approx} \mathscr{P}(X) \xleftarrow{\approx} K(\mathscr{R}_n)$.

PROOF. Let $f_i : \mathscr{P}_{n+1} \to \mathscr{P}_n$ by $\mathscr{F} \mapsto \mathscr{F}(i)^{\binom{r+1}{i}}$ for $i \geq 1$. Twisting the Koszul sequence (2) by $r + 1$ gives

$$(5) \qquad 0 \to \mathscr{F} \to \mathscr{F}(1)^{r+1} \to \cdots \to \mathscr{F}(i)^{\binom{r+1}{i}} \to \cdots \to \mathscr{F}(r+1) \to 0.$$

Just as in the proof of Lemma 9.12 we see that $\sum_{i \geq 1} (-1)^{i-1} f_{i*} : K(\mathscr{P}_{n+1}) \to K(\mathscr{P}_n)$ is inverse to $K(\mathscr{P}_n) \to K(\mathscr{P}_{n+1})$. The same argument applies to $\mathscr{R}_n \subset \mathscr{R}_{n+1}$. Note that all kernels in (5) are locally free and lie in the correct categories by Corollary 11.4.

Define $u_i : \mathscr{P}(k) \to \mathscr{P}(\mathbb{P}^r)$ for $i = 0, \ldots, r$ by $u_i(M) = \mathscr{O}_{\mathscr{P}}(-i) \otimes M$.

THEOREM 11.13. $(u_0, \ldots, u_r) : K(\mathscr{P}(k))^{r+1} \xrightarrow{\approx} K(\mathscr{P}(\mathbb{P}_k^r))$.

PROOF. By Lemma 11.1, $u_i(M) \in \mathscr{P}_0$ for $0 \leq i \leq r$. Define $v_i : \mathscr{P}_0 \to \mathscr{P}(k)$ by $v_i(\mathscr{F}) = \Gamma(\mathscr{F}(i))$. These are exact functors on $\mathscr{P}_0$ since $H^1(\mathbb{P}^r, \mathscr{F}(i)) = 0$ for $i \geq 0$ if $\mathscr{F} \in \mathscr{P}_0$. By Lemma 11.1,

$$v_n u_m(M) = \Gamma(\mathscr{O}_{\mathbb{P}}(n - m) \otimes M) = \begin{cases} M & \text{if } n = m, \\ 0 & \text{if } n < m. \end{cases}$$

Therefore the composition $K(\mathscr{P}(k))^{r+1} \xrightarrow{(u_i)} K(\mathscr{P}(\mathbb{P}^r)) \xrightarrow{(v_i)} K(\mathscr{P}(k))^{r+1}$ is given by a matrix of the form $\begin{pmatrix} 1 & & * \\ & \ddots & \\ 0 & & 1 \end{pmatrix}$ and so is an isomorphism. This shows that (u_i) is injective.

The functors $T_i : \mathscr{R}_0 \to \mathscr{P}(k)$ of Lemma 11.10 are exact and $T_i = 0$ for $i > r$. By the canonical resolution and the additivity property, $\sum (-1)^i u_{i*} T_{i*} : K(\mathscr{R}_0) \to K(\mathscr{P}(\mathbb{P}^r))$ is the isomorphism induced by $\mathscr{R}_0 \hookrightarrow \mathscr{P}(\mathbb{P}^r)$. This shows that (u_i) is onto, proving the theorem.

12. Severi-Brauer varieties

In [Q, 8.4] Quillen shows how to generalize the calculation in §11 to Severi-Brauer varieties. A Severi-Brauer variety over a field k is a projective variety X over k such that for some finite galois extension K/k we have $K \otimes_k X \cong \mathbb{P}_K^r$. We say that K splits X.

If $G = \text{Gal}(K/k)$, we can recover X from $K \otimes_k X$ by collapsing under the action of G. Therefore we can classify Severi-Brauer varieties by classifying actions of G on $\mathbb{P}^r_K$ compatible with the G-action on K. This is done as follows. Choose an isomorphism $\varphi : K \otimes_k X \cong \mathbb{P}^r_K = K \otimes_k \mathbb{P}^r_k$. For each $\sigma \in G$ the composition

$$K \otimes_k \mathbb{P}^r_k \xrightarrow{\sigma^{-1} \otimes 1} K \otimes_k \mathbb{P}^r_k \underset{\varphi}{\xleftarrow{\cong}} K \otimes_k X \xrightarrow{\sigma \otimes 1} K \otimes_k X \underset{\varphi}{\xrightarrow{\cong}} K \otimes_k \mathbb{P}^r_k$$

gives an automorphism $c(\sigma)$ of $\mathbb{P}^r_K = K \otimes_k \mathbb{P}^r_k$ over K. Note that $\mathrm{Aut}_K(\mathbb{P}^r_K)$ $= \mathrm{PGL}_{r+1}(K)$. It is easy to check that c is a 1-cocycle i.e. $c(\sigma\tau) = c(\sigma)\sigma(c(\tau))$ where the action of σ on $\mathrm{PGL}_{r+1}(K)$ is induced by its action on K. A new choice of σ will differ from the old one by an automorphism $a \in \mathrm{PGL}_{r+1}(K)$ of $\mathbb{P}^r_K$ and c will be replaced by $c'(\sigma) = a^{-1} c(\sigma)\sigma(a)$, a cohomologous cocycle. Therefore the isomorphism classes of Severi-Brauer varieties split by K are in 1-1 correspondence with the elements of $H^1(G, \mathrm{PGL}_{r+1}(K))$. This set is also in 1-1 correspondence with the isomorphism classes of Azumaya algebras A of rank $(r+1)^2$ over k split by K, i.e. such that $K \otimes_k A \approx \mathcal{M}_{r+1}(K)$. This follows by the same argument and the fact that $\mathrm{Aut}_{K_{\mathrm{alg}}}(\mathcal{M}_{r+1}(K)) = \mathrm{PGL}_{r+1}(K)$.

If a group G acts on a ring A, let $\mathcal{M}od_G(A)$ be the category of A-modules with compatible G-action, in other words, A-modules M with an action of G on M which satisfies $\sigma(x+y) = \sigma(x) + \sigma(y)$ and $\sigma(ax) = \sigma(a)\sigma(x)$ for $a \in A$, $x \in M$. Morphisms are G-equivariant A-homomorphisms.

LEMMA 12.1 (Speiser). *If $G = \mathrm{Gal}(K/k)$ for a finite galois extension K/k, then $\mathcal{M}od(k) \xrightarrow{\approx} \mathcal{M}od_G(K)$ by $V \mapsto K \otimes_k V$ with G acting on K. The inverse sends W to W^G, the submodule fixed by G.*

PROOF. Clearly $(K \otimes_k V)^G = V$. We have to show that $K \otimes_k W_G \xrightarrow{\approx} W$. For this we can extend the ground field k so that K splits. This reduces the verification to the trivial case where $K = \prod_G k$ with G permuting the factors.

COROLLARY 12.2. *Let K/k be a finite galois extension with $G = \mathrm{Gal}(K/k)$. Let A be a k-algebra. Then $\mathcal{M}od(A) \approx \mathcal{M}od_G(K \otimes_k A)$. Also $\mathscr{P}(A) \approx \mathscr{P}_G(K \otimes_k A)$, the category of finitely generated projective $K \otimes_k A$-modules with compatible G-action.*

The last statement follows from the fact that $K \otimes_k A = A'$ is faithfully flat over A so $M \in \mathcal{M}od(A)$ is finitely generated projective if and only if $A' \otimes_A M$ is.

Let X be a scheme over k and let $X' = K \otimes_k X$ where K/k is a finite galois extension. Let $G = \mathrm{Gal}(K/k)$. Let $q\mathscr{C}oh_G(X')$ be the category of quasicoherent sheaves on X' with compatible G-action and let $\mathscr{P}_G(X')$ be the category of vector bundles on X' with compatible G-action.

COROLLARY 12.3. $q\mathscr{C}oh(X) \xrightarrow[\pi^*]{\approx} q\mathscr{C}oh_G(X')$ *and* $\mathscr{P}(X) \xrightarrow[\pi^*]{\approx} \mathscr{P}_G(X')$

where $\pi : X' \to X$. *Also* $\pi_* \pi^*(\mathscr{F}) \approx K \otimes_k \mathscr{F}$ *and* $\mathrm{Hom}(\pi^* \mathscr{F}, \pi^* \mathscr{G}) = K \otimes_k \mathrm{Hom}(\mathscr{F}, \mathscr{G})$

PROOF. If $X = \mathrm{Spec}\, A$ this is just Corollary 12.2. Let $\pi : X' \to X$ and let $X = \bigcup U_\alpha$ with U_α open and affine. If $U_\alpha = \mathrm{Spec}\, A_\alpha$ then $\pi^{-1}(U_\alpha) = \mathrm{Spec}(K \otimes_k A_\alpha)$. A sheaf on X can be specified by giving sheaves on the U_α with isomorphisms over $U_\alpha \cap U_\beta$ satisfying standard well-known compatibility conditions [H]. The same is true on X' using the $\pi^{-1}(U_\alpha)$. It is trivial to verify that this is compatible with the isomorphisms of Corollary 12.2. The last two statements follow from the corresponding results for modules.

Now let X be a Severi-Brauer variety and choose $\varphi : K \otimes_k X \cong \mathbb{P}_K^r$ as above. For $\sigma \in G = \mathrm{Gal}(K/k)$, let $\sigma' = \sigma \otimes 1$ on $K \otimes_k \mathbb{P}_k^r = \mathbb{P}_K^r$ and let σ'' be the automorphism of $\mathbb{P}_K^r$ corresponding to $\sigma \otimes 1$ on $K \otimes_k X$ so that $\sigma'' = \varphi \sigma' \varphi^{-1}$. Then $c(\sigma) = \sigma'' \sigma'^{-1}$ As in [Ka], we can lift σ'' to an automorphism $f(\sigma)$ of the homogeneous coordinate ring $S = K[x_0, \dots, x_r]$ of $\mathbb{P}_K^r$ as follows. We lift $c(\sigma) \in \mathrm{PGL}_{r+1}(K)$ to $c'(\sigma) \in \mathrm{GL}_{r+1}(K)$. Let this act linearly on $\sum K x_i$ and extend it to a K-algebra automorphism of S. Then set $f(\sigma) = c'(\sigma)\bar{\sigma}$ where $\bar{\sigma}$ acts on S by σ on K with $\bar{\sigma}(x_i) = x_i$. This $f(\sigma)$ is defined up to multiplication by an element of $K^* = K - \{0\}$. The $f(\sigma)$ satisfy $f(\sigma \tau) = \lambda_{\sigma, \tau} f(\sigma) f(\tau)$ on $S_1 = \sum K x_i$ where $\lambda_{\sigma, \tau} \in K^*$. On S_n, the forms of degree n, we therefore have $f(\sigma \tau) = \lambda_{\sigma, \tau}^n f(\sigma) f(\tau)$.

Suppose M is a graded S-module with automorphisms $g(\sigma)$ for $\sigma \in G$ such that $g(\sigma)(sm) = (f(\sigma)s)(g(\sigma)m)$ for $s \in S$, $m \in M$ and such that on M_n we have $g(\sigma \tau) = \lambda_{\sigma, \tau}^n g(\sigma) g(\tau)$. The sections of the associated sheaf $\tilde{M}$ locally have the form m/s where $m \in M$, $s \in S$ are homogeneous of the same degree. It follows that the automorphisms $h(\sigma)$ induced on $\tilde{M}$ satisfy $h(\sigma)h(\tau) = h(\sigma \tau)$. Therefore by Corollary 12.3, $\tilde{M} = \pi^* \mathscr{F}$ for a quasicoherent sheaf $\mathscr{F}$ on X where $\pi : \mathbb{P}_K^r = K \otimes_k X \to X$.

Applying this construction to the modules $S(-1) \otimes S_1 \to S$ (where $g(\sigma) = f(\sigma) \otimes f(\sigma)$ on the left) gives a map of sheaves $\mathscr{J} \to \mathscr{O}_X$ on X such that $\pi^* \mathscr{J} \to \pi^* \mathscr{O}_X$ is $\mathscr{O}_\mathbb{P}(-1)^{r+1} \to \mathscr{O}_\mathbb{P}$. Extend $\mathscr{J} \to \mathscr{O}_X$ to a derivation of the exterior algebra $\bigwedge^* \mathscr{J}$. This gives a Koszul resolution

$$(1) \qquad 0 \to \bigwedge^{r+1} \mathscr{J} \to \bigwedge^r \mathscr{J} \to \cdots \to \bigwedge^2 \mathscr{J} \to \mathscr{J} \to \mathscr{O}_X \to 0$$

which maps under π^* to the usual Koszul resolution

$$(2) \qquad 0 \to \mathscr{O}_\mathbb{P}(-r-1) \to \mathscr{O}_\mathbb{P}(-r)^{r+1} \to \cdots \to \mathscr{O}_\mathbb{P}(-1)^{r+1} \to \mathscr{O}_\mathbb{P} \to 0$$

on $\mathbb{P}_K^r$. In particular (1) is exact since (2) is and π is faithfully flat.

LEMMA 12.4. *If* $\mathscr{F}$ *is quasicoherent on* X *then* $H^i(\mathbb{P}^r, \pi^* \mathscr{F}) = K \otimes_k H^i(X, \mathscr{F})$.

PROOF. If $\mathscr{G}$ is quasicoherent on $\mathbb{P}^r$, the Leray spectral sequence shows that $H^i(\mathbb{P}^r, \mathscr{G}) = H^i(X, \pi_* \mathscr{G})$. Apply this to $\mathscr{G} = \pi^* \mathscr{F}$ and use the fact that $\pi_* \pi^*(\mathscr{F}) \approx K \otimes_k \mathscr{F}$ by Corollary 12.3.

We say that a coherent sheaf $\mathcal{F}$ on X is regular if $\pi^*\mathcal{F}$ is regular on $\mathbb{P}^r$. If $\mathcal{F}$ is regular define $T_0(\mathcal{F}) = \Gamma(\mathcal{F})$.

LEMMA 12.5. *If $\mathcal{F}$ is regular on X then $\mathcal{O}_X \otimes_k T_0(\mathcal{F}) \to \mathcal{F}$ is an epimorphism.*

PROOF. Since π^* is faithfully flat it is sufficient to show that π^* of this map is an epimorphism. Now $\pi^*(\mathcal{O}_X \otimes_k T_0(\mathcal{F})) = \pi^*\mathcal{O}_X \otimes_k T_0(\mathcal{F}) = \mathcal{O}_\mathbb{P} \otimes_k H^0(X, \mathcal{F}) = \mathcal{O}_\mathbb{P} \otimes_K (K \otimes_k H^0(X, \mathcal{F})) = \mathcal{O}_\mathbb{P} \otimes_K H^0(\mathbb{P}^r, \pi^*\mathcal{F})$ which maps onto $\pi^*\mathcal{F}$ since $\pi^*\mathcal{F}$ is regular.

Define $\mathcal{Z}_0(\mathcal{F})$ to be the kernel in $0 \to \mathcal{Z}_0(\mathcal{F}) \to \mathcal{O}_X \otimes_k T_0(\mathcal{F}) \to \mathcal{F} \to 0$. The definition of the canonical resolution must be modified here since in general there is no $\mathcal{O}_X(-n)$ with $\pi^*\mathcal{O}_X(-n) = \mathcal{O}_\mathbb{P}(-n)$. Instead we use the sheaves $\mathcal{J}_n = \mathcal{J}^{\otimes n}$ which have $\pi^*(\mathcal{J}_n) = (\mathcal{O}_\mathbb{P}(-1)^{r+1})^{\otimes n} = \mathcal{O}_\mathbb{P}(-n)^{(r+1)^n}$. If $\mathcal{Z}_{n-1} = \mathcal{Z}_{n-1}(\mathcal{F})$ is defined, let $T_n(\mathcal{F}) = \mathrm{Hom}_{\mathcal{O}_X}(\mathcal{J}_n, \mathcal{Z}_{n-1})$. This is a module over $\Lambda_n = \mathrm{End}_{\mathcal{O}_X}(\mathcal{J}_n)$. Let $\Lambda = \Lambda_1 = \mathrm{End}_{\mathcal{O}_X}(\mathcal{J})$. Then $K \otimes_k \Lambda = \mathrm{End}_{\mathcal{O}_\mathbb{P}}(\pi^*\mathcal{J}) = \mathrm{End}_{\mathcal{O}_\mathbb{P}}(\mathcal{O}_\mathbb{P}(-1)^{r+1}) = \mathcal{M}_{r+1}(K)$ showing that Λ is an Azumaya algebra over k. An easy calculation shows that it is the Azumaya algebra corresponding to the same cocycle as X. The map $\Lambda^{\otimes n} \to \mathrm{End}_{\mathcal{O}_X}(\mathcal{J}^{\otimes n}) = \Lambda_n$ is an isomorphism since under π^* it becomes $\mathcal{M}_{r+1}(K)^{\otimes n} \xrightarrow{\approx} \mathcal{M}_{(r+1)^n}(K)$. The evaluation map $\mathcal{J}_n \otimes T_n(\mathcal{F}) \to \mathcal{Z}_{n-1}$ factors through $\mathcal{J}_n \otimes_{\Lambda_n} T_n(\mathcal{F})$. Let $\mathcal{Z}_n$ be the kernel in

$$(3) \qquad 0 \to \mathcal{Z}_n \to \mathcal{J}_n \otimes_{\Lambda_n} T_n(\mathcal{F}) \to \mathcal{Z}_{n-1} \to 0.$$

This sequence is exact since under π^* it becomes

$$0 \to \pi^*\mathcal{Z}_n \to \pi^*\mathcal{J}_n \otimes_{\Lambda_n} T_n(\mathcal{F}) \to \pi^*\mathcal{Z}_{n-1} \to 0.$$

Since $\pi^*\mathcal{J}_n = \mathcal{O}_\mathbb{P}(-n)^{(r+1)^n}$ is a sheaf of K-modules, we can rewrite the middle term as

$$\mathcal{O}_\mathbb{P}(-n)^{(r+1)^n} \otimes_{K \otimes \Lambda_n} (K \otimes \mathrm{Hom}_{\mathcal{O}_X}(\mathcal{J}_n, \mathcal{Z}_{n-1}))$$

$$= \mathcal{O}_\mathbb{P}(-n)^{(r+1)^n} \otimes_{\mathcal{M}_{(r+1)^n}(K)} \mathrm{Hom}_{\mathcal{O}_\mathbb{P}}(\mathcal{O}_\mathbb{P}(-n)^{(r+1)^n}, \pi^*\mathcal{Z}_{n-1})$$

$$= \mathcal{O}_\mathbb{P}(-n) \otimes_K \mathrm{Hom}_{\mathcal{O}_\mathbb{P}}(\mathcal{O}_\mathbb{P}(-n), \pi^*\mathcal{Z}_{n-1}) = \mathcal{O}_\mathbb{P}(-n) \otimes_K \Gamma(\pi^*\mathcal{Z}_{n-1}(n)).$$

Therefore π^* sends (3) to

$$(4) \qquad 0 \to \pi^*\mathcal{Z}_n \to \mathcal{O}_\mathbb{P}(-n) \otimes_K \Gamma(\pi^*\mathcal{Z}_{n-1}(n)) \to \pi^*\mathcal{Z}_{n-1}(n) \to 0.$$

The sequences (4) splice together to give the canonical resolution of $\pi^*\mathcal{F}$. Therefore we can splice the sequences (3) getting a canonical resolution

$$0 \to \mathcal{J}_r \otimes_{\Lambda_r} T_r(\mathcal{F}) \to \cdots \to \mathcal{J}_1 \otimes_{\Lambda_1} T_1(\mathcal{F}) \to \mathcal{O}_X \otimes_k T_0(\mathcal{F}) \to \mathcal{F} \to 0$$

on X which maps under π^* to the canonical resolution of $\pi^*\mathcal{F}$ considered in Lemma 11.10. Lemma 11.11 shows that this sequence stops at $\mathcal{J}_r \otimes_{\Lambda_r} T_r(\mathcal{F})$.

Define $u_i : \mathscr{P}(\Lambda_i) \to \mathscr{P}(X)$ by $u_i(M) = \mathscr{J}_i \otimes_{\Lambda_i} M$ where we set $\mathscr{J}_0 = \mathscr{O}_X$ and $\Lambda_0 = k$. Let the functor K satisfy the conditions (1), (2), (3) of §11, e.g. $K = K_i$ for any i.

THEOREM 12.6.

$$\bigoplus_{i=0}^{r} K(\mathscr{P}(\Lambda_i)) \xrightarrow{\approx} K(\mathscr{P}(X)).$$

To prove this we imitate the proof of Theorem 11.13. Let $\mathscr{P}_n$ be the full subcategory of $\mathscr{P}(X)$ of all $\mathscr{F}$ satisfying $H^i(\mathbb{P}^r, \pi^*\mathscr{F}(p)) = 0$ for $i \neq 0$, $p \geq n$. Let $\mathscr{R}_n$ be the full subcategory of $\mathscr{F}$ such that $\mathscr{F}(n)$ is regular. By tensoring the sequence (1) with $\mathscr{F} \otimes (\bigwedge^{r+1} \mathscr{J})^{\vee}$ where $(\bigwedge^{r+1} \mathscr{J})^{\vee} = \underset{\sim}{hom}_{\mathscr{O}_X}(\bigwedge^{r+1} \mathscr{J}, \mathscr{O}_X)$, we get a sequence which maps under π^* to the sequence (5) of §11. We use this sequence as in the proof of Lemma 11.12 to show that $\mathscr{P}_n \hookrightarrow \mathscr{P}_{n+1}$ and $\mathscr{R}_n \hookrightarrow \mathscr{R}_{n+1}$ induce isomorphisms $K(\mathscr{P}_n) \xrightarrow{\approx} K(\mathscr{P}_{n+1})$ and $K(\mathscr{R}_n) \xrightarrow{\approx} K(\mathscr{R}_{n+1})$.

Define $v_i : \mathscr{P}_0 \to \mathscr{P}(\Lambda_i)$ by $v_i(\mathscr{F}) = \mathrm{Hom}_{\mathscr{O}_X}(\mathscr{J}_i, \mathscr{F})$. Then $v_n u_n(M) = \mathrm{Hom}_{\mathscr{O}_X}(\mathscr{J}_n, \mathscr{J}_n \otimes_{\Lambda_n} M) = \mathrm{Hom}_{\mathscr{O}_X}(\mathscr{J}_n, \mathscr{J}_n) \otimes_{\Lambda_n} M = \Lambda_n \otimes_{\Lambda_n} M = M$ while $v_n u_m(M) = 0$ for $n < m$ as we see by applying π^*. As in §11 we conclude that $(u_i) : \bigoplus_0^r K(\mathscr{P}(\Lambda_i)) \to K(\mathscr{P}(X))$ is injective.

Define $T_i : \mathscr{R}_0 \to \mathscr{P}(\Lambda_i)$ to be the functors used in the canonical resolution. Just as in §11 we see that $\sum (-1)^i u_{i*} T_{i*} : K(\mathscr{R}_0) \to K(\mathscr{P}(X))$ is an isomorphism showing that (u_i) is onto.

13. Quadric hypersurfaces

Let $X \subset \mathbb{P}_k^r$ be a quadric hypersurface defined by $f = 0$ where f is a nonsingular quadratic form. By Corollary 11.8, $\mathscr{O}_X(1)$ is regular so we can construct a canonical resolution

(1)
$$\cdots \to \mathscr{O}_X(-n) \otimes T_n(\mathscr{F}) \to \cdots \to \mathscr{O}_X(-1) \otimes T_1(\mathscr{F}) \to \mathscr{O}_X \otimes T_0(\mathscr{F}) \to \mathscr{F} \to 0$$

for regular sheaves $\mathscr{F}$. However the proof of Lemma 11.11 breaks down here because $\mathscr{O}_X$ is not regular and, in general, the canonical resolution will be of infinite length.

Following [SQ] we will construct a truncated canonical resolution which can be used to calculate $K_i(X)$. As in §11 we denote the kernels in (1) by $\mathscr{Z}_n(\mathscr{F})$ so that we have $\mathscr{Z}_{-1}(\mathscr{F}) = \mathscr{F}$ and

(2) $$0 \to \mathscr{Z}_n(\mathscr{F}) \to \mathscr{O}_X(-n) \otimes T_n(\mathscr{F}) \to \mathscr{Z}_{n-1}(\mathscr{F}) \to 0.$$

Let $d = r - 1 = \dim X$.

LEMMA 13.1. *If $\mathscr{F}$ is a regular vector bundle and if $\mathscr{G}(d)$ is regular then* $\mathrm{Ext}_{\mathscr{O}_X}^q(\mathscr{Z}_{n-1}(\mathscr{F}), \mathscr{G}) = 0$ *for $q > 0$.*

PROOF. Apply $\text{Ext}^*_{\mathcal{O}_X}(-, \mathcal{G})$ to (2). We have $\text{Ext}^q(\mathcal{O}(-n) \otimes T_n, \mathcal{G}) = T_n^* \otimes \text{Ext}^q(\mathcal{O}(-n), \mathcal{G}) = T_n^* \otimes H^q(X, \mathcal{G}(n)) = 0$ for $q > 0$, $q \geq d - n$. The exact Ext sequence now shows that $\text{Ext}^q(\mathcal{Z}_{n-1}, \mathcal{G}) \xrightarrow{\approx} \text{Ext}^{q+1}(\mathcal{Z}_n, \mathcal{G})$ for $q > 0$, $n + q \geq d$. Therefore $\text{Ext}^q(\mathcal{Z}_{d-1}, \mathcal{G}) \approx \text{Ext}^{q+1}(\mathcal{Z}_d, \mathcal{G}) \approx \cdots \approx \text{Ext}^{q+d}(\mathcal{Z}_{-1}, \mathcal{G}) = \text{Ext}^{q+d}(\mathcal{F}, \mathcal{G}) = H^{q+d}(X, \underline{\text{hom}}(\mathcal{F}, \mathcal{G})) = 0$ for $q > 0$.

COROLLARY 13.2. *If $0 \to \mathcal{F}' \to \mathcal{F} \to \mathcal{F}'' \to 0$ is a short exact sequence of regular sheaves then the exact sequence $0 \to \mathcal{Z}_{d-1}(\mathcal{F}') \to \mathcal{Z}_{d-1}(\mathcal{F}) \to \mathcal{Z}_{d-1}(\mathcal{F}'') \to 0$ splits*

PROOF. The construction of the canonical resolution in §11 shows that $\mathcal{Z}_{d-1}(\mathcal{F}')$ is regular so $\text{Ext}^1(\mathcal{Z}_{d-1}(\mathcal{F}''), \mathcal{Z}_{d-1}(\mathcal{F}')) = 0$ by the lemma.

Now suppose $\mathcal{F}(-1)$ is regular. The canonical resolution gives $0 \to \mathcal{Z}_0 \to \mathcal{O} \otimes T_0(\mathcal{F}) \to \mathcal{F}(-1) \to 0$ with $\mathcal{G} = \mathcal{Z}_0(1)$ regular. Let $m = \dim T_0(\mathcal{F})$. Then we have $0 \to \mathcal{G} \to \mathcal{O}(1)^m \to \mathcal{F}(-1) \to 0$ so, by Corollary 13.2, $\mathcal{Z}_{d-1}(\mathcal{F}) \oplus \mathcal{Z}_{d-1}(\mathcal{G}) \approx \mathcal{Z}_{d-1}(\mathcal{O}(1)^m)$.

Let $\mathcal{V} = \mathcal{Z}_{d-1}(\mathcal{O}(1))$ and let $E = \text{End}_{\mathcal{O}_X}(\mathcal{V})$ acting on $\mathcal{V}$ from the right. If $\mathcal{W} \oplus \mathcal{W}' \approx \mathcal{V}^m$ then $\mathcal{V} \otimes_E \text{Hom}_{\mathcal{O}_X}(\mathcal{V}, \mathcal{W}) \xrightarrow{\approx} \mathcal{W}$ since this is true for $\mathcal{W} = \mathcal{V}$ and both sides preserve direct sums. For the same reason $\text{Hom}_{\mathcal{O}_X}(\mathcal{V}, \mathcal{W}) \in \mathcal{P}(E)$. Define $T(\mathcal{F}) = \text{Hom}_{\mathcal{O}_X}(\mathcal{V}, \mathcal{Z}_{d-1}(\mathcal{F}))$.

LEMMA 13.3. *If $\mathcal{F}(-1)$ is regular we have a truncated canonical resolution*

$$0 \to \mathcal{V} \otimes_E T(\mathcal{F}) \to \mathcal{O}_X(1-d) \otimes T_{d-1}(\mathcal{F}) \to \cdots \to \mathcal{O}_X \otimes T_0(\mathcal{F}) \to \mathcal{F} \to 0.$$

LEMMA 13.4. *$E = C_0(\mathcal{F})$, the even part of the Clifford algebra of the quadratic form f.*

The proof requires a long calculation. I will give some indication below of how this is done. Full details are given in [SQ]. In particular, we see that E is semisimple since f is nonsingular.

Define $u_i : \mathcal{P}(k) \to \mathcal{P}(X)$ by $u_i(M) = \mathcal{O}_X(-i) \otimes M$ and define $u : \mathcal{P}(E) \to \mathcal{P}(X)$ by $u(M) = \mathcal{V} \otimes_E M$. Let the functor K satisfy the conditions (1), (2), (3) of §11, e.g $K = K_i$ for any i.

THEOREM 13.5. *Let X be a nonsingular quadric hypersurface of dimension d. Then*

$$(u_{0*}, \ldots, u_{d-1*}, u_*) : \bigoplus_0^{d-1} K(\mathcal{P}(k)) \oplus K(\mathcal{P}(E)) \xrightarrow{\approx} K(\mathcal{P}(X)).$$

PROOF. Define $\mathcal{P}_n$, $\mathcal{R}_n$ as in §11. Note that Lemma 11.12 applies. By Lemma 11.7, $u_i : \mathcal{P}(k) \to \mathcal{P}_0$ for $0 \leq i \leq d - 1$. Also $u : \mathcal{P}(E) \to \mathcal{P}_0$ by the following lemma.

LEMMA 13.6. *If $\mathcal{F}$ is regular then $H^q(X, \mathcal{Z}_i(\mathcal{F})) = 0$ for $0 \leq i \leq d - 1$ and all q.*

PROOF. By the remark before Lemma 11.9, $H^0(X, \mathscr{Z}_0) = 0$. By Lemma 11.9, $\mathscr{Z}_0(1)$ is regular so $H^q(X, \mathscr{Z}_0) = 0$ for $q > 0$. By Lemma 11.7, $H^q(X, \mathscr{O}(-i)) = 0$ for $0 < i < d$ and all q so the cohomology sequence of $0 \to \mathscr{Z}_i \to \mathscr{O}(-i) \otimes T_i \to \mathscr{Z}_{i-1} \to 0$ shows that the lemma for $\mathscr{Z}_{i-1}$ implies the lemma for $\mathscr{Z}_i$ as long as $0 < i < d$.

Define $v_n : \mathscr{P}_0 \to \mathscr{P}(k)$ by $v_n(\mathscr{F}) = \Gamma(\mathscr{F}(n))$ for $0 \leq n \leq d-1$, and define $v : \mathscr{P}_0 \to \mathscr{P}(E)$ by $v(\mathscr{F}) = \mathrm{Hom}_{\mathscr{O}_X}(\mathscr{V}, \mathscr{F})$. Note that v is an exact functor by Lemma 13.1 because $\mathscr{F} \in \mathscr{P}_0$ implies that $\mathscr{F}(d)$ is regular. In fact $H^q(X, \mathscr{F}(d-q)) = 0$ for $q \leq d$ but $H^q(X, -) = 0$ if $q > d = \dim X$.

Now

$$v_n u_m(M) = \Gamma(\mathscr{O}_X(n-m) \otimes M) = \begin{cases} M & \text{if } n = m, \\ 0 & \text{if } n < m, \end{cases}$$

and $vu(M) = \mathrm{Hom}_{\mathscr{O}_X}(\mathscr{V}, \mathscr{V} \otimes_E M) = \mathrm{Hom}_{\mathscr{O}_X}(\mathscr{V}, \mathscr{V}) \otimes_E M = E \otimes_E M = M$.

LEMMA 13.7. $v_i u(M) = 0$ for $0 \leq i \leq d-1$.

PROOF. $v_i u(M) = \Gamma((\mathscr{V} \otimes_E M)(i)) = \Gamma(\mathscr{V}(i)) \otimes_E M$. Since $\mathscr{V} = \mathscr{Z}_{d-1}(\mathscr{O}(1))$ we have $\Gamma(\mathscr{V}(d-1)) = 0$. If $i < d-1$ the canonical resolution for $\mathscr{O}(1)$ gives $0 \to \mathscr{Z}_{d-1}(\mathscr{O}(1)) \to \mathscr{O}(1-d) \otimes T_{d-1}(\mathscr{O}(1))$ so $\Gamma(\mathscr{V}(i)) \hookrightarrow \Gamma(\mathscr{O}(i+1-d)) \otimes T_{d-1} = 0$.

It follows that the composition

$$\overset{d-1}{\underset{0}{\bigoplus}} K(\mathscr{P}(k)) \oplus K(\mathscr{P}(E)) \xrightarrow{(u_i, u)_*} K(\mathscr{P}(X)) \xrightarrow{(v_i, v)_*} \overset{d-1}{\underset{0}{\bigoplus}} K(\mathscr{P}(k)) \oplus K(\mathscr{P}(E))$$

is given by a matrix of the form $\begin{pmatrix} 1 & & * \\ & \ddots & \\ 0 & & 1 \end{pmatrix}$ showing that $(u_i, u)_*$ is injective.

Let $T_i : \mathscr{R}_{-1} \to \mathscr{P}(k)$, $T : \mathscr{R}_{-1} \to \mathscr{P}(E)$ be the functors occurring in Lemma 13.3. As in §11 we see that $u_{0*}T_{0*} - u_{1*}T_{1*} + \cdots \pm u_* T_*$ is the isomorphism $K(\mathscr{R}_{-1}) \to K(\mathscr{P}(X))$ showing that $(u_i, u)_*$ is onto. This proves the theorem.

All that remains is to prove Lemma 13.4 and to identify the vector bundle $\mathscr{V}$.

Suppose that $\mathscr{F}$ is regular and let $M_n = 0$ for $n < 0$, $M_n = \Gamma(\mathscr{F}(n))$ for $n \geq 0$. Let $M = \oplus M_n$. Then $\tilde{M} = \mathscr{F}$. In the proof of Lemma 11.6 we saw that $S_n \otimes_k \Gamma(\mathscr{F}) \twoheadrightarrow \Gamma(\mathscr{F}(n))$ is onto and it follows that M is generated by M_0 as a graded A-module where $A = S/(f)$. The sequence $0 \to N \to A \otimes M_0 \to M \to 0$ is the start of a minimal resolution of M over A and induces $0 \to \mathscr{Z}_0 \to \mathscr{O} \otimes \Gamma(\mathscr{F}) \to \mathscr{F} \to 0$, the start of the canonical resolution of $\mathscr{F}$. We have $N_0 = 0$ and the diagram

$$0 \longrightarrow N_n \longrightarrow A_n \otimes M_0 \longrightarrow M_n \longrightarrow 0$$

$$0 \longrightarrow \Gamma(\mathscr{Z}_0(n)) \longrightarrow \Gamma(\mathscr{O}_X(n)) \otimes M_0 \longrightarrow \Gamma(\mathscr{F}(n)) \longrightarrow 0$$

shows that $N_n \xrightarrow{\approx} \Gamma(\mathscr{Z}_0(n))$ so $N(1)_n \approx \Gamma(\mathscr{Z}_0(1)(n))$ and N has the same relation to $\mathscr{Z}_0(1)$ that M had to $\mathscr{F}$. We can therefore iterate the construction getting a minimal resolution

$$\cdots \to A(-1) \otimes N_1 \to A \otimes M_0 \to M \to 0$$

of M over A and the above remarks show that this induces the canonical resolution of $\mathscr{F}$.

Since all minimal resolutions of M are isomorphic it will suffice to construct such a resolution in any way. We want to apply this to $\mathscr{F} = \mathscr{O}(1)$ for which $M_n = A(1)_n$ for $n \neq 1$ while $M_{-1} = 0$, $A(1)_{-1} = A_0 = k$. Therefore $M = A^+(1)$ where A^+ is the kernel in $0 \to A^+ \to A \to k \to 0$. This is the start of a minimal resolution for k so we can construct the canonical resolution of $\mathscr{O}(1)$ by taking a minimal resolution for k and lopping off the lowest term. Such a minimal resolution was constructed by Tate [T].

We have $A = k[x_0, \ldots, x_{d+1}]/(f)$. The Koszul complex $K_\bullet = K_\bullet(x_0, \ldots, x_{d+1})$ over A has K_1 free on elements $e_0, \ldots, e_{d+1}$ with $\partial e_i = x_i$. Write $f = \sum \lambda_i x_i$ where the λ_i are linear forms in the x_i. Then $\partial \sum \lambda_i e_i = 0$ so $\gamma = \sum \lambda_i e_i$ is a cycle of $K_\bullet$. We kill off γ by adjoining a divided power algebra $\Gamma[v]$ with $\deg v = 2$, $\partial v = \gamma$. Tate shows that $K_\bullet \otimes \Gamma[v]$ is the required minimal resolution.

In dimensions $\geq d$, this resolution can be obtained by the following construction used by Atiyah, Bott, and Shapiro [ABS]. Let $C = C(f)$ be the Clifford algebra of f. It has a mod 2 grading $C = C_0 \oplus C_1$. C is generated by elements $e_0, \ldots, e_{d+1}$ such that $\left(\sum_0^{d+1} x_i e_i\right)^2 = f(x)$. If $M = M_0 \oplus M_1$ is a graded C-module, define $\mathscr{O}(n) \otimes M_0 \to \mathscr{O}(n+1) \otimes M_1$ by the map $\sum x_i \otimes e_i$ where $x_i : \mathscr{O}(n) \to \mathscr{O}(n+1)$ is the map of sheaves induced by $x_i : A(n) \to A(n+1)$ and $e_i : M_0 \to M_1$ by the C-module structure. In the same way define $\mathscr{O}(n) \otimes M_1 \to \mathscr{O}(n+1) \otimes M_0$.

Define a complex $\mathrm{Cliff}_\bullet(M)$ by $\mathrm{Cliff}_n(M) = \mathscr{O}(-n) \otimes M_{n+d+1}$ with the map $\mathrm{Cliff}_n(M) \to \mathrm{Cliff}_{n-1}(M)$ given by $\sum x_i \otimes e_i$. This is a complex because $(\sum x_i \otimes e_i)(\sum x_i \otimes e_i) = f(x) \in A_2$ but $f(x) = 0$ in A.

LEMMA 13.8. *The canonical resolution of $\mathscr{O}(1)$ in dimensions $\geq d$ is isomorphic to* $\mathrm{Cliff}_\bullet(C(q))$ *in dimensions $\geq d$.*

This is proved by an explicit calculation comparing Tate's resolution to the complex $\mathrm{Cliff}_\bullet(C(q))$. The details may be found in [SQ, §8].

Define $\mathscr{V}(M) = \mathrm{ckr}[\mathrm{Cliff}_{d+1}(M) \to \mathrm{Cliff}_d(M)]$. It follows that $\mathscr{V} = \mathscr{V}(C(q))$.

LEMMA 13.9. *Let M and N be graded $C(q)$-modules. Then*

$$\mathrm{Hom}_{C(q)}(M, N)_0 \xrightarrow{\approx} \mathrm{Hom}_{\mathscr{O}_X}(\mathscr{V}(M), \mathscr{V}(N)).$$

The subscript 0 here denotes maps of degree zero.

PROOF. The complex $\mathrm{Cliff}_\bullet(M)$ is periodic with period 2 and it follows from Lemma 13.8 that it is exact. We have $\mathscr{O}(-d-1) \otimes M_0 \to \mathscr{O}(-d) \otimes M_1 \to \mathscr{V}(M) \to 0$ and $0 \to \mathscr{V}(N) \to \mathscr{O}(1-d) \otimes N_0 \to \mathscr{O}(2-d) \otimes N_1$ so $\mathrm{Hom}_{\mathscr{O}_X}(\mathscr{V}(M), \mathscr{V}(N))$ can be identified with the set of maps $\mathscr{O}(-d) \otimes M_1 \to \mathscr{O}(1-d) \otimes N_0$ such that both compositions in

$$\mathscr{O}(-d-1) \otimes M_0 \to \mathscr{O}(-d) \otimes M_1 \to \mathscr{O}(1-d) \otimes N_0 \to \mathscr{O}(2-d) \otimes N_1$$

are zero. Since $\mathrm{Hom}_{\mathscr{O}_X}(\mathscr{O}(n), \mathscr{O}(n+1)) = A_1$ this reduces the problem to linear algebra. The details may be found in [**SQ**, §8].

Lemma 13.4 now follows by setting $M = N = C(q)$. We also see that $\mathscr{V}$ can be constructed explicitly by the method of [**ABS**].

REFERENCES

[ABS] M. F. Atiyah, R. Bott, and A. Shapiro, *Clifford modules*, Topology **3** (1964), 3–38.

[B] H. Bass, *Algebraic K-theory*, Benjamin, New York, 1968.

[Bo] A. Borel, *Stable real cohomology of arithmetic groups*, Ann. Sci. École Norm. Sup. (4) **7** (1974), 235–272.

[CE] H. Cartan and S. Eilenberg, *Homological algebra*, Princeton Univ. Press, Princeton, NJ, 1956.

[F] P. Freyd, *Abelian categories*, Harper and Row, New York, 1964.

[Ga] P. Gabriel, *Des categories abéliennes*, Bull. Soc. Math. France **90** (1962), 323–448.

[G] D. Grayson, *Products in K-theory and intersecting algebraic cycles*, Invent. Math. **47** (1978), 71–83.

[GQ] ——, Higher algebraic K-theory. II (*after D. Quillen*) *Algebraic K-Theory*, Lecture Notes in Math., vol. 551, Springer-Verlag, Berlin, 1976, pp. 217–240.

[H] R. Hartshorne, *Algebraic geometry*, Springer-Verlag, Berlin, 1977.

[K] M. C. Kang, *Construction of Brauer-Severi varieties and norm hypersurfaces*, Canad. J. Math. **42** (1990), 230–238.

[Ma] J. P. May, *Simplicial objects in algebraic topology*, van Nostrand, Princeton, NJ, 1967.

[MaE] J. P. May, E^∞ *ring spaces and* E^∞ *ring spectra*, Lecture Notes in Math., vol. 657, Springer-Verlag, Berlin, 1977.

[Mi] J. Milnor, *Introduction to algebraic K-theory*, Ann. of Math. Stud., vol. 72, Princeton Univ. Press, Princeton, NJ, 1972.

[Mit] B. Mitchell, *Theory of categories*, Academic Press, New York, 1965.

[Q] D. Quillen, *Higher algebraic K-theory. I*, *Algebraic K-Theory* I, Lecture Notes in Math., vol. 341, Springer-Verlag, Berlin, 1973, pp. 85–147.

[QC] D. Quillen, *On the cohomology and K-theory of the general linear groups over a finite field*, Ann. of Math. **96** (1972), 552–586.

[QF] ——, *Finite generation of the groups K_i of rings of algebraic integers*, *Algebraic K-Theory* I, Lecture Notes in Math., vol. 341, Springer-Verlag, Berlin, 1973, pp. 179–214.

[ST] H. Seifert and W. Threlfall, *Lehrbuch der Topologie*, Teubner, Leipzig-Berlin, 1934.

[S] V. Srinivas, *Algebraic K-theory*, Birkhäuser, Boston, MA, 1991.

[SK] R. G. Swan, *Algebraic K-theory*, Lecture Notes in Math., vol. 76, Springer-Verlag, Berlin, 1968.

[SQ] ——, *K-theory of quadric hypersurfaces*, Ann. of Math. (2) **122** (1985), 113–153.

[T] J. Tate, *Homology of Noetherian rings and local rings*, Illinois J. Math. **1** (1957), 14–25.

[W] F. Waldhausen, *Algebraic K-theory of generalized free products*, Ann. of Math. (2) **108** (1978), 135–256.

[We]   C. A. Weibel, A survey of products in K-theory, *Algebraic K-Theory*, Evanston 1980, Lecture Notes in Math., vol. 854,, Springer-Verlag, Berlin, 1981, pp. 494–517.

DEPARTMENT OF MATHEMATICS THE UNIVERSITY OF CHICAGO, 5734 UNIVERSITY AVENUE CHICAGO, ILLINOIS 60637

E-mail address: swan@zaphod.uchicago.edu

ISBN 0-8218-1498-2 (set)
ISBN 0-8218-0339-5 (part 1)

9 780821 803394